APR 0 3 1999

APR 2 1 1999

DEC 1 6 1999

NOV 1 4 2000

May 5/2001

OCT 2 1 2003

DEC 9 2004

FEB 2 3 2006

DEC 0 4 2007

EVOLUTION OF SOCIAL BEHAVIOUR PATTERNS IN PRIMATES AND MAN

PROCEEDINGS OF THE BRITISH ACADEMY · 88

EVOLUTION OF SOCIAL BEHAVIOUR PATTERNS IN PRIMATES AND MAN

A Joint Discussion Meeting of
the Royal Society and the British Academy

Edited by
W. G. RUNCIMAN, JOHN MAYNARD SMITH
& R. I. M. DUNBAR

Published for THE BRITISH ACADEMY
by OXFORD UNIVERSITY PRESS

Oxford University Press, Walton Street, Oxford OX2 6DP
Oxford New York
Athens Auckland Bangkok Bombay
Calcutta Cape Town Dar es Salaam Delhi
Florence Hong Kong Istanbul Karachi
Kuala Lumpur Madras Madrid Melbourne
Mexico City Nairobi Paris Singapore
Taipei Tokyo Toronto
and associated companies in
Berlin Ibadan

© The British Academy, 1996

All rights reserved. No part of this publication may be reproduced,
stored in a retrieval system, or transmitted, in any form or by any means,
without the prior permission in writing of the British Academy

British Library Cataloguing in Publication Data
Data available

ISBN 0–19–726164–7
ISSN 0068–1202

Phototypeset by Wyvern Typesetting Ltd, Bristol
Printed in Great Britain
on acid-free paper by
Bookcraft (Bath) Ltd
Midsomer Norton, Avon

Contents

Introduction   1
W. G. RUNCIMAN

Social Evolution in Primates: The Role of Ecological Factors
and Male Behaviour   9
CAREL P. VAN SCHAIK

Determinants of Group Size in Primates: A General Model   33
R. I. M. DUNBAR

Function and Intention in the Calls of Non-Human Primates   59
DOROTHY L. CHENEY & ROBERT M. SEYFARTH

Why Culture is Common, but Cultural Evolution is Rare   77
ROBERT BOYD & PETER J. RICHERSON

An Evolutionary and Chronological Framework for
Human Social Behaviour   95
ROBERT A. FOLEY

Friendship and the Banker's Paradox: Other Pathways to the
Evolution of Adaptations for Altruism   119
JOHN TOOBY & LEDA COSMIDES

The Early Prehistory of Human Social Behaviour:
Issues of Archaeological Inference and Cognitive Evolution   145
STEVEN MITHEN

The Emergence of Biologically Modern Populations in Europe:
A Social and Cognitive 'Revolution'?   179
PAUL MELLARS

Responses to Environmental Novelty: Changes in Men's
Marriage Strategies in a Rural Kenyan Community   203
MONIQUE BORGERHOFF MULDER

Genetic Language Impairment: Unruly Grammars   223
M. GOPNIK, J. DALALAKIS, S. E. FUKUDA, S. FUKUDA & E. KEHAYIA

The Emergence of Cultures among Wild Chimpanzees   251
CHRISTOPHE BOESCH

Terrestriality, Bipedalism and the Origin of Language 269
    LESLIE C. AIELLO

Conclusions 291
    JOHN MAYNARD SMITH

Introduction

W. G. RUNCIMAN
Trinity College, Cambridge, CB2 1TQ
Fellow of the British Academy

THE IDEA for the Royal Society/British Academy discussion meeting (held in April 1995) at which the papers in this volume were presented goes back to a similar joint meeting held as long ago as 1966 on the theme of ritualization of behaviour. That meeting was organized by Julian Huxley, who in his introduction paid tribute to Darwin for having, as he put it, 'paved the way for a unified psycho-physiological approach in the study of the behaviour of not only animals but man' (Huxley 1966: 250). Since then, an avowedly Darwinian approach to the study of social behaviour has come to embrace a range of disciplines extending well beyond the frontier which conventionally separates the concerns of the British Academy from those of the Royal Society. At the same time, a growing understanding of the behavioural repertoire of primates has modified many earlier assumptions about the distinctiveness of human behaviour while re-emphasizing the mental capacities which do indeed mark off *Homo sapiens sapiens* from other species. And a growing body of research on the longstanding question of human origins has yielded an increasingly convincing reconstruction of where and when (if not exactly how) it occurred.

It is only to be expected that the closer to present-day human societies a Darwinian approach is brought to bear, the more hostility it will continue to generate (Stove 1994). In anthropology, moreover, the difference in approach between social and biological (or 'sociobiological') anthropologists has widened to the point that although, as a recent article in the journal *Science* points out, anthropologists are all trained to bridge the gap between cultures, 'today many American anthropologists find themselves divided by one of those very gaps—and are having a tough time spanning the chasm' (Holden 1993: 1641). But rival approaches to common topics are not always mutually exclusive to the extent that the protagonists may make them appear. Not only do different kinds of question call for different kinds of answer (Foley 1995: 70, citing Tinbergen 1963), but the understandable propensity of researchers with new ideas to claim too much for them is, for all the controversy generated thereby, an encouragement to their correction by further research. Thus the signal inability of behaviourist psychology, despite its success in accounting for certain limited aspects of both animal and human behaviour in terms of operant conditioning, to deal with language-learning (Chomsky 1959) has helped to move the study of language in directions which have proved significantly more rewarding. Similarly the claim that language (if not, indeed, the whole complex of 'culture' as Tylor originally defined it) is 'socially constructed', although it derives from a presupposition which is palpably fallacious (Noble & Davidson 1991: 234–5), has nevertheless served to emphasize the need to reconcile the unquestionable diversity of human societies with the behavioural universals underlying it (Brown 1991).

It is, moreover, common ground that the operation of natural selection over the five or more million years since we diverged from our nearest genetic relatives continues to bear directly on contemporary human behaviour, whatever may or may not be the changes bound up with the transition from nature to culture. We can say with confidence that, among other things, young adult males are predisposed to homicide relative both to older males and to coeval females across the range of human societies (Daly & Wilson 1988; cf. Cronin 1991: 331); that certain facial expressions correspond to certain emotional states independently of variations in culture (Ekman 1973); and that all humans are genetically endowed with a common set of domain-specific, content-dependent, functionally specialized psychological modules and algorithms (Tooby & Cosmides 1992), including a 'language instinct' (Pinker 1994). It remains, inevitably, debatable how much of the variation in social behaviour patterns can be explained by hypotheses directly derived from models of natural selection. Even where the data show a consistent fit with what could be predicted in terms of maximization of inclusive reproductive fitness, there may be an equally good fit with what could be predicted by derivation from a model of autonomous cultural evolution. Moreover, the more complex the pattern of behaviour, the more difficult it is likely to be to disentangle the effects of genetic, ecological, demographic, and cultural variables. But in social, no less than in biological, anthropology, it is a matter of finding evidence which will enable competing hypotheses to be tested against one another.

It follows that reductionist hypotheses ought not to be dismissed a priori, as some social anthropologists and sociologists have been inclined to do. But the speed of change and breadth of range of social behaviour patterns which can be observed once the potential of our distinctive mental capacities begins to be realized is such that a reductionist approach becomes increasingly difficult to sustain. For example, the claim that (even) monastic celibacy is explicable in terms of maximization of inclusive reproductive fitness (Alexander 1979: 80) can hardly fail to invite a comment to the effect that, as it is cautiously phrased by Boyd & Richardson (1985: 203), 'it seems more plausible to us that other mechanisms are at work'. Once units of information affecting phenotype are being transmitted by genuine imitation and learning from the mind (or brain) of one member of a conspecific population to that of another, the diversity of tastes, fashions, beliefs, rituals, styles, and techniques can be accommodated convincingly within 'gene-culture coevolutionary theory' only if the possibility of autonomous cultural evolution as well as selection for genetic fitness is allowed for and the 'adaptationist' issue left open (Laland *et al.* 1995). Whether the units of selection are taken to be discrete particles of information ('memes') or 'instructions' (Tschauner 1994: 79), or 'bundles' (Durham 1990: 203), or what many anthropologists still prefer to call 'traits' can likewise be left open, so long as it is recognized that 'the smallest units of phenotypic consequence need not be equal to the functional unit of transmission' (Durham 1991: 422). What matters, once again, is recognition of the possibility that episodes of innovation and diffusion of social behaviour *can* come about through 'rapid and multivariate patterns of purely internal cultural change' (Mellars 1992: 17).

To say this, however, is not to assume that culture is unique to the human species. This issue has sometimes been bedevilled by definitional disputes over what is to count as 'culture'. But culture is unique to humans only under a definition so restrictive that the distinction between genetic and exosomatic transfer of information is itself lost. Rare as genuine imitation and learning as opposed to (mere) stimulus enhancement appears to be in other species (Boyd & Richerson, this volume), there is no doubt that it exists, and not only in our closest genetic relatives. Nor can it any longer be disputed that chimpanzees, in particular, can not only make, use and reuse tools but transmit different styles of tool use independently of population density or environmental parameters (Boesch *et al.* 1994; Boesch, this volume). Moreover, the capacities of individual animals such as the bonobo Kanzi (Savage-Rumbaugh & Rumbaugh 1993), who are studied in non-natural laboratory or domestic environments where they learn or imitate the behaviour of humans, have to some degree diminished what was previously taken to be the divide between human and primate intelligence. Nevertheless, the recent conclusion of a primatologist whose own work has done much to bridge the hominid/pongid gap is a firm restatement that 'chimpanzees do not have human culture, material or otherwise' (McGrew 1992: 230); and the most significant difference between 'them' and 'us' from the viewpoint of biology and the social sciences alike is the innate capacity of three-year-old human infants to construct novel sentences whose lexical and syntactic complexity is far beyond even the most carefully trained adult chimpanzee.

When, how, and why this capacity evolved is perhaps the most intriguing puzzle confronting researchers concerned with the question 'what is being selected for what in the transition from nature to culture?' (Aiello, this volume). The advantages to a species endowed with language are evident. Speech-users can maintain social cohesion in extended groups, explore unfamiliar and inhospitable territory, distinguish friends from enemies, and plan the acquisition and storage of material resources far more effectively in consequence. But to list the advantages is not to reconstruct the evolution, or explain its uniqueness to the human species. Nor does it yield the answers to such questions as how linguistic competence relates to other mental abilities, or what inferences (if any) can be drawn about language from archaeological evidence for increasing sophistication in technology or art, or when the necessary neuroanatomical changes (whatever they were) took place, or why humans resemble birds more than they do primates in their vocal development, or whether gestural rather than verbal communication may hold the key to the origin of language. But it seems safe to suppose not only that the evolution of the capacity for language was both gradual and complex but that as it is further elucidated the implications for the relation between natural and cultural selection will be increasingly revealing.

Meanwhile, it is as well to remember that although language cannot be said to be necessary for the exosomatic transmission even of complex information (Bloch 1991), once it is fully developed it brings about a further change of equal importance for the evolution of social behaviour patterns. As it is put by Whallon (1989: 438), other people come to be identified not merely as individuals but 'in terms of social categories, among which mutual rights and obligations are defined by the system'. At this point, the term 'role' takes on a meaning beyond its meaning in the theory of natural selection, where it denotes differences such as male and female, or owner and intruder, which are perceived by contestants in asymmetric contests for territory or resources and thereby cause them to adopt one strategy rather than another (Maynard Smith 1982: 204). Roles are now positions in social space whose incumbents share expectations and beliefs about their capacity to influence each other's behaviour because of the roles that they occupy. Moreover, whereas in cultural evolution the units of selection are replicated by transmission from one person's mind (or brain) to another's, in social evolution they are units of *reciprocal* action—with, of course, the possibility of mutation without which evolution of any kind comes to a standstill. At this level, the equivalents of the genes of natural selection (and the 'traits', 'bundles', or 'memes' of cultural selection) are the practices by which roles are defined as such (Runciman 1986). In consequence, just as it is bound to be difficult to disentangle the effects of natural from those of cultural selection, so is it bound to be difficult to disentangle the effects of cultural from those of social selection.

Let me give one brief example taken from Sperber (1985: 86) where he says, in talking about pre-literate cultures, that 'in an oral tradition, cultural representations which are hard to remember are forgotten'. This is not the circular proposition which it may look like at first glance, since it is a question to be settled empirically whether formulae which can be shown to be easier to remember are in fact transmitted either vertically or horizontally within the population under study more frequently than others which can be shown to be more difficult to remember. But suppose we apply it to the much-studied transmission of oral epic poetry in Archaic Greece. It is well known that the replication and diffusion of the *Iliad* and *Odyssey* was facilitated not only by their hexametric form but by their use of stock epithets, recurring themes, conventional sequences, and repeated lines. But the mnemonic function of devices like these is far from furnishing an adequate explanation of their remarkable survival and spread throughout the Greek-speaking world. For whatever the powers of memory, enhanced by these devices, of non-literate populations, the transmission of long oral epics required training and application on the part of semi-professional singers who had to earn at least part of their living from patrons in whose courts or households they performed when called on to do so (Kirk 1965). Thus, for all that oral traditions in pre-literate societies are a textbook example of exosomatic transmission from one person's mind (or brain) to another's, they at the same time take us into the 'civilized' world of roles, ranks, and institutional relationships of power and patronage which is as familiar to archaeologists reconstructing the social structure of non-literate peoples from their grave-goods and the spatial distribution of their material remains as it is to social anthropologists studying them directly in the field, or sociologists and historians studying them through documentary records.

This particular example leads into topics outside the concerns of the contributors to this volume. But on any topic where the possibility of cultural selection is present, we have to address the difference between behaviour which is and is not, in the words of one social anthropologist, 'activity devoid of control by a knowing subject' (Ingold 1983: 5) or, in the words of another, the indisputable fact that humans are 'intrinsically active and choice-making beings in pursuit of particular goals and objectives' (Robarchek 1989: 909). Well—yes. But who ever thought otherwise? The question which needs to be asked is 'so what?'—or, more circumspectly, 'what follows for the study of social behaviour patterns in animals and man?' One thing which immediately follows is that the question 'what is it like to be the way they are?' arises in a new and different form. That question is pointless when addressed to termite workers which can be said to 'know' that they have reached the boundary of a ground plan only in a sense similar to that in which cells at the boundary of a liver 'know' that they are not in the middle of it (Dawkins 1982: 204), fascinating but ineluctably opaque when addressed to vervet monkeys at the foot of Mt Kilimanjaro alerting one another to the presence of predators without, however, attributing to each other minds like their own (Cheney & Seyfarth 1990, and this volume), and deceptively straightforward when applied to present-day humans interviewed by survey researchers about their tastes, life-styles, and political opinions. It is, however, a question to be categorically distinguished from explanatory questions, whether motivational, functional, or genetic. It calls for different methods; it appeals to different criteria; it leaves the researcher with discretion in answering it of a kind which does not arise in the formulation and test of explanatory hypotheses; and it may even be as well answered by a work of fiction as by an ethnographic monograph in the manner of Malinowski or Evans-Pritchard or a slice of 'thick description' in the manner of Clifford Geertz (1973: Chapter 1).

What, then, is the difference which it makes when explanatory hypotheses *are* being formulated and tested? In one sense, the answer is: none. The question 'what is it about the antecedent history and present environment of this population which makes its patterns of social behaviour what they are?' is the same where the history and environment cause its members to have purposes and goals in consequence of which they behave as they do as it is where the history and environment cause their behaviour to be an immediate and instinctive response to the stimuli which evoke it. The members of !Kung San foraging bands are aware of the function of meat-sharing in generating a sense of mutual obligation, just as the members of university departments of anthropology are aware of the latent social functions of the seminar and the coffee-break. But the functions would be the same even if they weren't. It is not the fact of self-awareness which explains the pattern of social behaviour of which the behaving selves are aware. Conversely, it is a commonplace in non-Freudian as much as in Freudian psychology that even people who are intensely self-aware may fail to understand (in the explanatory, as opposed to the descriptive, what-it-feels-like sense) the motives which impel them to behave as they do. The point is, rather, that since something different *is* going on 'inside "their" heads' from the vervets' heads, and the vervets' from the termites' heads, we can hardly fail to be better placed to explain both human and animal social behaviour patterns if we can succeed in finding out how it works and what it does. Thus, the potential value of models of rational human action (however 'rational' is defined) lies in their capacity to account for the behaviour observed by relating it to the modules and algorithms inside 'their' heads whether or not 'they' are aware of it and whether or not 'we' have an empathetic sense of what it is like to be one of 'them' (cf. e.g. Cohen & Machalek 1988, 1994). And as further progress in this direction comes to be made, the 'chasm' separating biological from social anthropology will more and more effectively be spanned.

This will surely involve further interdisciplinary co-operation of the kind which has been growing steadily over the last thirty years and is, indeed, exemplified by the papers in this volume and the discussions which followed their presentation at the joint meeting. In saying this, I am well aware that for both Popperian and other reasons it is impossible to forecast the future course of scientific enquiry. But I do not think it foolishly optimistic to suggest that we are witnessing an integration of approaches which will turn out to be as important as the old 'new synthesis' which integrated evolutionary theory with population genetics. Indeed, I am prepared to venture the prediction that this new 'new synthesis', which has already enlisted primatologists, anthropologists, archaeologists, psychologists, demographers, and linguists, will increasingly involve even general sociologists hitherto as remote from biology as myself.

REFERENCES

Alexander, R.D. 1979: *Darwinism and Human Affairs.* Seattle: University of Washington Press.
Boesch, C., Marchesi, P., Marchesi, N., Fruth, B. & Joulian, F. 1994: Is nut cracking in wild chimpanzees a cultural behaviour? *Journal of Human Evolution* 26, 325–338.
Bloch, M. 1991: Language, anthropology and cognitive science. *Man* n.s. 26, 183–198.
Boyd, R. & Richerson, P. J. 1985: *Culture and the Evolutionary Process.* Chicago: University of Chicago Press.
Brown, D.E. 1991: *Human Universals.* New York: McGraw Hill.
Cheney, D.L. & Seyfarth, R.M. 1990: *How Monkeys See the World: Inside the mind of another species.* Chicago: University of Chicago Press.
Chomsky, N. 1959: Review of Skinner's 'Verbal Behaviour'. *Language* 35, 26–58.
Cohen, L.E. & Machalek, R. 1988: A general theory of expropriative crime: an evolutionary ecological approach. *American Journal of Sociology* 94, 465–501.
Cohen, L.E. & Machalek, R. 1994: The normalcy of crime: from Durkheim to evolutionary ecology. *Rationality and Society* 6, 286–308.
Cronin, H. 1991. *The Ant and the Peacock.* Cambridge: Cambridge University Press.
Daly, M. & Wilson, M. 1988: *Homicide.* New York: Aldine de Gruyter.
Dawkins, R. 1982: *The Extended Phenotype.* San Francisco: W.H. Freeman.
Durham, W. 1990: Advances in evolutionary culture theory. *Annual Review of Anthropology* 19, 187–210.
Durham, W. 1991: *Coevolution: Genes, Culture and Human Diversity.* Stanford: Stanford University Press.
Ekman, P. 1973: Cross-cultural studies of facial expression. In *Darwin and Facial Expression: A century of research in review* (ed. P. Ekman), pp. 169–222. New York: Academic Press.

Foley, R. 1995: Causes and consequences in human evolution. *Journal of the Royal Anthropological Institute* n.s. 1, 67–86.
Geertz, C. 1973: *The Interpretation of Cultures.* New York: Basic Books.
Holden, C. 1993: Failing to cross the biology-culture gap. *Science* 262, 1641–2.
Huxley, J. 1966: Introduction. *Philosophical Transactions of the Royal Society of London* Series B 251, 249–271.
Ingold, T. 1983: The architect and the bee: reflections on the work of animals and men. *Man* n.s. 18, 1–20.
Kirk, G. 1965: *Homer and the Epic.* Cambridge: Cambridge University Press.
Laland, K.N., Kumm, J. & Feldman, M.W. 1995: Gene-culture coevolutionary theory: a test case. *Current Anthropology* 36, 131–156.
Maynard Smith, J. 1982: *Evolution and the Theory of Games.* Cambridge: Cambridge University Press.
McGrew, W.C. 1992: *Chimpanzee Material Culture: Implications for human evolution.* Cambridge: Cambridge University Press.
Mellars, P. 1992: Archaeology and modern human origins in Europe. *Proceedings of the British Academy* 82, 1–35.
Noble, W. & Davidson, I. 1991: The evolutionary emergence of modern human behaviour: language and its archaeology. *Man* n.s. 26, 223–253.
Pinker, S. 1994: *The Language Instinct.* Harmondsworth: Penguin.
Robarchek, C.A. 1989: Primitive warfare and the ratomorphic image of mankind. *American Anthropologist* 91, 903–920.
Runciman, W.G. 1986: On the tendency of human societies to form varieties. *Proceedings of the British Academy* 72, 149–165.
Savage-Rumbaugh, E.S. & Rumbaugh, D.M. 1993: The emergence of language. In *Tools, Language and Cognition in Human Evolution* (ed. K.R. Gibson & T. Ingold), pp. 86–108. Cambridge: Cambridge University Press.
Sperber, D. 1985: Anthropology and psychology: towards an epidemiology of representations. *Man* n.s. 20, 73–89.
Stove, D. 1994: So you think you are a Darwinian? *Philosophy* 69, 267–277.
Tinbergen, N. 1963: On aims and methods in ethology. *Zeitschrift für Tierpsychologie* 20, 410–433.
Tooby, J. & Cosmides, L. 1992: The psychological foundations of culture. In *The Adapted Mind: Evolutionary psychology and the generation of culture* (ed. J. Barkow, L. Cosmides & J. Tooby), pp. 19–136. New York: Oxford University Press.
Tschauner, H. 1994: Archaeological systematics and cultural evolution: retrieving the honour of cultural history. *Man* n.s. 29, 77–93.
Whallon, R. 1989: Elements of cultural change in the later palaeolithic. In *The Human Revolution: Behavioural and biological perspectives on the origins of modern humans* (ed. P. Mellars & C. Stringer), pp. 433–454. Edinburgh: Edinburgh University Press.

Social Evolution in Primates: The Role of Ecological Factors and Male Behaviour

CAREL P. VAN SCHAIK

Department of Biological Anthropology and Anatomy, Duke University, Box 93083, Durham, NC 27708, USA

Keywords: socio-ecology; social strategies; competition, scramble, contest; predation; infanticide.

Summary. In order to explain the variation in primate social systems, socio-ecology has focussed on the role of ecological factors to explain female associations and relationships and on the spatio-temporal distribution of mating opportunities to explain male associations and relationships. While this approach has been quite successful, it ignores male–female associations and relationships and ignores the possibility that male behaviour modifies other aspects of the social system. In this paper, the ecological approach is complemented by consideration of a social factor found to limit fitness, namely infanticide by males. Infanticide risk is proposed to have selected for male–female associations and relationships, and to have modified female–female relationships in some cases. It is also hypothesized to have selected for the unusual male bonding by species such as chimpanzees. Finally, its possible impact on between-group relations is examined. The findings suggest that infanticide is of equal importance to ecological factors, with which it may interact in sometimes complex ways, in shaping primate social systems.

INTRODUCTION

SOCIAL SYSTEMS AMONG PRIMATES vary widely from species to species (Smuts *et al.* 1987). This variation concerns both patterns of membership

© The British Academy 1996.

and spatial distribution (i.e. the associations of individuals), and the nature of the social relationships among the members of social units. Ever since the extent of this variation became apparent, attempts to explain it considered social behaviour an adaptation produced by natural selection (Crook & Gartlan 1966). Since the selective pressures were thought to be mainly ecological, this endeavour has become known as socio-ecology. The aim of this paper is to provide an overview of recent developments in primate socio-ecology.

We should be careful in defining the object of inquiry. Social systems arise through behavioural interactions between individuals. Hence, they are not adaptations; only the social strategies of individuals are. Unfortunately social strategies cannot be directly observed, but must be deduced. Deducing the social strategies requires an iterative approach, because the primary rules used by the relevant players will be modified to deal with the social and demographic context they themselves have produced. Thus, the ideal socio-ecological model merely specifies the behavioural rules used by individuals, and the social system emerges from their interactions. despite some promising starts (e.g. te Boekhorst & Hogeweg 1994), this is still a distant ideal.

Socio-ecology has dealt with this complexity by developing a priori arguments. If social strategies are adaptations, then those factors that exert the strongest limitation on lifetime reproductive success should provide the strongest selection pressures toward the evolution of social strategies, i.e. the spatial associations and social relationships that individuals engage in. Primate populations are often limited by food and predation, as shown by deliberate and accidental experiments (Mori 1979; Richard 1985). Thus, socio-ecology has focused on the role of the abundance and distribution of food and predators in shaping social strategies (Crook & Gartlan 1966; Wrangham 1980; Dunbar 1988).

It is also clear that the two sexes tend to be limited by different factors (Trivers 1972). Again, an a priori argument suggests that female social strategies mainly serve to reduce the impact of predation and feeding competition. In contrast, variation in male fitness is often largely due to differences in the number of infants sired, and we should therefore expect the associations and relationships formed by adult males to increase access to mates.

This deductive approach has engendered the fundamental paradigm of socio-ecology (Emlen & Oring 1977): the spatial distribution and social relationships among females are thought to reflect ecological conditions, in particular distribution of risks and food, whereas the distribution of males, and the social relationships among them, are determined by the spatio-temporal distribution of mating opportunities (Figure 1). This approach has been verified experimentally in small mammals (Ims 1988; Ostfeld 1990), and has also been widely applied in primates. The first section of this paper

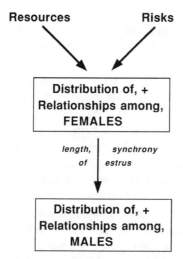

Figure 1. Ecology of social systems. The classic socio-ecological paradigm, which links female associations and relationships to ecological factors and male associations and relationships to the spatio-temporal distribution of mating opportunities.

will briefly review primate work that examines the impact of ecological factors on female distribution (spatial associations) and social relationships and on the male response to this.

However, while this approach is elegant, it inevitably oversimplifies. First, it is incomplete, in that it ignores the common existence of significant associations or social relationships between the sexes, which, as we shall see, are quite important in the order primates. Second, it may also be wrong. In particular, it is conceivable that male–female interactions might alter spatial associations and relationships among females (Wrangham 1979), those among males, and even the relations between groups. Both additions and modifications arise from the increased appreciation of the action of a different factor that limits fitness; this factor is a social one, namely infanticide by males (Watts 1990; van Schaik & Dunbar 1990; Hiraiwa-Hasegawa & Hasegawa 1994). Hence, the remainder of this paper will focus on male–female associations and relationships, and on their impact on other features of social systems.

CLASSIC SOCIO-ECOLOGY

The ecological model for female associations and relationships

Permanent gregariousness is quite common among female primates. While none of the nocturnal taxa shows female association, 79% of diurnal taxa does (42 of the 53 diurnal taxa, where a taxon is a genus or a species or group of species within a genus that is homogeneous for the relevant social variables). Female within-group coalitionary relationships (alliances) are also fairly common (26% of 34 gregarious taxa for which this information is available). It is more difficult to assess the presence of between-group alliances among females, because they may occur with low frequency and are often conditional. An estimated 23% of 31 gregarious taxa for which this is known has more than occasional between-group conflicts by females.

Briefly, the ecological model for females is as follows. Females form spatial associations because it reduces predation risk (Janson 1992). The feeding competition that inevitably arises acts as a countervailing force, leading to groups of some intermediate size (van Schaik 1983; Dunbar 1988). The social relationships among these gregarious females will depend on the competitive regime (van Schaik 1989). Where access to limiting resources is not defensible or not worth defending, competition will be by scramble, and variation in power cannot be translated into variation in access, so social behaviour is of no use to improve access to limiting resources. Thus, females should not show frequent aggression over food, should lack formal submission signals, and they should not form alliances. Female fitness will depend on group size, and the easiest way to manipulate group size is by moving to other groups or starting new ones. Females are therefore expected to migrate freely between groups, whenever ecological, social or reproductive considerations (e.g. inbreeding) make such moves advantageous. This prediction assumes that diurnal primates, being mainly gregarious, face relatively few ecological constraints on dispersal (cf. Watts 1990), although there may be serious social constraints at high densities. This type is referred to as non-female-bonded, following Wrangham's (1980) terminology.

In contrast, where the limiting resources that females compete for are monopolizable, competition is by contest. Then, power differences give rise to differences in access (e.g. net food intake or safety), and aggression is selected for. Provided certain assumptions are met (Figure 2), we therefore expect that females form decided dominance relationships, with formal submission signals that go in one direction within a dyad (bared teeth among many cercopithecines, pant grunts among chimps, spat calls in lemurs, etc.: de Waal 1986; Pereira & Kappeler, in press). They should form alliances, either because coalitionary aggression is needed to achieve access to the limiting resources (cf. Wrangham 1980) or because they can benefit through kin selection by improving the agonistic power of their relatives. Alliances with relatives are also more stable. Hence, association with relatives is expected, which can be achieved by female philopatry. When emigration occurs, it is in the form of subgroups budding off and striking

Figure 2. Diagram of the social consequences of contest competition in animals, to illustrate the derivation of the links between decided dominance relationships, alliances and philopatry of the contesting sex.

out on their own. This suite of characters is called female-bonded (Wrangham 1980; van Schaik 1989).

This basic dichotomy is complicated by the possibility that contest competition is also possible between groups rather than just between individuals (see van Schaik 1989). Strong between-group contest would change the predictions made above as follows. First, because relatives make the best allies in such between-group contests as well (especially since there is a possible collective action problem), females of non-female-bonded groups are expected to be philopatric when between-group contest is high. Second, because subordinates derive a source of power from their ability to withhold support to the large alliance, we should see a more tolerant form of dominance relations in the female-bonded groups. This leads to four types of female social structure (Table 1).

Before this model can be put to the test, it should be established that it accurately describes the situation in non-human primates. First, we should assess its internal consistency. Obviously, it was consistent with the evidence available when it was formulated, but the new descriptive material accumulated since then indicates that the association between decided dominance relationships on the one hand and female alliances on the other hand remains extremely strong. Likewise, there are no examples of taxa where decided dominance and alliances are accompanied by routine dispersal of the females (review: Sterck et al., in manuscript). Second, detailed long-term field studies allow evaluation of the effect of dominance

Table 1. The predicted effects of the nature of competition for limiting resources on female social relationships and dispersal (based on van Schaik 1989).

Competition		Social response		
WG^b contest	BG^C contest	Female philopatry	Female relationships (dominance) type	Designation[a]
Low	Low	No[d]	(none) egalitarian	Non-female-bonded
Low	High	Yes	(none) egalitarian	Female-resident
High	Low	Yes	(yes) despotic	Female-bonded
High[e]	High	Yes	(yes) tolerant	Tolerant female-bonded

[a] The definitions of these designations deviate from those originally used by Wrangham (1980).
[b] WG = within-group
[c] BG = between-group
[d] Dispersal is not compulsory, but likely to be the norm.
[e] WG contest is at least potentially high. In practice, tolerant dominance relations may lead to a relaxation of within-group contest for access to limiting resources or their reproductive consequences.

rank on energy budgets and reproduction. Crude ranks can often be recognized among females in non-female-bonded groups on the basis of displacements with no or mutual aggression. However, these ranks do not affect aspects of energy budgets or reproduction in two well-studied non-female-bonded species, gorillas (*Gorilla g. beringei*) and thomas langurs (*Presbytis thomasi*), whereas predictable dominance effects are commonly found among many female-bonded species (Silk 1993; review in Sterck *et al.*, in manuscript).

Actual tests of the model consider the relation between ecological conditions and female social relationships. Because of the limited phenotypic plasticity of a species' social behaviour, experimental manipulations of ecological variables need not always produce the predicted social changes. The best tests are probably comparisons that examine ecological differences between closely related but socially distinct species. One such test (Mitchell *et al.* 1991) concerns two squirrel monkeys, the Peruvian *Saimiri sciureus*, and the Costa Rican *S. oerstedi*. These two species are quite similar, in that groups are about the same size, face serious predation risk by raptors, and have similar diets and activity budgets. Yet, *S. sciureus* is clearly female-bonded (showing frequent resource-based aggression, dominance, alliances, and female philopatry), whereas *S. oerstedi* is non-female-bonded (showing 70 times lower aggression rates, no dominance, no alliances, and female breeding dispersal). An extreme ecological difference was found in the fruit trees, in which females have most of their conflicts: the Peruvian *S. sciureus* lives in a tropical rain forest where trees have the normal range of crown and crop sizes. The Costa Rican *S. oerstedi* live in forests with densely packed tiny trees with minute fruit crops that are exploited in a dispersed fashion.

Because of the low resolution of such semi-qualitative comparisons, many more such tests are needed for a proper evaluation of the model. The provisional results of other such comparisons, with baboons (*Papio spp.*), also support the model (R. Barton, personal communication; G. Cowlishaw, personal communication).

No non-ecological alternatives for the variation in female social relationships have been published. Those that can be developed (Sterck *et al.*, in manuscript) do not lead to the rejection of the ecological model. In conclusion, the ecological model provides for now the most satisfactory explanation for variation in female social relationships.

Male associations and relationships

The classic socio-ecological approach for males states that their distribution reflects the spatio-temporal distribution of mating opportunities (Emlen & Oring 1977). Thus, the associations and relationships formed by males should improve their ability to gain access to mates. Associations are less common among males than among females: 60% vs. 79% of diurnal taxa (using same conventions as above). Within-group alliances are also less common among males: they occur in 12% of the 25 diurnal taxa with male association whose social behaviour is well known, as opposed to 26% among females (only taxa with male association are included because spatial association is a precondition for alliance formation). Thus, primate males are both less likely to associate among themselves and to form alliances when associated.

The sex difference in association is straightforward. Males can derive the benefit of reduced predation risk by associating with females. On the other hand, they will derive strong mating benefits from excluding other males from access to females. Indeed, as in other mammals, the number of males in a group of primates is generally considered a function of the extent to which one male can monopolize sexual access to females (Clutton-Brock 1989; Altmann 1990).

Several factors may explain the reduced incidence of within-group alliances among males. First, males will benefit from excluding other males from mating, even if they cannot exclude these other males from being in the group. Second, male alliances tend to be less stable due to the faster rise and fall of a male's fighting power, and thus of his value as an ally. Third, there is a fundamental difference in the nature of the resources the two sexes compete for. Females compete for access to food, males for fertilizations. If two females compete for food, both will gain since both will obtain access to more food than they would obtain alone. In contrast, while collaborating males may each gain more matings, this does not mean that each gains more fertilizations: the total amount of fertilizations in a group is a constant quantity. Preliminary quantification of this argument showed that only a few pairs of mid-rankers showed the expected gains that would make it profitable for them to form an alliance (C. van Schaik & C. Nunn, in preparation), very similar to the pattern observed in baboons (e.g. Noë 1990).

Male alliances that function in between-group conflicts are probably more common than those within groups, although they remain opportunistic. This is not unexpected because the total amount of fertlizations is no longer a fixed quantity in between-group competition, as males can increase their tenure or even increase the number of females attracted by collaborating. Not easily reconciled with the classic approach is the occurrence of male bonding and philopatry where solitary females occupy indefensible ranges, such as in chimpanzees (*Pan troglodytes*), spider monkeys (*Ateles spp.*) and woolly spider monkeys (*Brachyeles arachnoides*) (Nishida & Hiraiwa-Hasegawa 1987; Wrangham 1987; Strier 1992). So far, no satisfactory explanation has been offered for this unusual situation, and I will return to it later.

MALE–FEMALE ASSOCIATIONS

As this brief survey shows, the classic socio-ecological framework explains many of the features of primate societies. However, it ignores male–female associations, and I will discuss these now.

Permanent male–female association is not self-evident. Given internal fertilization and lactation, i.e. obligatory female association with the zygote and the infant, desertion after fertilization is a viable male option. Hence, desertion is quite common, especially where mating is seasonal or at least predictable and punctuated by long periods of no mating. It is therefore not surprising that permanent male–female association is rather uncommon among mammals (see e.g. Wilson 1975). Curiously, primates are the order in which permanent male–female association is by far the most common; indeed, among diurnal species it is almost 100% (see Smuts *et al.* 1987). Mate guarding does not explain this pattern since even the most seasonal breeders have year-round male–female association.

Why is this? Other factors than ecological ones may also limit fitness, and may thus exert selective pressure towards the evolutions of social strategies. The major social problem for primate females is infanticide by males that have not mated with the female before (Hiraiwa-Hasegawa &

Hasegawa 1994). Infanticide is estimated to be responsible for 35% of infant mortality in hanuman langurs (*Presbytis entellus*; Sommer 1994), 37% in mountain gorillas (Watts 1990) and as much as 64% in red howler monkeys (*Alouatta seniculus*; Crockett & Sekulic 1984). While these numbers are likely to be lower for most other populations, they demonstrate that infanticide can exert a strong selective pressure on primate social strategies.

Although broad comparative data are still lacking, infanticide may well be more prevalent in primates than in most other mammals (but see Pusey & Packer 1994 on lions, *Panthera leo*). Several factors may conspire to make it particularly acute in primates. First, primates have very slow life histories (Harvey *et al.* 1987), making the period of vulnerability to male infanticide long. Second, primate infants are generally conspicuous (not hidden in nests or dens) and defenseless (unable to run away or fight back very effectively). Third, perhaps primate females show greater site tenacity than many other mammalian taxa, which increases the probability that an infanticidal male can mate with the female once she returns to oestrus, thus rendering infanticide a beneficial male strategy.

Various social strategies could evolve that would reduce a female's risk of infanticide. First, mating behaviour can be modified (along with physiological changes). For instance, females could actively pursue promiscuity when sexually receptive and show situation-dependent receptivity, as during pregnancy or post partum (Hardy & Whitten 1987). Second, females could migrate away from groups in which infanticide is likely (see below). Third, females could ally with effective protectors. Evidence suggests that males, specifically the possible sires of the females' infants, play a special role in preventing infanticide by unfamiliar males: males that were reproductively active are often associated with infants (van Schaik & Dunbar 1990), infanticide is most likely when male representation in the group changes (Hiraiwa-Hasegawa & Hasegawa 1994; Sommer 1994), and when males are removed (by accident or experimentally) infanticide is highly likely (e.g. Sugiyama 1966).

Association with the male may therefore be the optimal strategy for a female, despite the costs of feeding competition that this will usually entail. This hypothesis leads to one very strong prediction. If permanent association between males and females in primates serves to reduce the risk of infanticide, it should only be found where females are spatially associated with their infants, usually because they carry them. Conversely, where the infant is not with its mother, but parked or left in a nest, permanent male-female association is not expected. A comparative test of this hypothesis finds strong support for this prediction in primates (C. van Schaik & P. Kappeler, in preparation). There are also no plausible alternatives for the observed taxonomic distribution of male-female association. For instance, activity period or litter size, provide a poorer fit than the mode of infant care.

The permanent association of males and females allows for the evolution of a rich variety of male–female social relationships. However, there is little quantitative description of these relationships, and also no theorizing about them (Smuts 1987).

SOCIAL CONSEQUENCES OF INFANTICIDE AVOIDANCE

The patterns in male–female associations and relationships indicate that infanticide risk is likely to be a potent selective force in primate social evolution. The obvious next step is to explore how infanticide risk may have affected the associations and relationships among females, among males, and perhaps even among groups.

Infanticide and social relationships among females

Since females were found to be less effective in reducing the risk of infanticide than males, it is unlikely that infanticide risk could have selected for female association directly. The ecological model claims that female association (gregariousness) serves to reduce predation risk. These female groups are then joined by one or more males. The competitive regime determines the nature of the social relationships among the females (see Figure 3.a).

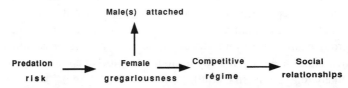

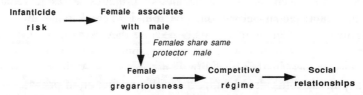

Figure 3. Female social relationships. Two alternative ways, ecological (a) or social (b), of deriving the non-female-bonded model of female social relationships in primates.

However, infanticide risk could indirectly lead to female association. Assume that females respond to the risk of infanticide by associating with a male. In fact, this is the hypothesized route to bonded monogamy in primates (van Schaik & Dunbar 1990). However, if the costs of female association are low, if females can transfer easily, if there is variation in the quality of males as protectors, and if females can share the anti-infanticide service of a male up to a point, then they could form groups around effective protectors (Wrangham 1979; Watts 1990). These groups are likely to be fairly small, because their size is set by the ratio of breeding females to able-bodied adult males. They are likely to be non-female-bonded because of the need for female migration and the low costs of association (Figure 3.b).

Female emigration decisions were studied in two non-female-bonded species, the mountain gorilla (Watts 1990) and thomas langur (Sterck & Steenbeek, in manuscript). In both species, females tend to transfer into smaller groups, consistent with the observed significant scramble component in their within-group competition. This pattern supports the original ecological model. However, female migration decisions are also clearly linked to infanticide risk: in gorillas females tend to transfer after their infants are killed, and in the langurs they tend to transfer during the brief periods when they are least vulnerable to infanticide. Furthermore, female transfer decisions seem to be guided primarily by the identity of the target male rather than the group of females.

The most plausible interpretation is that predation risk and infanticide risk operate simultaneously and both provide significant pressure toward the observed system in these two species. The relative importance of each factor is bound to vary with the ecological conditions. For instance, the langurs show various patterns compatible with the predation reduction function of grouping (small groups avoid the ground layer; males form all-male bands). And the social model can not explain large non-female-bonded groups in high-predation environments such as those of *Saimiri*. But it is possible that infanticide risk could be a significant contributing factor to non-female-bonded groups, and maybe even the predominant one in some species facing negligible predation risk.

Infanticide and male associations and relationships

Male bonding

At first sight, it is extremely unlikely that infanticide could have affected male social relationships. Males reduce the risk of losing their offspring to infanticidal rivals by forming associations and relationships with females. However, what are a male's options if females are solitary and if the costs of permanent male–female association are too high, as is observed among large arboreal frugivores such as spider monkeys, orangutans (*Pongo pygmaeus*) and chimpanzees?

In general, where a male cannot defend access to females directly, he could defend access to the range containing the female or females (cf. Clutton-Brock 1989). This may be effective in defence of mating access especially when female oestrus is brief and advertised. However, for infanticide prevention to be effective all male trespassing has to be minimized. I hypothesize that male alliances may provide this protection. The best studied example of male bonding is the chimpanzee, in which males form parties that patrol the boundaries of a communal range and respond in highly antagonistic fashion toward male strangers, sometimes with lethal consequences (Manson & Wrangham 1991). This should serve to make males very reluctant to enter unfamiliar territory, which thus provides a measure of safety to females. Females, moreover, tend to stay away from the boundary area when they have infants (Goodall 1986).

It is difficult to test this hypothesis, but no other plausible scenarios have been presented so far. It seems worthy of further investigation, especially since male bonding probably represents the ancestral hominid system (Foley & Lee 1989).

The number of males in a group

In female groups with permanent male representation, the number of males is an important determinant of social behaviour (e.g. Hamilton & Bulger 1992). The classic socio-ecological approach proposes that the number of males in a group is determined by the potential for monopolization of potentially fertile matings. However, in small groups, it may be in the females' interest to allow multiple males to be attached to the group despite the increase in feeding competition caused by this, because their presence may reduce the risk of predation or of infanticide. Females may exert some direct influence over male immigration (Smuts 1987). Furthermore, the monopolization potential is determined in part by female behaviour, such as the degree to which females invite promiscuity, and by the temporal clumping of female attractivity, which depends on the length and accuracy of ovulation signalling and the degree of synchrony among females. Natural selection can affect all these traits, and thus the monopolization potential. Females may therefore have some control over the number of males in their group.

Males may help to reduce the risk of predation to females and their offspring because of their higher vigilance levels and consequently greater ability to detect predators, and their tendency to face down predators (van

Schaik & van Noordwijk 1989). Thus, one could predict that where predation risk is particularly severe but feeding competition does not allow for large female groups, groups should be more likely to contain multiple adult males. This prediction was tested by van Schaik & Hörstermann (1994) in a controlled comparison of arboreal folivores in three different continents: American howler monkeys, African colobus (*Colobus spp.*) and Asian langurs. Large monkey-eating eagles are absent in the range of the Asian langurs and one population of African colobus, and, as predicted, in these populations average-sized female groups are most likely to contain a single adult male at a given group size (Figure 4). Alternative hypotheses did not produce this pattern. These findings are therefore consistent with the hypothesis that predation risk can affect male representation in primate groups.

The number of males in a group may also affect the risk that a female is subject to infanticide. Infanticide is expected to be less likely if a group contains multiple reproductively active males because male immigration is less likely to be in the form of violent take-overs of top dominance, and potentially infanticidal newcomers face several possible sires of the infants (of course this assumes promiscuous matings in multi-male groups, but this is a common phenomenon in the larger ones: Hrdy & Whitten 1987). Comparisons have shown that infanticide risk is lower in multi-male groups than in single-male groups in hanuman langurs (Newton 1986) and mountain gorillas (Robbins 1995). This argument would predict that it would generally be to the females' advantage to live in groups with multiple adult males unless the feeding competition that this produces is too severe or unless females have alternative means of reducing infanticide risk.

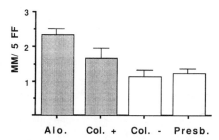

Figure 4. The mean (+ standard error) number of adult males in a group of five adult females as estimated from regression equations relating the number of males to the number of females in groups of different arboreal folivores. The comparison is between taxa inside the range of large monkey-eating eagles (Alo, Col+) and those outside the range of such eagles (Col−, Presb). Alo = *Alouatta*; Col = *Colobus*; Presb = arboreal Southeast Asian *Presbytis* (excluding *P. entellus*). The grand mean group size of these species is about 5 females. The differences among these estimated values are highly significant. From van Schaik & Hörstermann (1994).

It is difficult to make general predictions for the degree of feeding competition imposed by males, but females have another strategy to reduce infanticide risk, namely breeding dispersal or transfer. Females can reduce the risk of infanticide if they can transfer to other groups during times of reduced vulnerability before a situation develops in which take-over by another male is likely and they would become vulnerable to infanticide. A recent comparative review of Asian langurs confirms this: where females could no longer move freely between groups, they were twice as likely to lose an infant to infanticide than where they could (Sterck, in manuscript).

This leads to a modified prediction. Having multiple males in the group or being able to disperse are complementary female social options to reduce infanticide risk. While each may have a cost, the absence of both is unlikely

Table 2. The relationship between the female breeding dispersal (absent in female-boned and female-resident taxa) and the representation of adult males in primate groups for taxa in which both social features are known[a]

Female breeding philopatry	Single-male groups[b]	Multi-male groups[b]
Present[c]	Cercopithecus non-aethiops Erythrocebus Theropithecus	Lemur catta Cebus Saimiri sciureus Cercocebus (?) Cercopithecus aethiops most Papio Macaca
Absent[d,e]	Nasalis (?) Rhinopithecus (?) Colobus badius p.p. Presbytis/Trachypithecus Papio hamadryas Gorilla g. beringei	Propithecus (?) Eulemur fulvus Saimiri oerstedi Alouatta (?) Brachyteles Ateles Colobus badius Papio ursinus p.p. Pan troglodytes

[a] Social designations compiled from various sources.
[b] A taxon is considered multi-male if many groups contain multiple males or if there is a strong positive relationships between number of females and number of males (cf. van Schaik & Hörstermann 1994).
[c] Includes female-bonded, tolerant female-bonded, and female-resident (see Table 1).
[d] Includes non-female-bonded (see Table 1).
[e] It is assumed that females in non-female-bonded show breeding dispersal in addition to the much better known natal dispersal (this assumption is known to be correct for most non-female-bonded taxa with single-male groups). *Presbytis entellus* is not included because of the possible human impact on female dispersal (Sterck, in press).

if infanticide reduction is an important objective of female social strategies. Thus, the combination of female philopatry and single-male groups should be rare.

At first sight, comparative data on primate social systems do not provide strong support for this hypothesis (Table 2). The combination of female philopatry and single adult males is found in several taxa. However, in the gelada (*Theropithecus gelada*) a second male, the 'follower', may reside in the group (Dunbar 1984), which is likely to reduce infanticide risk. In the other two taxa, guenons, *Cercopithecus* non-*aethiops*, and patas, *Erythrocebus patas*, influxes of many males during the mating season are common (Cords 1988). This phenomenon, too, may be seen as an anti-infanticide strategy, in which paternity is confused to some extent, with the effect that infanticide by new resident males is less likely. Depending on one's inclination, these phenomena may be regarded as a refutation of the initial hypothesis, or as indicating that alternative tactics can be adopted to minimize the impact of infanticide risk where the use of the common strategies is precluded. Finally, it might be argued that multi-male groups are surprisingly common among primates in general, but again the proper comparisons with representatives of other mammalian orders have not been undertaken.

In conclusion, there is some evidence in support of the notion that the number of males associated with a group of females is governed in part by the need for male protection against infanticidal males or predators. However, much more work is needed before this conclusion can be accepted unequivocally.

Infanticide risk and between-group relations

Finally, could infanticide risk have affected the nature of between-group relations? At the outset, it is interesting to note that in most species and situations, between-group relations in primates are tantamount to between-group antagonism. Theorizing to date about the possible functions of between-group antagonism has exclusively focused on defence of resources and of mates, and there is evidence for both (Cheney 1987; Kinnaird 1992; van Schaik *et al.* 1992; Cowlishaw 1995), although a thorough review of the evidence in favour of these functions is long overdue.

An additional function is possible. There is a surprising number of reports that indicate that infanticide can also occur during between-group encounters: in savanna baboons (*Papio anubis*: Shopland 1982; Collins *et al.* 1984), vervets (*Cercopithecus aethiops*: Cheney *et al.* 1988), ringtailed lemurs (*Lemur catta*: Hood 1994), hanuman langurs (Sommer 1994), gorillas (Watts 1990) and thomas langurs (R. Steenbeek, personal communication).

In fact, in some cases, such as thomas langurs, males that are already attached to a group of females make violent sneak-attacks on other groups. Such attacks may be selectively advantageous, if females are more likely to transfer into the group of the infanticidal male (because he is the best protector against future attacks). All this suggests that infanticide risk may affect between-group relations in primates.

How can this suggestion be tested? Some simple predictions can be made for the behaviour of individuals or classes during group conflicts. First, when a group contains no infants its members should be more likely to actively engage other groups. Second, females with infants are expected to hang back during encounters. These predictions have not yet been tested systematically, but preliminary support for them comes from van Schaik & Dunbar's (1990) analysis of Mitani's (1987) gibbon experiments.

Predictions on the rate and nature of between-group conflicts can also be developed. The critical prediction made by the infanticide prevention hypothesis is that the adults and infants of primate groups should avoid intermingling even when no contested resources are present and when there is no mating activity. However, developing further predictions is beset by problems. First, it is often difficult to separate mate defence and infant defence functions in group-level phenomena. Second, groups are not homogeneous units, but are composed of a multiplicity of players of both sexes, who are simultaneously pursuing different, partly incompatible, objectives. Third, between-group relations are affected by the collective action problem (see Hawkes 1992). Individuals produce, at some cost to themselves, a benefit (e.g. they acquire a resource such as food, mates, or safety for infants) to which all group members will subsequently have access. If some beneficiaries do not assist in this process, and thus do not share the cost, the best course of action of the producers of the benefit may be to stop providing the benefit, unless these free-riders are close relatives. In between-group relations, this translates into avoidance of conflicts (and perhaps the adoption of 'bourgeois'-like solutions).

A compilation of primate studies indicates that groups containing multiple adult males are indeed less likely to have conflicts with their neighbours (Figure 5), indicating between-group avoidance in multi-male groups. There are also reports of vocally mediated avoidance of close contact between groups in populations where all groups contain multiple males (e.g. Waser 1976). Given this effect it is not surprising that group composition also affects range overlap (Figure 6). This effect is retained in each of the four radiations represented in the data set. Range overlap is not correlated with the defensibility of the range, as measured by Mitani & Rodman's (1979) D-index ($R = -0.085$, $n = 24$). This result indicates that

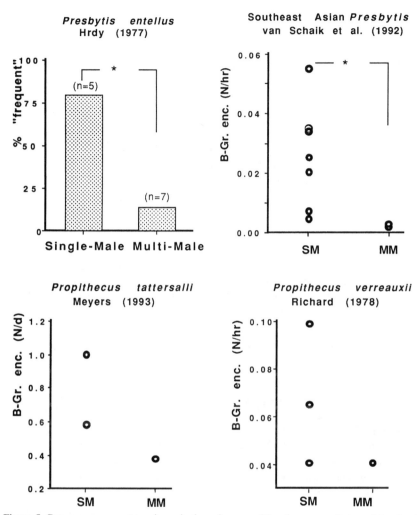

Figure 5. Between-group antagonism: single-male vs. multi-male groups. Rates of (predominantly antagonistic) encounters between groups for groups containing multiple adult males from species or populations containing groups of both kinds, in langurs and sifakas (*Propithecus spp.*). Asterisks indicate significance at the $P = 0.05$ level.

social factors, probably the collective action problem, may prevent the expression of the expected functions of between-group relations, such as resource defence. On the other hand, between-group avoidance facilitates infant defence.

In conclusion, at this time there is no solid evidence that the nature of between-group relations serves to reduce infanticide risk, but I believe the suggestion deserves serious scrutiny in the future.

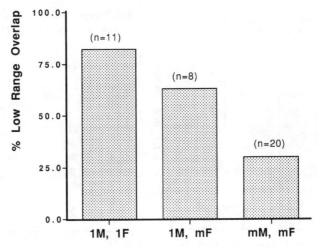

Figure 6. Group composition and range overlap. The percentage of studies with low home range overlap in relation to group composition (M = male, F = female; m = multiple). Low overlap is defined here as less than 30% qualitative overlap, corresponding to approximately the highest overlap shown by behaviourally territorial groups of gibbons (van Schaik & Dunbar 1990). The data are taken from the two most extensive published compilations (Mitani & Rodman 1979; Cheney 1987).

DISCUSSION

The assumptions of socio-ecology

Beginning with Crook & Gartlan (1966), socio-ecology has adopted an adaptationist stance, and has focused on the effect of ecological factors, in particular the abundance and distribution of food and predators, on female distribution and relationships, and on the effect of the spatio-temporal distribution of mating opportunities in determining male distribution and relationships. More recently, the role of social forces is increasingly highlighted, especially harassment of females and killing of infants by males (e.g. Wrangham 1979; Smuts & Smuts 1993). This paper has attempted to integrate the impacts of social and ecological forces on social features. While it is still early, there seems to be increasing evidence that infanticide risk is a selective force similar in strength to ecological factors, with which it may interact in complex ways.

It has often been stressed that social structure and organization are correlated with phylogeny (Struhsaker 1969; Di Fiore & Rendall 1994). The influence of phylogeny can be subsumed under the socio-ecological approach, if it is thought to reflect its correlation with features of morphology, physiology, life style and life history, all of which mediate the impact of external factors (cf. Harvey & Pagel 1991). For instance, dispersal opportunities will be severely curtailed in animals that rely on elaborate dwellings that require a major collective effort to build (e.g. Waser 1988). Likewise, variation in altriciality or litter size constrain the social options of adults. More subtly, phylogenetic position may affect cognitive capabilities, and thus the possible complexity of social relationships (Cheney & Seyfarth 1990).

The true alternative to the adaptationist approach of socio-ecology is the long-standing critical undercurrent that essentially considers social behaviour to express adaptively neutral variation whose correlation with phylogeny is entirely due to common descent (Rowell 1979; Di Fiore & Rendall 1994), or assumes that social inventions, while adaptive, are so rare that only a small and arbitrary set of taxa will have them (Thierry 1990). The adaptive approach is adopted here especially because it is more heuristic in that testable hypotheses are more easily framed and tested. To the extent that these hypotheses are supported by empirical data, this vindicates the adaptive approach. However, as shown by the discussion of female philopatry and male representation in primate groups, one of the greatest obstacles in testing these hypotheses is the occurrence of alternative strategies and tactics which may have arisen where the common strategy was less effective or too costly. This seemingly endless list of functionally equivalent alternative strategies introduces an element of faith into the adoption of the approach.

Testing socio-ecological models

Socio-ecological hypotheses are evolutionary models. Some tests of these models have employed experimental manipulations of ecological conditions (e.g. Gore 1993). For such manipulations to work, the predicted social change must be within the norm of reaction of the species. The increased appreciation of the existence of alternative strategies (Dunbar 1983) and the resulting flexibility (e.g. Hamilton et al. 1976) has led to the assumption of near-infinite behavioural flexibility (Dunbar 1989). However, many species maintain a similar social system in captivity, despite wide variation in conditions; and studies of hybrid baboons indicate that some social behaviour cannot easily be modified by short-term experience (e.g. Nagel 1973). An additional technical problem with experiments is that animals may not be able to interpret the modified conditions correctly when they do not last long enough or alternate with other conditions (cf. Berger 1988).

Lack of phenotypic plasticity is less likely to plague experiments that manipulate the social or demographic context because variation in these variables is more likely at ecological time scales. There is surprisingly little documentation of the social consequences of intraspecific variation in group composition, group density, etc.

Regardless, comparisons between taxa remain one of the most powerful ways to test these models, and the preliminary comparisons presented here attest to their suitability. However, the limited phenotypic plasticity referred to above means that the populations compared must live in undisturbed habitats, so as to ensure we study the impact of ecological and demographic conditions that prevailed during history rather than the impact of unintended and undocumented recent experimental alterations. This caveat is not a gratuitous one: habitat disturbance and modification and fragmentation, leading to unbalanced ecological communities and sometimes to hyperabundance of primates where they are protected, are beginning to affect many of the field sites where the data are collected that are used in these comparisons. For instance, some of the observed behavioural differences between the chimpanzee populations of Gombe and Mahale are likely due to the small size and lack of dispersal opportunities at Gombe relative to the more natural situation at Mahale (Nishida *et al.* 1990).

Note. I thank Robert Barton, Tim Clutton-Brock, Guy Cowlishaw, Robin Dunbar, Beth Fox, Charles Nunn, Romy Steenbeek, Liesbeth Sterck, Jan van Hooff, Maria van Noordwijk, and Frances White for useful discussion and personal communication of unpublished material.

REFERENCES

Altmann, J. 1990: Primate males go where the females are. *Animal Behaviour* 39, 193–195.
Berger, J. 1988: Social systems, resources, and phylogenetic inertia: an experimental test and its limitations. In *The Ecology of Social Behavior* (ed. C. N. Slobodchikoff), pp. 157–186. San Diego: Academic Press.
Cheney, D.L. 1987: Interactions and relationships between groups. In *Primate Societies* (ed. B.B. Smuts, D.L. Cheney, R.M. Seyfarth, R.W. Wrangham & T.T. Struhsaker), pp. 267–281. Chicago: University of Chicago Press.
Cheney, D.L. & Seyfarth, R.M. 1990: *How Monkeys See the World*. Chicago: Chicago University Press.
Cheney, D.L., Seyfarth, R.M., Andelman, S.J. & Lee, P.C. 1988: Reproductive success in vervet monkeys. In *Reproductive Success* (ed. T.H. Clutton-Brock), pp. 384–402. Chicago: The University of Chicago Press.
Clutton-Brock, T.H. 1989: Mammalian mating systems. *Proceedings of the Royal Society of London, B* 236, 339–372.
Collins, D.A., Busse, C.D. & Goodall, J. 1984: Infanticide in two populations of savanna baboons. In *Infanticide: comparative and evolutionary perspectives* (ed. G. Hausfater & S.B. Hrdy), pp. 193–215. New York: Aldina Publ. Co.
Cords, M. 1988: Mating systems of forest guenons: a preliminary review. In *A Primate Radiation: evolutionary biology of the African guenons* (ed. A. Gauthier-Hion, F. Bourliere, J.P. Gauthier & J. Kingdon), pp. 323–339. Cambridge: Cambridge University Press.

Cowlishaw, G. 1995: Behavioural patterns in baboon group encounters: the role of resource competition and male reproductive strategies. *Behaviour* 132, 75–86.

Crockett, C.M. & Sekulic, R. 1984: Infanticide in red howler monkeys (*Alouatta seniculus*). In *Infanticide: comparative and evolutionary perspectives* (ed. G. Hausfater & S.B. Hrdy), pp. 173–191. New York: Aldine Publ. Co.

Crook, J.H. & Gartlan, J.C. 1966: Evolution of primate societies. *Nature* 210, 1200–1203.

de Waal, F.B.M. 1986: The integration of dominance and social bonding in primates. *Quarterly Review of Biology* 61, 459–479.

Di Fiore, A. & Rendall, D. 1994: Evolution of social organization: a reappraisal for primates by using phylogenetic methods. *Proceedings of the National Academy of Sciences* 91, 9941–9945.

Dunbar, R.I.M. 1983: Life history tactics and alternative strategies of reproduction. In *Mate Choice* (ed. P.P.G. Bateson), pp. 423–434. Cambridge: Cambridge University Press.

Dunbar, R.I.M. 1984: *Reproductive Decisions—an economic analysis of gelada baboon social strategies*. Princeton: Princeton University Press.

Dunbar, R.I.M. 1988: *Primate Social Systems*. Ithaca: Cornell University Press.

Dunbar, R.I.M. 1989: Social systems as optimal strategy sets: the costs and benefits of sociality. In *Comparative Socioecology* (ed. V. Standen & R.A. Foley), pp. 131–150. Oxford: Blackwell.

Emlen, S.T. & Oring, L.W. 1977: Ecology, sexual selection, and the evolution of mating systems. *Science* 197, 215–223.

Foley, R.A. & Lee, P.C. 1989: Finite social space, evolutionary pathways, and reconstructing hominid behavior. *Science* 243, 901–906.

Goodall, J. 1986: *The Chimpanzees of Gombe*. Cambridge, MA: Harvard University Press.

Gore, M. 1993: Effects of food distribution on foraging competition in rhesus monkeys, *Macaca mulatta*, and hamadryas baboons, *Papio hamadryas*. *Animal Behaviour* 45, 773–786.

Hamilton, W.J. & Bulger, J. 1992: Facultative expression of behavioral differences between one-male and multimale savanna baboon groups. *American Journal of Primatology* 28, 61–71.

Hamilton, W.J., Buskirk, R.E.R. & Buskirk, W.H. 1976: Defense of space and resources by chacma (*Papio ursinus*) baboon troops in an African desert and swamp. *Ecology* 57, 1264–1272.

Harvey, P. & Pagel, M. 1991: *The Comparative Method in Evolutionary Biology*. Oxford: Oxford University Press.

Harvey, P., Martin, R.D. & Clutton-Brock, T.H. 1987: Life histories in comparative perspective. In *Primate Societies* (ed. B.B. Smuts, D.L. Cheney, R.M. Seyfarth, R.W. Wrangham & T.T. Struhsaker), pp. 181–196. Chicago: University of Chicago Press.

Hawkes, K. 1992: Sharing and collective action. In *Evolutionary Ecology and Human Behavior* (ed. E.A. Smith & B. Winterhalder), pp. 269–300. New York: Aldine de Gruyter.

Hiraiwa-Hasegawa, M. & Hasegawa, T. 1994: Infanticide in nonhuman primates: sexual selection and local resource competition. In *Infanticide and Parental Care* (ed. S. Parmigiani & F.S. vom Saal), pp. 137–154. London: Harwood Academic Publishers.

Hood, L.C. 1994: Infanticide among ringtailed lemurs (*Lemur catta*) at Berenty reserve, Madagascar. *American Journal of Primatology* 33, 65–69.

Hrdy, S.B. & Whitten, P.L. 1987: Patterning of sexual activity. In *Primate Societies* (ed. B.B. Smuts, D.L. Cheney, R.M. Seyfarth, R.W. Wrangham & T.T. Struhsaker), pp. 370–384. Chicago: University of Chicago Press.

Ims, R.A. 1988: Spatial clumping of sexually receptive females induces space sharing among male voles. *Nature* 335, 541–543.

Janson, C.H. 1992: Evolutionary ecology of primate social structure. In *Evolutionary Ecology and Human Behavior* (ed. E.A. Smith & B. Winterhalder), pp. 95–130. New York: Aldine de Gruyter.

Kinnaird, M.F. 1992: Variable resource defense by the Tana River crested managabey. *Behavioral Ecology and Sociobiology* 31, 115–122.

Manson, J. & Wrangham, R.W. 1991: Intergroup aggression in chimpanzees and humans. *Current Anthropology* 32, 369–390.

Mitani, J.C. 1987: Territoriality and monogamy among agile gibbons (*Hylobates agilis*). *Behavioral Ecology and Sociobiology* 20, 265–269.

Mitani, J.C. & Rodman, P.S. 1979: Territoriality: the relation of ranging pattern and home range size to defendability, with an analysis of territoriality among primate species. *Behavioral Ecology and Sociobiology* 5, 241–251.

Mitchell, C.L., Boinski, S. & van Schaik, C.P. 1991: Competitive regimes and female bonding in two species of squirrel monkeys (*Saimiri oerstedi* and *S. sciureus*). *Behavioral Ecology and Sociobiology* 28, 55–60.

Mori, A. 1979: Analysis of population changes by measurement of body weight in the Koshima troop of Japanese monkeys. *Primates* 20, 371–397.

Nagel, U. 1973: A comparison of anubis baboons, hamadryas baboons, and their hybrids at a species border in Ethiopia. *Folia Primatologica* 19, 104–165.

Newton, P.N. 1986: Infanticide in an undisturbed forest population of hanuman langurs, *Presbytis entellus*. *Animal Behaviour* 34, 785–789.

Nishida, T. & Hiraiwa-Hasegawa, M. 1987: Chimpanzees and bonobos: cooperative relationships among males. In *Primate Societies* (ed. B.B. Smuts, D.L. Cheney, R.M. Seyfarth, R.W. Wrangham & T.T. Struhsaker), pp. 165–177. Chicago: University of Chicago Press.

Nishida, T., Takasaki, H. & Takahata, Y. 1990: Demographic and reproductive profiles. In *The Chimpanzees of the Mahale Mountains: sexual and life history strategies* (ed. T. Nishida), pp. 63–97. Tokyo: University of Tokyo Press.

Noë, R. 1990: A veto game played by baboons: a challenge to the use of the prisoners's dilemma as a paradigm for reciprocity and cooperation. *Animal Behaviour* 39, 78–90.

Ostfeld, R.S. 1990: The ecology of territoriality in small mammals. *Trends in Ecology and Evolution* 5, 411–415.

Pereira, M.E. & Kappeler, P.M. (In press). Divergent systems of agonistic relationship in lemurid primates. *Behaviour*.

Pusey, A.E. & Packer, C. 1994: Infanticide in lions: consequences and counterstrategies. In *Infanticide and Parental Care* (ed. S. Parmigiani & F.S. vom Saal). London: Harwood Academic Publishers.

Richard, A.F. 1985: *Primates in Nature*. New York: Freeman and Co.

Robbins, M.M. 1995: A demographic analysis of male life history and social structure of mountain gorillas. *Behaviour* 132, 21–47.

Rowell, T.E. 1979: How would we know if social organization were not adaptive? In *Primate Ecology and Social Organization* (ed. I.S. Bernstein & E.O. Smith), pp. 1–22. New York: Garland.

Shopland, J.M. 1982: An intergroup encounter with fatal consequences in yellow baboons. *American Journal of Primatology* 3, 263–266.

Silk, J.B. 1993: The evolution of social conflict among female primates. In *Primate Social Conflict* (ed. W.A. Mason & S.P. Mendoza), pp. 49–83. New York: State University of New York.

Smuts, B.B. 1987: Sexual competition and mate choice. In *Primate Societies* (ed. B.B. Smuts, D.L. Cheney, R.M. Seyfarth, R.W. Wrangham & T.T. Struhsaker), pp. 385–399. Chicago: University of Chicago Press.

Smuts, B.B., Cheney, D.L., Seyfarth, R.M., Wrangham, R.W. & Struhsaker, T.T. (eds.). 1987: *Primate Societies*. Chicago: Chicago University Press.

Smuts, B.B. & Smuts, R.W. 1993: Male aggression and sexual coercion of females in nonhuman primates and other mammals: Evidence and theoretical implications. *Advances in the Study of Behavior* 22, 1–63.

Sommer, V. 1994: Infanticide among the langurs of Jodhpur: testing the sexual selection hypothesis with a long-term record. In *Infanticide and Parental Care* (ed. S. Parmigiani & F.S. vom Saal), pp. 155–198. London: Harwood Academic Publishers.

Sterck, E.H.M. (In press) Determinants of female transfer in Thomas langurs. *American Journal of Primatology.*

Sterck, E.H.M., Watts, D.P. & van Schaik, C.P. (In manuscript) The evolution of female social relationships in nonhuman primates.

Strier, K.B. 1992: *Faces in the Forest: the endangered muriqui monkeys of Brazil.* New York: Oxford University Press.

Struhsaker, T.T. 1969: Correlates of ecology and social organization among African cercopithecines. *Folia Primatologia* 11, 80–118.

Sugiyama, Y. 1966: An artificial social change in a hanuman langur troop (*Presbytis entellus*). *Primates* 7, 41–72.

te Boekhorst, I.J.A. & Hogeweg, P. 1984: Self-structuring in artificial "CHIMPS" offers new hypotheses for male grouping in chimpanzees. *Behaviour* 130, 229–252.

Thierry, B. 1990: Feedback loop between kinship and dominance: the macaque model. *Journal of Theoretical Biology* 145, 511–521.

Trivers, R.L. 1972: Parental investment and sexual selection. In *Sexual Selection and the Descent of Man* (ed. B. Campbell), pp. 136–179. Chicago: Aldine.

van Schaik, C.P. 1983: Why are diurnal primates living in groups? *Behaviour* 87, 120–144.

van Schaik, C.P. 1989: The ecology of social relationships amongst female primates. In *Comparative Socioecology* (ed. V. Standen & R.A. Foley), pp. 195–218. Oxford, Blackwell.

van Schaik, C.P., Assink, P.R. & Salafsky, N. 1992: Territorial behavior in Southeast Asian langurs: resource defense or mate defense? *American Journal of Primatology* 26, 233–242.

van Schaik, C.P. & Dunbar, R.I.M. 1990: The evolution of monogamy in large primates: a new hypothesis and some crucial tests. *Behaviour* 115, 30–62.

van Schaik, C.P. & Hörstermann, M. 1994: Predation risk and the number of adult males in a primate group: A comparative test. *Behavioral Ecology and Sociobiology* 35, 261–272.

van Schaik, C.P. & van Noordwijk, M.A. 1989: The special role of male *Cebus* monkeys in predation avoidance and its effect on group composition. *Behavioral Ecology and Sociobiology* 24, 265–276.

Waser, P.M. 1976: *Cercocebus albigena*: site attachment, avoidance, and intergroup spacing. *American Naturalist* 110, 911–935.

Waser, P.M. 1988: Resources, philopatry, and social interactions among mammals. In *The Ecology of Social Behavior* (ed. C.N. Slobodchikoff), pp. 109–130. San Diego: Academic Press.

Watts, D.P. 1990: Ecology of gorillas and its relation to female transfer in mountain gorillas. *International Journal of Primatology* 11, 21–45.

Wilson, E.O. 1975: *Sociobiology.* Cambridge: Belknap Press.

Wrangham, R.W. 1979: On the evolution of ape social systems. *Social Sciences Information* 18, 334–368.

Wrangham, R.W. 1980: An ecological model of female-bonded primate groups. *Behaviour* 75, 262–300.

Wrangham, R.W. 1987: Evolution of social structure. In *Primate Societies* (ed. B.B. Smuts, D.L. Cheney, R.M. Seyfarth, R.W. Wrangham & T.T. Struhsaker), pp. 282–297. Chicago: University of Chicago Press.

Determinants of Group Size in Primates: A General Model

R. I. M. DUNBAR

*Department of Psychology, University of Liverpool,
PO Box 147, Liverpool, L69 3BX*

Keywords: group size; systems model; time budgets; baboons; chimpanzees.

Summary. Significant constraints are placed on group size by local habitat conditions as a consequence of both the selection pressures that act on the animals and the design of their physiological systems. I use a linear programming approach to develop a model of habitat-specific minimum and maximum group sizes for baboons. Three main variables define the state space of realizable group sizes. These are the maximum group size within which the animals can still balance their time budgets (the maximum ecologically tolerable group size), the minimum group size that reduces predation risk to some (undefined) acceptable level (the minimum permissible group size) and the maximum group size that animals' neocortex size will allow them to maintain as a coherent stable social entity (the cognitive group size). Similar models have also been developed for gelada and chimpanzees. Once group size can be determined for a particular habitat, a number of other behavioural patterns can be determined as a consequence of well-understood general principles. I illustrate this with the example of male mating strategies.

OVER THE LAST TWO DECADES, our understanding of primate behaviour, and the selection forces acting on it, has grown spectacularly, thanks largely to the shift in emphasis generated by sociobiology during the early 1970s.

© The British Academy 1996.

Being able to view behavioural interactions from a strategic (or goal-directed) perspective opened up layers of complexity in animal behaviour that the traditional stimulus-response analyses of classical ethology had been unable to tap.

What we have tended to overlook, however, is the fact that many animals live in groups. The size, composition and dispersion of groups imposes limits on the range of options open to any given individual. These demographic factors are, in turn, largely a consequence of the local ecology interacting with the species' ecological adaptations.

It is important to understand that the optimal group size is habitat-specific: it is a consequence of the way in which the particular environmental and climatic variables characteristic of a given habitat influence the behavioural ecology of the animal in question. Hitherto, attempts to explain social evolution have too often tended to view mating systems and grouping patterns as species-specific phenomena. Variance around the species typical value has often been viewed as little more than inevitable biological error. The assumption adopted here is that variations in group size are a direct consequence of optimisation decisions by the animals. Indeed, it is precisely the variation in group size across habitats and, within habitats, through time that we are trying to explain.

Group size is, of course, a consequence of decisions made by animals about the optimal size for groups in a given habitat. The costs and benefits on which this decision is based are a function of local environmental conditions. Strictly speaking, of course, the choice of optimal group size may itself be a consequence of the costs and benefits of the behavioural options open to an animal in terms of mating and parenting. Group size may, for example, influence fertility rates (van Schaik 1983), and so alter the anticipated gains of different mating strategies. Nonetheless, there is a useful sense in which we can see decisions about group size as antecedent to the decisions that an animal makes about mate choice and parenting effort. In effect, these strictly behavioural decisions are made in the context of prior decisions about grouping patterns (Dunbar 1988).

In this paper, I summarize our attempts to build functional models of primate socio-ecological systems designed to explore these issues. Unlike the micro-economic models characteristic of much of behavioural ecology over the past 30 years, these models owe more to the macro-economic approach favoured by the systems ecologists of the 1960s. Their principal purpose is to allow us to explore the relationship between environmental parameters and demographic variables. While we understand that the relationships involved are in fact mediated by conventional optimality decisions, our interest lies not in the optimization processes themselves (though these must ultimately be part of the story) but in the *consequences* of these decisions. Once we can understand these, we will have a much clearer idea of the systemic constraints that act on individuals' choices at the strategic level.

The model I outline here is based on studies of baboons. We are, however, also building similar models for several other taxa. I shall allude to an earlier model developed for gelada, as well as to one currently being developed for chimpanzees by Daisy Williamson (1996). In addition, we have also started work on a similar model for gibbons (Sear 1994). In each case, I conceive the core problem as identifying the determinants of group size. Once group size is determined, a number of fairly straightforward lifehistory considerations dictate the composition and reproductive characteristics of the group.

A LINEAR PROGRAMMING MODEL OF GROUP SIZE

We can approach the problem of optimal group sizes by considering it as a linear programming model. This assumes that the optimal group size is the intersection of a set of benefit and cost equations together with a number of constraints. These create a region of possible group sizes (the range of realizable group sizes) within the state space created by the range of conceivable group sizes: the zone of realizable group sizes must lie above the line generated by the benefit equation(s), and below those generated by the cost and constraint equations.

The basic model is shown in Figure 1. Here, group size is plotted against a notional environmental variable (e.g. rainfall). Strictly speaking, this should be a three-dimensional graph with fitness (or lifetime reproductive success) emerging out of the page at right angles to the other two dimensions. Each of the cost, benefit and constraint curves can then be represented more correctly as a surface in three-dimensions. Since this is difficult to show, it is simpler to illustrate the main points when the surfaces are represented by their projections in two dimensions. All points on a line are thus isometric with respect to fitness: each line represents the point at which fitness drops below some minimally acceptable level. Fitness increases above the benefit curve and it also increases below the cost and constraint curves.

The curves themselves need some interpretation. I assume that the benefit curve represents the selective advantage(s) of group size. For primates, these are usually understood to be either protection from predation risk or defence of resources (van Schaik 1983; Wrangham 1980, 1987; Dunbar 1988). Irrespective of which of these hypotheses is in fact correct, fitness increases with increasing group size. This is because both the risk of predation and the defendability of a resource increase monotonically

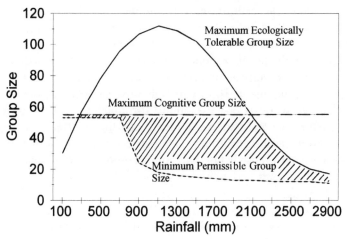

Figure 1. Linear programming model of group size. Group size is plotted against a notional environmental variable (in this case, mean annual rainfall). The range of realizable group sizes (*hatched area*) is defined by the minima set by the benefit variables and the maxima set by the cost and constraint variables. In this model, only one graph is shown for each type of variable. The plotted values for the maximum and minimum group sizes are those predicted by the baboon model at 25°C (approximate centre of the taxon's preferred thermal zone). The cognitive constraint is that set by the relationship between observed *mean* group size and relative neocortex volume, and is almost certainly an underestimate of the true maximum cognitive group size.

with the size of the group. However, predation risk will always set a lower limit on the minimum group size than the demands of territorial defence, because the latter will inevitably tend to force the population into an arms race in which minimum group sizes are driven upwards towards the maximum tolerable by the spiralling effects of between-group competition. I define the size of group that reduces predation risk to some constant tolerable level as the *minimum permissible group size*. This is shown as the lower broken line in Figure 1.

However, the benefits that derive from grouping are necessarily offset by the costs of grouping. These come in two distinct kinds: direct (those due to competition and harassment, often reflected in reduced fecundity) and indirect (those that arise from the additional costs of servicing larger groups, such as increased day journey lengths and the additional feeding time required to fuel these, as well as the social time required to ensure the social cohesion of the group). Taken together, these will impose an upper limit on the size of the group that can remain together. This will arise partly from the marginal cost of reduced fecundity that females are unwilling to bear when group size exceeds some critical threshold. However, an important constraint may be imposed by the fact that the increased feeding, moving and social time requirements demanded by large groups may exceed the total time the animals have available during the day. The limiting value on group size at which all spare resting time (the only category containing free or 'convertible' time) has been assigned to feeding, moving and social interaction is defined as the *maximum ecologically tolerable group size* (see Dunbar 1992a). It is graphed as the upper (solid) line in Figure 1.

Finally, since we are dealing with the proximate mechanisms of the system, there is likely to be a cognitive constraint on the number of relationships that any one individual can maintain, and this in turn will impose an upper limit on group size. This appears to be a consequence of the size of the neocortex (Dunbar 1992b). From an evolutionary point of view, of course, there is no cognitive constraint: species will presumably evolve brains that are big enough to ensure cohesion in the size of group that their ecological circumstances typically demand. However, seen from a particular animals' point of view, brain size is not a variable that can be manipulated in the here-and-now, and it must therefore be taken as a constraint. The cognitive limit on group size is shown as the horizontal dashed line in Figure 1.

Our task, then is to put the flesh on the bones of this model by determining exactly what the relevant relationships are. I have approached this problem largely using forward stepwise regression. This is partly because we are searching for the set of equations that provide the best predictive power. However, in doing so, I have also remained mindful of biological plausability. In other words, I have asked of each equation: does it make biological sense? Can we provide an explanation in terms of biological first principles for why such a relationship should exist?

The principal analysis has been carried out on *Papio* baboons. The choice of taxon was dictated mainly by the fact that this genus had been subjected to more intensive quantitative study over the past three decades than any other primate taxon. However, a parallel study was run on the gelada (*Theropithecus gelada*). Although this species has been studied at only three sites, the fact that it is the most closely related taxon to the *Papio* baboons meant that it was reasonable to assume that it responded physiologically in rather similar ways. At the same time, the fact that gelada are grazers, whereas baboons are principally frugivores, provided a unique basis for examining the impact of dietary niche on the various system components.

In the meantime, we have begun work on two more taxa, the gibbons (*Hylobates* spp.) and the chimpanzees (*Pan* spp.). These were selected partly because of the number of field studies that are available and partly because they represent interesting extremes of ecological and lifehistory adaptation. Chimpanzees live in large polygamous fission-fusion social systems, but are dietetically much more restricted than baboons (essentially ripe fruit frugivores); as such, they provide a further insight into the influence of dietary niche. Gibbons are medium-sized strictly arboreal frugivores that live in small (monogamous) groups, and thus provide an insight into the constraints imposed by arboreality. Gibbons also contrast with the other species in being non-African, thereby allowing us to examine the impact of different kinds of forest environment on behavioural ecology and grouping patterns.

In all these analyses, we have ignored taxonomic differences at the species level. Species of the same primate genus differ principally in terms of their body weight and fine details of social behaviour. Differences in reproductive parameters and ecological niche are invariably minimal or non-existent, with variance in dietary and other behavioural variables being much greater between populations of the same species than between species (Dunbar 1992a). Moreover, in many cases (notably the baboons), species often seem to constitute a geographical cline rather than good biological species. Indeed, there is some evidence to suggest that the genetic distances between species of the same genus for many catarrhine primates is of a magnitude that would warrant only subspecific status in other Orders (Shotake *et al.* 1977, Kawamuto *et al.* 1982).

CLIMATIC VARIABLES

The basic model assumes that group size, day journey length, activity patterns and various environmental variables are all inter-related in a complex web of cause-effect relationships. Although the density and dispersion (patchiness) of vegetation (especially food species) are likely to be important environmental variables driving behaviour, these data are rarely available in the literature. However, these aspects of plant biology are themselves determined by climatic variables such as temperature and rainfall, and these variables are widely available (either for the study sites themselves or from nearby weather stations). It consequently seemed reasonable to bypass the intermediate steps and relate behavioural variables directly to the climatic variables.

My original analyses (Dunbar 1992a) were based on the use of four key climatic variables: mean ambient temperature, total annual rainfall and two measures of rainfall dispersion (an evenness index for monthly rainfall and the number of months in the year that received less than 50 mm of rainfall). I chose these variables mainly because they were widely available or easy to calculate given the data available. Daisy Williamson (1996) has since undertaken a very detailed analysis of data for 218 weather stations randomly chosen throughout sub-Saharan Africa from Wernstedt's (1972) *World Climatic Data*. She carried out a principal components analysis of nine weather variables and found that they clustered on just three key dimensions: mean annual temperature, total annual rainfall and rainfall dispersion (seasonality). All three indices are important determinants of the growing conditions for plants, and hence of primary productivity. Although other variables (including soil type and aspect, temperature variation, relative humidity, evapo-transpiration) are known to be important determinants of vegetation growth, very few sites provide enough information to include these in a comparative analysis. Moreover, Williamson's analysis, combined with the broad success of the models based on just these three key parameters, suggests that the net gain from increasing the level of environmental data is likely to be marginal.

Only with respect to one variable is there a real problem, and that is the quantity of standing water (or water table level). It is clear that, whenever permanent water provides sufficient vegetation at the micro-habitat level, baboons can survive in extreme habitats (e.g. the Namib desert) that would otherwise be incapable of supporting them. Although in principle it would be possible to include standing water as a factor in the equations (its effect would be the equivalent of raising the value for rainfall: see Dunbar 1993a), in practice we do not yet have an easy way of assessing its impact.

MAXIMUM ECOLOGICALLY TOLERABLE GROUP SIZE

I assume that the main habitat-dependent constraint on grouping is imposed by the inelasticity of the time budget. In effect, this is equivalent to the indirect costs of grouping (i.e. the marginal moving, feeding and social time costs required to sustain an additional increment in group size). Although these are strictly speaking costs of grouping, they can conveniently be thought of as imposing an upper limit on group size. This upper limit occurs when all spare time has been allocated to those activities needed to enable group size to be increased (the upper line in Figure 1).

Approximately 95% of an animal's waking time is devoted to just four categories of activity (feeding, moving, social interaction and resting). Other activities (e.g. territorial defence, drinking, monitoring the environment) occupy a negligible proportion of the time budget and can be ignored for present purposes.

I assume that feeding time is largely determined by the animal's body weight (following from Kleiber's Law), environmental variables that determine the costs of thermoregulation, nutrient availability in plants and the energy costs of travel. Time spent moving is assumed to be a function of vegetation dispersion (patchiness), ambient temperature and day journey length. Because social time is associated with the maintenance of social bonds (and hence the cohesion of groups), I assume that the amount of time devoted to social interaction (principally social grooming) is a monotonic (but not necessarily linear) function of group size.

Finally, I assume that resting time acts as a reserve of uncommitted time that animals can draw on when they need to increase the time allocation to any of the other three categories. Analyses of time budgets for different populations (Dunbar & Sharman 1984, Dunbar 1992a), for seasonal differences within habitats (Dunbar 1992a) and for mothers responding to the escalating energy demands of growing infants (e.g. Dunbar & Dunbar 1988) demonstrate that additional feeding time requirements are invariably taken first from resting time. Only once resting time reaches some minimum threshold is additional feeding time taken from elsewhere (normally social time: see also Altmann 1980). However, moving time can also provide some capacity in this respect in that savings of time may be achieved by travelling faster. There is evidence to suggest that baboons do travel faster as the environment deteriorates: as a result, moving time remains more or less constant across habitats despite changes in group size, even though day journey varies across populations by an order of magnitude (Dunbar 1992a). In the present analyses, this form of time-saving is in fact already incorporated into the data on moving time: moving time as we observe it is the net value *after* the animals have made all the adjustments they want to make.

In the original model (Dunbar 1992a), I determined linear regression equations for day journey length and time budget variables from a set of 14 study sites for which data were available, with an additional four sites providing data on day journey length but not time budgets. (Note that 18 sites were used for day journey length, not 21 as implied in Dunbar 1992a.) The effects of body weight were incorporated into the analyses, since it has been shown that baboon body weights vary systematically with rainfall and temperature (Dunbar 1990).

These equations were checked by using them to predict the time budgets of four other study sites not used in the original regression analyses: the mean difference between observed and predicted values was $z = 0.44$ (with only one of 16 values having $P < 0.05$: if this value is omitted, the mean difference between the remaining 15 observed and predicted values is only $z = 0.28$).

We have since been able to improve the data on climatic variables (notably for the Amboseli and Ruaha sites), and the equations were rerun to check for a better fit in each case. The new equation for feeding time is as follows:

$$\ln(F) = 6.866 + 4.077 \ln(Z) - 0.750 \ln(T)$$
$$-0.390 \ln(V) + 0.155 \ln(J)$$

where F is the percentage of time devoted to feeding, Z is Simpson's index of evenness for monthly rainfall across the year, V is the number of dry months (i.e. months with less than 50 mm of rainfall) and J is the length of the day journey (in km). The moving time equation remains unchanged from that given in Dunbar (1992a).

The final step is to determine the maximum ecologically tolerable group size for any given habitat. This was done by determining the group size at which all available spare resting time has been allocated to feeding, moving and social activity (i.e. resting time is at the minimum value specific for that habitat).

In doing this, I used different equations for resting and social time to those derived from the stepwise analysis of time budgets. I argued that these two variables are of lower ecological priority than feeding and moving. Whereas feeding and moving time requirements are dictated in a rather strict way by environmental and demographic variables (and are thus beyond the control of the animals), the animals have rather more control over whether or not they invest in resting and social time. Hence, the observed values for resting and social time are likely to represent the compromise values *after* the animals have evaluated the difference between what they *ought* to do and what they think they can get away with in order to spare more time for feeding and moving. In marginal habitats, the ability to compromise on the strict demands for resting and social activity may mean the difference between being able to survive in that habitat and not being able to do so.

I assumed that the primary constraint on resting time is the need to seek shelter when ambient temperatures rise above a crucial threshold around the middle of the day. In order to estimate this, I reran the stepwise regression for resting time with all time budget and demographic variables excluded. This yielded a best-fit equation in the two indices of the seasonality of rainfall, which I interpret as reflecting seasonal temperature load (the rainfall diversity index is largely a function of ambient temperature) and the availability of cover (length of dry season is a key determinant of bush cover). I use this equation as an attempt to identify the environmentally determined minimum resting requirement. The new equation for resting time using the updated database is:

$$\ln(R) = 0.97 - 7.923 \ln(Z) + 0.601 \ln(V)$$

where R is the percentage of time devoted to resting during the day. As in Dunbar (1992a), resting time is subject to a minimum value of 5%.

In the case of social time, I assumed that the primary concern is the amount of time required to service relationships. Previous analyses of grooming time allocations by Old World monkeys and apes (Dunbar 1991) suggested that social time increases with group size (perhaps reflecting the need to service proportionately more relationships as group size increases). In the original model (Dunbar 1992a), I set a linear regression to the data on species grooming time allocations given in Dunbar (1991). However, the data suggest that a nonlinear equation may be more appropriate. For the present version of the model, I therefore reanalysed the data for all catarrhine primates for which data are given by Dunbar (1991). The following quadratic equation provided the best fit:

$$\ln(S) = -2.275 + 1.32 \ln(N) - 0.0445(\ln(N))^2$$

($r^2 = 0.997$, $N = 13$ generic means for Catarrhine primates).

In addition, a number of additional changes were introduced that improves the biological validity of the model compared to the original version given in Dunbar (1992a). Jeanne Altmann has pointed out to me that the original model generates impossible values of Z (the index of rainfall diversity): an upper limit of $Z = 0.9167$ (the maximum possible value for a set of 12 months) was therefore imposed. In addition, it was possible for significant values of maximum group size to be obtained at temperatures in excess of 40°C. In practice, non-fossorial animals cannot physically survive in habitats where mean ambient temperatures exceed about 35°C (Peter Wheeler, personal communication). An upper limit on survival was therefore placed at 35°C. Finally, the original equation for feeding time incorporated a negative relationship between feeding time and ambient temperature (reflecting the costs of thermoregulation in low temperature environments). Strictly speaking, energy consumption does not decrease indefinitely as temperatures rise, but rather starts to *increase* again once ambient temperature exceeds 30°C (Mount 1979). I therefore amended the feeding time equation so that it was symmetrical about the 30°C point, with an absolute cut-off at an ambient temperature of 35°C. For temperatures exceeding 30°C, the feeding time equation was modified to reverse its slope against temperature as follows:

$$\ln(F) = 1.768 + 4.077 \ln(Z) + 0.750 \ln(T) - 0.390 \ln(V)$$
$$+ 0.155 \ln(J)$$

This equation simply sets a new intercept at $T = 30°C$ and then reverses the sign of the slope parameter for T.

The simulation resulting from this analysis yields a maximum ecologically tolerable group size that is habitat-specific. For ease of presentation, these are given against just two habitat variables, total annual rainfall and mean ambient temperature, in Table 1. In order to do this, it was necessary to reduce the original four independent climatic variables to two. This was possible

Table 1. Maximum ecologically tolerable group size, N_{max}, for baboons under different climatic conditions.

Rainfall (mm)	Annual temperature (°C)							
	0	5	10	15	20	25	30	35
100	0	0	9	23	31	30	22	7
300	0	0	15	39	54	58	52	57
500	0	1	18	48	70	79	79	57
700	0	1	19	52	80	96	101	80
900	0	1	19	53	84	107	119	100
1100	0	0	16	51	84	112	131	114
1300	0	0	12	45	79	109	136	121
1500	0	0	8	35	70	102	131	118
1700	0	0	4	24	56	89	118	105
1900	0	0	2	14	40	71	101	85
2100	0	0	0	7	26	53	81	63
2300	0	0	0	4	16	37	62	44
2500	0	0	0	2	10	26	48	31
2700	0	0	0	1	7	20	39	24
2900	0	0	0	1	6	17	35	21

Note: Values predicted by new version of model based on more realistic climatic constraints (see text).

because both the evenness of rainfall (Z) and the number of dry months (V) turn out to be weakly related to rainfall (P) and temperature (T) by the following equations (based on Williamson's [1995] analysis of 218 weather stations distributed throughout sub-Saharan Africa):

$$V = 11.49 - 0.0078P + 1.5 * 10^{-6}P^2 \qquad (r^2 = 0.714)$$
$$Z = 1.04 - 0.0122V - 0.003T \qquad (r^2 = 0.475) \qquad (1)$$

The resulting distribution for N_{max}, the maximum ecologically tolerable group size, shown in Table 1 differs only in detail from that generated by the original version of the model given in Dunbar (1992a). Indeed, we have run a number of versions using slightly different forms for the key equations, and these produce essentially similar results. The main message is that there is a limit to the range of habitats that baboons can occupy. By and large, baboons cannot survive in very hot or very dry habitats, or in cooler climates. Secondly, there is clearly very considerable variation in the maximum tolerable group sizes that baboons can maintain over the range of habitats that they can occupy. In some hotter/drier habitats, maximum tolerable group sizes may be as low as 10–15 animals; in some wetter/warmer habitats, it may be as high as 135. However, as rainfall increases above about 1500 mm per year, baboons find it increasingly difficult to maintain groups of any significant size.

It is clear, however, that rainfall seasonality (reflected in the model by the two variables Z and V) does have a significant effect on baboon time budgets, and therefore on the maximum ecologically tolerable group size. I therefore reran the simulation model to produce an output in three dimensions with V, the number of dry months, as the third independent variable (with Z calculated from equation [1] as before). Since V is related to P, it was necessary to impose some constraints on the range of possible values for V. Williamson (1995) obtained the following limits for V from her analyses of the data for 218 sub-Saharan weather stations:

$$V_{min} = 13.219 - 0.0073P \qquad (r^2 = 0.916)$$

$$V_{max} = 13.065 - 0.0066P + 1.2 \times 10^{-6} P^2 \qquad (r^2 = 0.671)$$

Figure 2 shows the combination of rainfall, temperature and dry months at which the model predicts that baboons can maintain groups of at least 15 individuals. I take 15 to be the minimum viable group size since the observed mean minimum group size for the 28 populations in the whole sample was 22.5 (range 7–51), with a distinct cluster of data points in the region 12–17. These results suggest that, within any given rainfall and temperature regime, baboons do rather better in habitats that are more seasonal.

We can test the validity of these predictions by comparing the predicted values obtained for maximum group sizes against those observed in real populations. The strongest test would come from showing that maximum group sizes predict the presence or absence of baboons in a geographically limited area. I have undertaken such tests for two areas in Ethiopia, neither of which contributes to the systems model presented here. In the Simen Mountains in northern Ethiopia, the 3000 m high escarpment provides a

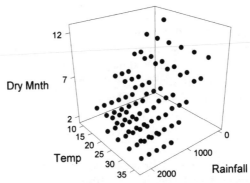

Figure 2. Zone of ecological survival for baboons: each point represents a specific combination of annual rainfall (mm), rainfall seasonality (indexed as the number of dry months: those with less than 50 mm rainfall) and mean annual temperature (°C) under which baboons would be able to maintain a maximum ecologically tolerable group size of at least 15 animals.

striking temperature gradient combined with a marked east-west rain shadow along a transect that is only about 150 km in length. We were able to determine the presence or absence of baboons at six sites within this area. In addition, we were able to obtain similar data from two sites on Mt Menegasha, some 500 km to the south of the Simen. We can use the simulation model to predict maximum group sizes for each site, given its observed annual rainfall, rainfall seasonality and temperature. Figure 3 shows the results, with an $N_{max} = 15$ again being taken as the minimum viable mean group size. The model appears to be able to predict the presence/absence of baboons in these two very different habitats extremely well. Group counts are available only for Menegasha: here the observed group size (20 animals) was well within the low maximum ecologically tolerable group size predicted by the model (38).

An alternative way to test the model is to compare observed and predicted group sizes for baboon populations throughout sub-Saharan Africa. Figure 4 plots observed mean group sizes for individual populations against the maximum predicted for each habitat by its specific rainfall and temperature characteristics (using all four climatic variables). (The Simen and Menegasha sites are not included here since we do not have adequate population demographic data for any of the populations in these two samples.) Populations which were used in determining the regression equations are shown as open circles; other populations not used in these analyses are shown as solid circles. With respect to the first group of

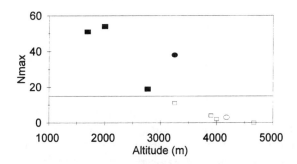

Figure 3. Maximum ecologically tolerable group sizes predicted by the model for two series of sites in Ethiopia (Simen Mountains in northern Ethiopia and Mt Menegasha in central Ethiopia), plotted against site altitude. The horizontal line (at $N_{max} = 15$) represents the minimum group size for a viable population. (In fact, minimum permissible group sizes predicted by the model given below vary from 11–54, but are always *less than* the predicted N_{max} for those populations where baboons actually occur, and are always *greater* than N_{max} at all sites where baboons do not occur.) N_{max} is estimated using site-specific values for all four climatic variables. Symbols: *squares*, Simen Mts; *circles*, Mt Menegasha. *Filled symbols*, sites at which baboons are observed; *open symbols*, sites where baboons do not occur.

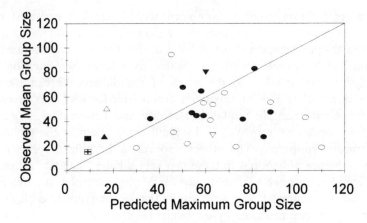

Figure 4. Observed mean group size for individual baboon populations, plotted against the maximum ecologically tolerable group size, N_{max}, for that population (calculated using all four climatic variables). *Open symbols*, sites used in the regression analyses for the model; *filled symbols*, other independent sites. Two sites used in the regression analysis (Gilgil 1984, Ruaha) were omitted because only the size of the study group is known. Data for Giant's Castle are split into 'low' and 'high' altitude sub-populations. Comparison of sites before and after population collapse: *square*, Kuiseb (1975 vs 1988); *upward triangle*, Amboseli (1969 vs 1975); *downward triangle*, Mikumi (1976 vs 1991). (Source: Dunbar 1992a, tables 2 and 7. Additional sources: Brain 1990; D. Hawkins & G. Norton personal communication.)

populations, it should be noted that the values shown in Figure 4 are population means. Since the regression equations were obtained from data for only a single group in each population and since the size of groups chosen for study are not always a random sample of the group sizes available in a population (Sharman & Dunbar 1982), these populations in fact also constitute a legitimate test of the model.

Figure 4 suggests that mean group size rarely exceeds the predicted maximum group size (and then only by a relatively small quantity). Of the eight sites whose means lie above the main diagonal (the line at which observed and predicted values are equal), three lie very close to the line and are well within the margin of error around the estimate of N_{max}. A further three (Mikumi, Amboseli and Kuiseb, indicated by separate symbols) concern sites where the population crashed shortly after the census was taken: in each case, the mean group size after the population collapse was close to or below the new predicted maximum group size. The remaining two deviant cases (*Papio papio* at Mt Assirik, Senegal, and *P. ursinus* at Suikerbosrand, South Africa) both involve populations where groups habitually fragmented into small unstable foraging parties that often slept and ranged alone.

These two tests thus provide compelling evidence for the validity of the model.

On balance, then, it seems that baboon populations do not normally exceed their ecologically maximum tolerable group size, and that when they do they either crash or are forced to fragment during foraging. These results also imply that when individual groups undergo fission, they do so because they have overshot the maximum tolerable size.

In addition to this baboon model, we have now run similar analyses for two other species, gelada baboons (grazers) and chimpanzees (forest-based frugivores that specialize on ripe fruit). The analyses for the gelada are given by Dunbar (1992c); those for the chimpanzees are available in Williamson (1996). I want to make only two observations based on these non-baboon models.

First, if we plot the geographical limits for the various species (defined as the range of habitats in which they could maintain a minimum group size of at least 15 animals) on the same graph, we find that these three species partition out the niche-space quite neatly (Figure 5). In plotting the data for baboons, I have plotted only those locations at which the maximum tolerable group size exceeds the predicted minimum permissible group size (given in Table 2 below). In addition, the distribution for gelada has been truncated at a rainfall value of 900 mm on the grounds that although rainfall does not feature in the

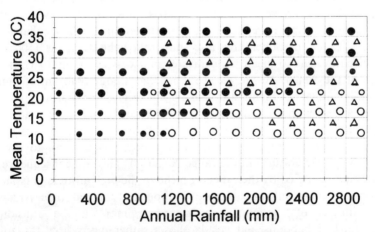

Figure 5. Geographical distributions for *Papio* baboons, gelada and chimpanzees, based on the range of habitats in which they could maintain a maximum ecologically tolerable group size, N_{max}, of at least 15. *Open circle*, gelada; *filled circle*, baboons; *triangle*, chimpanzees. For baboons, a small circle indicates habitats where the minimum permissible group size (see Table 2) would in fact exceed the maximum tolerable group size (see Table 1). The distribution for gelada has been curtailed for habitats with less than 900mm of rainfall because such habitats usually have temperatures in excess of $T = 20°C$; a small circle for gelada indicates that $15 < N_{max} < 20$ and is probably less than the minimum permissible group size.

Table 2. Minimum ecologically permissable group size, N_{max}, for baboons under different climatic conditions.

	Annual temperature (°C)							
Rainfall (mm)	0	5	10	15	20	25	30	35
100	7	29	38	44	49	53	57	60
300	7	29	38	44	49	53	57	60
500	7	29	38	44	49	53	57	60
700	7	29	38	44	49	53	57	60
900	3	13	17	20	22	24	26	27
1100	2	10	13	15	17	18	19	20
1300	2	9	11	13	15	16	17	18
1500	2	8	10	12	13	15	16	17
1700	1	7	10	11	13	14	15	16
1900	1	7	9	11	12	13	14	15
2100	1	7	9	10	12	13	14	14
2300	1	7	9	10	11	12	13	14
2500	1	6	8	10	11	12	13	13
2700	1	6	8	9	11	12	12	13
2900	1	6	8	9	10	11	12	13

Note: Values predicted by new version of model based on more realistic climatic constraints (see text).

gelada model, there is a relationship between mean temperature and minimum rainfall in the African weather station database: low rainfall values are only found in high temperature habitats. More importantly, data collated by Hurni (1982) for the Simen area also show that habitats below 1500 m in altitude (equivalent to mean temperatures in excess of about 19°C at the latitude of the Simen) do not receive more than about 1000 mm of rain a year, while habitats at higher altitudes do not receive less than this.

The data show that gelada occur only in cooler habitats (mean ambient temperatures of around 10–15°C), as a result of which there is relatively little overlap in geographical range with baboons (who tend to favour habitats with temperatures in the range 20–30°C). This is a consequence of gelada being restricted by their dietary niche to the high altitude grasslands that currently occur only in habitats over about 1500 m in altitude. Analysis of the impact of changing temperature regimes on the altitudinal distributions of baboons and gelada shows rather nicely how the zonal distributions of these two taxa moves up and down the altitudinal gradient as global temperatures rise and fall (Dunbar 1992d). Similarly, the chimpanzee distribution is a more or less mirror image of that for baboons, but with rainfall being the main factor separating the two taxa. This apparently reflects the chimps' preference for tree-based feeding sites in contrast to the baboons' preference for feeding sites in the shrub/bush layer.

Second, the niche separation between the three taxa can be traced back to the dietary differences between them and the way in which their preferred dietary sources respond to climatic variables. The easiest way to show this is in respect of the gelada and the baboon feeding time equations. When these are reduced to the first two independent variables, they have the form:

$$\ln(F_{Gel}) = 5.9 - 0.6 \ln(T) - 0.9 \ln(Q) \tag{2}$$

$$\ln(F_{Bab}) = 6.4 - 0.6 \ln(T) + 5.7 \ln(Z) \tag{3}$$

where F_{Gel} and F_{Bab} are the percentages of time devoted to feeding by the gelada and the baboon respectively, T is mean ambient temperature, Q is the protein content of grass (% protein by weight) and Z is Simpson's index of the diversity of rainfall across the months of the year. Now, it turns out that, for this set of baboon study sites, both Q and Z are quadratic functions of T:

$$\ln(Q) = -26.7 + 23.9 \ln(T) - 4.8(\ln(T))^2 \tag{4}$$

($r_2 = 0.97$) and

$$\ln(Z) = -4.9 + 3.2 \ln(T) - 0.6(\ln(T))^2 \tag{5}$$

($r_2 = 0.65$).

Substituting equations (4) and (5) into equations (2) and (3) yields:

$$\ln(F_{Gel}) = 14.2 - 8.0 \ln(T) + 1.5(\ln(T))^2$$

$$\ln(F_{Bab}) = -22.2 + 17.5 \ln(T) - 3.1(\ln(T))^2$$

which are virtual mirror images of each other: in habitats where baboons have to feed a lot, gelada have to feed relatively little, and vice versa. These turn out to have this form because the two taxa's primary food sources (grass for the gelada, the bush layer vegetation for baboons) respond in diametrically opposite ways to temperature. Grasses (at least of the kind on which gelada feed) are common at low temperatures, whereas bush level cover is common at higher temperatures. Time spent moving behaves similarly due to the fact that inter-patch distance for each vegetation layer is inversely related to vegetation density.

MINIMUM PERMISSIBLE GROUP SIZE

I assume that the minimum permissible group size is determined by the level of predation risk in a given habitat. Primates in general use group size as a key means of deterring predators (van Schaik 1983; Dunbar 1988). In trying to determine how minimum group size relates to environmental variables, we need to identify the key problems that animals encounter with respect to predation.

The first point to note is that mortality *per se* is not necessarily a good guide to the problem that animals face. If group size is the animals' response to predation risk, then mortality rates should be constant across habitats (except where animals are prepared to trade up predation risk against other variables in order to be able to survive at all). I therefore assume that animals adjust the minimum size of group they are prepared to live in so that predation risk is equilibrated (to some roughly constant low level) across habitats.

Cowlishaw (1993) found that, in a Namibian baboon population, the degree of cover was the most important factor influencing both the risk of exposure to predator attack and the animals' nervousness. Similarly, Rasmussen (1983) found that baboons in southern Tanzania were more likely to bunch and to act nervously during travel at those times of year when the level of vegetation cover made it difficult for them to see stalking predators, while Altmann & Altmann (1970) found that both alarms and actual predator attacks in their Kenyan baboon population were concentrated in wooded areas. Equally, however, baboons may be nervous in very open areas when they are far from trees and other suitable refuges (baboons: Byrne 1981; gelada: Dunbar 1989). This suggests that, for baboons at least, predation risk may be positively related to the density of low level cover (e.g. ground and bush layers), but negatively related to the density of large trees that can function as refuges.

Usable data on tree and bush level cover are not available for most of these sites. I therefore used my own data on the percentage of ground surface with tree and bush level cover at nine sites in eastern Africa (see Dunbar 1992a, with an additional Ugandan forest site sampled by Louise Barrett) and derived regression equations relating each of these two variables to fundamental climatic variables. The best-fit equations are:

$$\ln(B) = -2.072 + 1.811 \ln(T) \qquad (r^2 = 0.36)$$

$$E = 86.28 - 14.078 V \qquad (r^2 = 0.85)$$

where B is the percentage of ground covered by bush/shrub layer vegetation and E is the percentage of ground covered by tree layer vegetation.

I used these equations to calculate tree and bush cover indices for each of the habitats in the sample, and then regressed the minimum observed group size for each population against these values. This yielded the following best-fit equation:

$$\ln(N_{min}) = 2.67 - 0.23 \ln(E) + 0.202 \ln(B) \tag{6}$$

($r^2 = 0.516$, $N = 33$, $P < 0.05$). The distribution in Figure 6 shows quite clearly that the minimum group size gets larger as the level of bush cover

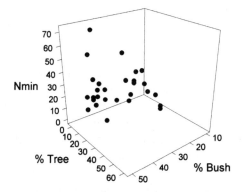

Figure 6. Minimum group sizes for individual baboon populations plotted against percentage of tree and bush level cover for each habitat. Group size increases as the density of bush cover increases and decreases as the density of tree cover increases.

increases (increased risk of unseen predator attack) and the level of tree cover declines (reduced availability of refuges).

If we use equation (6) to determine minimum permissible group sizes, we obtain the distribution given in Table 2. (Once again, I show the data for combinations of rainfall and temperature only, and use these variables to estimate habitat-specific values of Z, V, B and E.) Comparison of Tables 1 and 2 reveals that the minimum permissible group size exceeds the maximum tolerable group size in some habitats, especially those characterized by less than about 400 mm of rainfall and temperatures below about 15°C. Baboons are prevented from occupying these habitats except where micro-habitat conditions (e.g. riverine forest) allow them to reduce the minimum group size or increase the maximum.

COGNITIVE CONSTRAINTS

The final component of the linear programming model is the constraint imposed on group size by the species' cognitive abilities. This constraint is derived directly from the 'Machiavellian Intelligence' hypothesis for primate brain size (Byrne & Whiten 1989). This argues that the principal selection pressure promoting the evolution of large brain size in primates has been the need to integrate and function effectively within increasingly large and tightly bonded social groups. The issue here is not simply creating large loosely structured aggregations. Primate groups are very tightly bonded structures whose coherence through time depends on the acquisition and exploitation of social knowledge about other group members. As a result, primate social groups differ in a number of important ways from those of other species. One of these is the complexity of their coalitions. Harcourt (1989; Harcourt & de Waal 1992) has argued that the coalitions of higher primates differ from those of other species in both their use of third parties and their temporal structure (primate coalitions are commonly established well in advance of the circumstances under which they are needed). These observations suggest that primates are able to acquire and make use of social knowledge about how other individuals behave that is based on deeper insights into others' mental worlds.

There are two corollaries to this claim. One is that the level of social complexity that one might expect from a species will increase with increasing brain size. This has in fact been shown to be true, both for the use of tactical deception (Byrne 1993) and the use of tactics that undermine the effects of linear dominance hierarchies (Pawlowski & Dunbar 1996). The second is that as the number of individuals that an animal has to keep track of increases, so the cognitive demands on it will increase, and thus demand proportionately larger brains with which to compute social spaces and animals' trajectories within them.

Note that the issue here is not simply remembering who-is-who in a large group, but rather remembering who-is-friends-with-who, constantly updating this knowledge as friendships change through time and, finally, using this knowledge in building and servicing one's own friendships and alliances. Prima facie evidence in support of this claim comes from the finding that, in primates, group size does correlate with brain size (Dunbar 1992d, 1995; Barton & Purvis 1995). In both these cases, it seems that it is neocortex size that is crucial (see Barton & Purvis 1995). This is not too surprising, since it is the disproportionate growth of the neocortex that has largely been responsible for the increases in total brain size during primate evolution (Stephan 1972; Passingham 1982).

These findings thus suggest that there is ultimately a cognitive limit to the size of group that a particular species can exhibit. The issue is not that large groups are impossible but that they are much harder to hold together, so that any centrifugal forces that might exist (foraging competition, patch size constraints, harassment) will tend to encourage their dispersal. Such groups will thus tend to fragment rather easily. We do not at present know exactly what the cognitive constraint is. We know only that there is a relationship between *mean* group size and brain size in primates. Presumably, the information-processing constraint imposed by neocortex size must act via *maximum* group size. However, we cannot at present identify what this is, since the observed maximum group sizes are likely to be confounded by ecological considerations.

There is, in addition, a further reason for supposing that the relationship between brain size and group size is more complex. Kudo et al. (in preparation) have examined grooming clique sizes in anthropoid primates and shown that these also correlate with relative neocortex size. Grooming cliques are the foundations for coalitions in primate groups, and we suspect that at least part of the constraint may reflect the number of coalitions partners whose conflicting interests can be managed simultaneously. Grooming clique size turns out to be a very tight linear function of group size in anthropoid primates. One plausible interpretation of this is that, as group size increases, so proportionately larger coalitions are needed to buffer individuals against the stresses and strains of living in such large groups.

DISCUSSION

The burden of these analyses is to show that we can define a set of equations that constrain quite tightly the range of group sizes that a given primate species can occupy. These group sizes are habitat-specific and reflect an individual species' ecological niche adaptations. Figure 1 plots the actual values generated by the baboon systems model for populations living in habitats with a mean annual temperature of 25°C (approximately the centre of the baboon's thermal zone). I have plotted the cognitive group size as that used in the analysis of neocortex size by Dunbar (1992b). It is therefore necessary to add the rider that this is almost certainly too low: it represents the *mean* group size not the cognitive *maximum* group size. It is important to be aware that this is very much a progress report. We are still very much engaged in developing the basic models and further refinements in parameter values and equations can be expected as we learn more about how the animals behave.

Nonetheless, given that we can proceed in this way, three important points follow.

One is that the model as I have presented it is underpinned by conventional optimality considerations. We should be able to identify the limiting processes in the animals' ecological world that give rise to these group size distributions. Indeed, these optimal solutions to conventional ecological and physiological problems are, in a very real sense, assumptions of the model. We should be able to test these directly by further field work. In many respects, the value of modelling exercises of this kind is to draw our attention to the key processes involved.

A second point arises from the fact that groups are the context in which animals play out their social and reproductive strategies. Their decisions on which strategies to pursue are influenced by the costs and benefits of the options available to them. Not only may the options themselves be determined by the size and composition of their groups (you can only choose to form an alliance with a sister if you have sisters living with you), but also the very costs and benefits that weight those strategies are a product of the demography of the group. Thus, once we know group size, a number of other things follow in a fairly straightforward way (Dunbar 1993b). I shall confine myself to just one example here.

It turns out that, for most primate species, the number of males in a group is a function of the number of females in the group (Dunbar 1988). In most cases, this is a direct consequence of an optimization problem being solved by the males in the population. The problem was originally identified by Emlen & Oring (1977) as one of monopolizing reproductive females. Dunbar (1988) has shown that, in primates at least, a male's ability to keep other males away from a group of females (so as to maintain a one-male group) depends on a combination of the size of the female group (a direct reflection of the ecological constraints on group size) and the reproductive synchrony among the females. (Srivastava & Dunbar [in press] have since been able to show that the distance males have to travel to find another female group [the search time in conventional optimal foraging terminology] is also an important consideration.) Thus, the choice between female-defence polygyny and conventional mate-defence promiscuity turns out to be a direct consequence of the way demographic and lifehistory variables weight the costs and benefits of mate defence for males.

Once males are in a multimale group, further consequences follow. Cowlishaw & Dunbar (1991) have shown that the dominant male's ability to monopolize matings within a multimale group is a negative function of the number of males in the group (and hence the pressure from rivals). In fact, there appears to be a crucial threshold at four males: with a small number of competitors in the group, the dominant male can command a disproportionate share of the matings (and, indeed, conceptions), but once there are more than four males it becomes increasingly difficult for him to do so.

In part, the dominant male's problem reflects the extent to which females can exert an influence on the situation through female choice. This, in turn, is partly a reflection of levels of sexual size dimorphism (something which varies considerably not just between species, but also *within* species in response to environmental parameters: for baboons, see Dunbar 1990). The larger males are relative to females, the more valuable they are as allies in any situation where power is a function of physical size. Although females have to balance a male's size-dependent value as an ally against the value of longer-lasting alliances formed with female relatives (Dunbar 1988, 1993b), there will inevitably be a point at which the power-asymmetry offered by large males outweighs the benefits offered by female relatives through kin selection.

As a result, group composition, male mating strategies and female social strategies should be a straightforward function of group size and other ecological factors. However, in contrast to the species-specific socio-ecological models of the 1970s, the important lesson from these analyses is that these variables may be expected to vary from one population to another *within* a species. Preliminary analyses (Dunbar 1993b) suggest that this simple deterministic model yields outcomes for these social variables that bear a rather complex relationship to climatic variables. Although the relationships involved are all causally deterministic, non-linear elements in the equations generate what appear superficially to be chaotic behaviour when carried through into further layers of equations.

Finally, given that we can define both demographic and behavioural aspects of a species' behaviour in this way, one obvious implication is that we can do the same for extinct species. The constraints lie only in the precision of our knowledge about the climatic parameter values for palaeoenvironments, body weights for extinct species and their dietary niches. Of these, only the latter remains genuinely problematic, although even in this respect tooth shape and wear patterns (Kay & Covert 1984) and the trace element content of fossil bone (e.g. Lee-Thorp *et al.* 1989) are beginning to elucidate matters. Significant advances have been made during the past decade in determining palaeoclimates from both faunal and vegetational assemblages and it is possible to specify with some degree of confidence the likely rainfall and temperature values by using modern habitats with comparable faunal or floral profiles (e.g. Vrba 1988). Similarly, it is now possible to estimate body weights with considerable accuracy from bone fragments: thanks mainly to general physical principles, the cross-sectional dimensions of weight-bearing bones provide an accurate estimate of the weight they carried in life (e.g. Martin 1990). Given this, it should be possible to say quite a lot about the population demography and behavioural ecology of individual populations of fossil taxa, at least so long as they belong to the same dietary grade as a living taxon. Preliminary attempts to do so have been carried out for fossil papionines (Dunbar 1991) and fossil theropithecines (Dunbar 1993a) with some success.

In addition to predicting how a taxon might have behaved at a given site, we can use the model in reverse to explore the likely reasons why a species went extinct. By using the model to predict group size at fossil sites where they are known not to have occurred or at different time horizons within a site, we may be able to show why a species went extinct.

In sum, the approach adopted here holds out significant hope for building a model that allows us both to predict the form and structure of species' socio-ecological systems and to explore aspects of that species' biogeography and evolutionary history postdictively as well as predictively.

In other words, not only may we be able to account for the species' past history, but we may also be able to predict what the consequences of major habitat or climatic change may be for the species' distribution and future survival. By simultaneously building top-down (i.e. systems models of the kind described here) and bottom-up (i.e. more conventional optimization models), we may be able to produce a very securely constructed general theory of primate social systems.

REFERENCES

Altmann, J. 1980: *Baboon Mothers and Infants.* Cambridge (MA): Harvard University Press.
Altmann, S.A. & Altmann, J. 1970: *Baboon Ecology.* Chicago: University of Chicago Press.
Barton, R.A. & Purvis, A.J. 1995: Primate brains and ecology: looking below the surface. In *Proceedings of XIVth Congress of the International Primatological Society, Strasburg.*
Brain, C. 1990: Spatial usage of a desert environment by baboons (*Papio ursinus*). *Journal of Arid Environment* 18, 67–73.
Byrne, R.W. 1981: Distance vocalisations of Guinea baboons (*Papio papio*) in Senegal: an analysis of function. *Behaviour* 78, 283–312.
Byrne, R. 1993: *The Thinking Ape.* Oxford: Oxford University Press.
Byrne, R.W. & Whiten, A. (eds) 1989: *Machiavellian Intelligence.* Oxford: Oxford University Press.
Cowlishaw, G. 1993: *Trade-offs Between Feeding Competition and Predation Risk in Baboons.* PhD thesis, University of London.
Cowlishaw, G. & Dunbar, R.I.M. 1991: Dominance rank and mating success in male primates. *Animal Behaviour* 41, 1045–1056.
Dunbar, R.I.M. 1988: *Primate Social Systems.* London: Chapman & Hall.
Dunbar, R.I.M. 1989: Social systems as optimal strategy sets: the costs and benefits of sociality. In *Comparative Socioecology* (ed. V. Standen & R. Foley), pp. 131–150. Oxford: Blackwell Scientific.
Dunbar, R.I.M. 1990: Environmental determinants of intraspecific variation in body weight in baboons (*Papio* spp.). *Journal of Zoology, London,* 220, 157–169.
Dunbar, R.I.M. 1991: Functional significance of social grooming in primates. *Folia Primatologica* 57, 121–131.
Dunbar, R.I.M. 1992a. Time: a hidden constraint on the behavioural ecology of baboons. *Behavioural Ecology and Sociobiology* 31, 35–49.
Dunbar, R.I.M. 1992b. Neocortex size as a constraint on group size in primates. *Journal of Human Evolution* 20, 469–493.
Dunbar, R.I.M. 1992c. A model of the gelada socioecological system. *Primates* 33, 69–83.
Dunbar, R.I.M. 1992d. Behavioural ecology of the extinct papionines. *Journal of Human Evolution* 22, 407–421.
Dunbar, R.I.M. 1993a. Socioecology of the extinct theropiths: a modelling approach. In *Theropithecus: The Rise and Fall of a Primate Genus* (ed. N.G. Jablonski), pp. 465–486. Cambridge: Cambridge University Press.
Dunbar, R.I.M. 1993b. Ecological constraints on group size in baboons. *Physiology & Ecology, Japan* 29, 221–236.
Dunbar, R.I.M. 1995: Neocortex size and group size in primates: a test of the hypothesis. *Journal of Human Evolution* 28, 287–296.
Dunbar, R.I.M. & Dunbar, P. 1988: Maternal time budgets of gelada baboons. *Animal Behaviour* 36, 970–980.

Dunbar, R.I.M. & Sharman, M. 1984: Is social grooming altruistic? *Zeitschrift für Tierpsychologie* 64, 163–173.

Emlen, S.T. & Oring, L. 1977: Ecology, sexual selection and the evolution of mating systems. *Science* 197, 215–223.

Harcourt, A.H. 1989: Social influences on competitive ability: alliances and their consequences. In *Comparative Socioecology* (ed. V. Standen & R. Foley), pp. 223–242. Oxford: Blackwell Scientific.

Harcourt, A.H. & de Waal, F. (eds) 1992: *Coalitions and Alliances in Humans and Other Animals*. Oxford: Oxford University Press.

Hurni, H. 1982: *Climate and the Dynamics of Altitudinal Belts from the Last Cold Period to the Present Day*. (*Simen Mountains*–Ethiopia, Vol. II) Bern: University of Bern Geographical Institute.

Kawamuto, Y., Shotake, T., & Nozawa, K. 1982: Genetic differentiation among three genera of Family Cercopithecidae. *Primates* 23, 272–286.

Kay, R.F. & Covert, B. 1984: Anatomy and behaviour of extinct primates. In *Food Acquisition and Processing in Primates* (ed. D.J.Chivers, B. Wood & A. Bilsborough), pp. 467–508. New York: Plenum Press.

Lee-Thorp, J.A., van der Merwe, N.J. & Brain, C.K. 1989: Isotopic evidence for dietary differences between two extinct baboon species from Swartkrans. *Journal of Human Evolution* 18, 183–190.

Martin, R.D. 1990: *Primate Origins and Evolution*. London: Chapman & Hall.

Mount, L.E. 1979: *Adaptation to Thermal Environment*. London: Arnold.

Passingham, R. 1982: *The Human Primate*. San Francisco: Freeman.

Pawlowski, B.B. & Dunbar, R.I.M. 1996: Neocortex size, social skills and mating success in male primates. (Submitted)

Rasmussen, D.R. 1983: Correlates of patterns of range use of a troop of yellow baboons (*Papio cynocephalus*). II. Spatial structure, cover density, food gathering, and individual behaviour patterns. *Animal Behaviour* 31, 834–856.

Sear, R. 1994: *A Quantitative Analysis of Gibbon behavioural Ecology*. MSc thesis, University College London.

Sharman, M. & Dunbar, R.I.M. 1982: Observer bias in selection of study group in baboon field studies. *Primates* 23, 567–573.

Srivastava, A. & Dunbar, R.I.M. (In press) The mating system of hanuman langurs: a problem in optimal foraging. *Behavioural Ecology and Sociobiology*.

Shotake, T., Nozawa, K., & Tanabe, Y. 1977: Blood protein variations in baboons. I. Gene exchange and genetic distance between *Papio anubis, Papio hamadryas* and their hybrids. *Japanese Journal of Genetics* 52, 223–237.

Stephan, H. 1972: Evolution of primate brains: a comparative anatomical investigation. In *Functional and Evolutionary Biology of Primates* (ed. R. Tuttle), pp. 155–174. Chicago: Aldine-Atherton.

van Schaik, C.P. 1983: Why are diurnal primates living in groups? *Behaviour* 87, 120–144.

Vbra, E.S. 1988: Late Pliocene climatic events and hominid evolution. In *Evolutionary History of the 'Robust' Australopithecines* (ed. F.Grine), pp. 183–238. Albany: SUNY Press.

Wernstedt, F.L. 1972: *World Climatic Data*. New York: Climatic Data Press.

Williamson, D. 1996: *Modelling the Socioecology of Early Hominids*. PhD thesis, University of London.

Wrangham, R.W. 1980: An ecological model of female-bonded primate groups. *Behaviour* 75, 262–300.

Wrangham, R.W. 1987: Evolution of social structure. In *Primate Societies* (eds. B.B.Smuts, D. Cheney, R. Seyfarth, R.W. Wrangham & T.T. Struhsaker), pp. 282–295. Chicago: Chicago University Press.

Function and Intention in the Calls of Non-Human Primates

DOROTHY L. CHENEY & ROBERT M. SEYFARTH

Departments of Biology and Psychology, University of Pennsylvania, Philadelphia, PA 19104, USA

Keywords: non-human primate; communication; mental state attribution; semantic signalling; reconciliation; contact calls.

Summary. Many of the vocalizations produced by non-human primates are functionally semantic, in the sense that they denote objects and events in the external world. Moreover, at least some monkey species appear to assess and compare calls on the basis of their meanings. In their social interactions, non-human primates also use their calls in ways that are functionally analogous to the ways that humans use language. The grunts given by free-ranging baboons, for example, serve to facilitate social interactions and to reconcile opponents following fights. The mental mechanisms underlying the vocalizations of non-human primates, however, appear to be fundamentally different from those that underlie human speech, because monkeys do not apparently call to one another with the intent of modifying or influencing each other's mental states. The alarm and contact calls of monkeys provide information about the signaller's current physical and mental states, but they are not deliberately given to inform or instruct others. Instead, listeners appear to extract relevant information about a call's function based on behavioural contingencies and their own experiences.

INTRODUCTION

DISCUSSIONS ABOUT THE EVOLUTION OF LANGUAGE typically focus on two apparently fundamental attributes of human speech. The first of these is

© The British Academy 1996.

semantics, which can be defined loosely as the meaning of words, or sounds. The second is syntax, defined equally loosely as a set of rules for assembling words into meaningful phrases or sentences (Jackendoff 1994; Pinker 1994).

Although the natural communication of monkeys and apes provides few examples of the kind of syntax found in human language (cf. Robinson 1984), there is evidence from a number of monkey species for what might be termed functionally semantic communication. Vervet monkeys (*Cercopithecus aethiops*), for example, use a variety of acoustically different alarm calls to denote predators that hunt in qualitatively different ways (Struhsaker 1967; Seyfarth *et al.* 1980a, 1980b). Each alarm call type elicits a different, apparently adaptive response from monkeys nearby. For instance, alarm calls given to leopards (*Panthera pardus*) cause vervets to run into trees, while alarm calls given to martial and crowned eagles (*Polemaetus bellicosus* and *Stephanoaetus coronatus*) cause vervets to look up or run into bushes. Playback experiments have demonstrated that alarm calls alone, even in the absence of an actual predator, elicit the same responses as do the predators themselves (Seyfarth *et al.* 1980b). The alarm calls, therefore, function as rudimentary semantic signals because each alarm call elicits the same response as would its referent, even when the referent is absent (Hockett 1960; Seyfarth & Cheney 1992).

The alarm calls of vervet monkeys are functionally semantic, but do they qualify as words? To answer this question we must consider not only how signals function in the animals' daily lives but also the proximate causal mechanisms that underlie their production and perception. Since the best studied mammalian communication system is our own, comparison with human language seems a reasonable place to begin.

It is often assumed that animals respond to vocal signals simply on the basis of the calls' physical features, or acoustic properties (e.g. Morton 1977). Humans, by contrast, make judgments about the similarity or difference between words on the basis of an abstraction, their meaning. For example, when asked to compare the words 'treachery' and 'deceit', we typically ignore the fact that the two words have different acoustic properties and describe them as similar because they have similar meanings. 'Treachery' and 'lechery', on the other hand, are judged as different because, despite their acoustic similarity, they mean different things. In making these judgments, we recognize the referential relation between words and the things for which they stand.

The 'ape language' projects provide a number of elegant cases in which chimpanzees (*Pan troglodytes*) have learned to assess and compare signs according to their meaning (e.g. Premack 1976; Matsuzawa 1985; Savage-Rumbaugh 1986). This ability, however, is not restricted to captive apes that have been trained in the use of artificial signs. Vervet monkeys also appear to have some mental representation of what their vocalizations stand for: when responding to calls, they seem to compare and assess them according to their meanings, and not just their acoustic properties.

When a vervet subject is repeatedly played a tape-recording of another individual's leopard alarm call when there is no leopard in the vicinity, she soon habituates to the call and ceases responding to it. If, however, the subject is then played the same individual's eagle alarm call, she responds strongly to it, in the same way that she would if an eagle had been sighted. Because the two calls have different referents, the subject does not transfer habituation across call types (Cheney & Seyfarth 1988).

Vervets do transfer habituation, however, between call types that have similar meanings. Vervet monkeys are hostile toward the members of neighbouring groups (Cheney 1981). When females encounter another group encroaching on their range, they often utter a loud, long, trilling call (termed a 'wrr'), which seems to function to alert other individuals of the encroachment. Roughly 45% of all inter-group encounters involve only the exchange of wrrs; others, however, escalate into aggressive chases and fights. When groups come together under these more aggressive conditions, females often give an acoustically different 'chutter'. The two calls, wrrs and chutters, therefore, are acoustically different but seem to share the same referent: another group. Moreover, vervets seem to treat the two calls as being, roughly speaking, synonymous. If a subject has habituated to repeated playback of another individual's inter-group wrr, she shows a similarly low level of response when played that individual's inter-group chutter. She transfers habituation from one call type to another, apparently because the two calls have the same general meaning despite their different acoustic properties (Cheney & Seyfarth 1988).

Vervet monkeys, therefore, appear to interpret their calls as sounds that represent, or denote, objects and events in the external world. When one vervet hears another calling, she forms a representation of what the call means. And if, shortly thereafter, she hears a second call, the two calls are compared on the basis of their meaning, and not just their acoustic properties.

INTENTIONAL COMMUNICATION

The comprehension of words by humans, however, involves more than just a recognition of the referential relation between sounds and the objects or events they denote. As listeners, we interpret words not just as signs for things but also as representations of the speaker's knowledge. We attribute mental states like knowledge and beliefs to others, and we recognize the causal relation between mental states and behaviour. We are, as a result, acutely sensitive to the relation between words and the mental states that underlie them. If, for example, we detect a mismatch between what another person says and what he thinks, we immediately consider the possibility that he is trying to deceive us.

H. P. Grice (1957) is one of many philosophers who have tried to clarify the distinction between human speech and simpler signalling systems that can nevertheless convey sophisticated, complex information. Grice distinguished the 'non-natural' meaning of linguistic phenomena, in which the speaker intends to modify both the behaviour and beliefs of his audience, from the 'natural' meaning of many other types of signs, in which, for example, thunder and lightning mean that it will soon rain (see also Bennett 1976; Tiles 1987). According to Grice's definition, truly linguistic communication does not occur unless both signaller and recipient take into account each other's states of mind—unless, in other words, both signaller and recipient take what the philosopher Dennett (1987) has called the 'intentional stance'.

All observations and experiments conducted to date suggest that monkeys do not attribute mental states different from their own to other individuals, though the evidence from chimpanzees is more equivocal (reviewed by Cheney & Seyfarth 1990b; Povinelli 1993; van Hooff 1994). Grice's definition of communication, therefore, may be completely irrelevant when applied to most cases of animal communication. Nevertheless, his definition is useful and provocative because it reminds us of precisely what is at stake when we compare non-human primate vocalizations with human language. Perhaps more important, it suggests that there can be communication systems that are complex and even semantic but that do not qualify as language because they fail to meet the criteria of language on intentional grounds.

It is this perspective, which compares a behavioural biologist's focus on function with a philosopher's focus on cognitive mechanisms, that we wish to consider in the remainder of this paper.

The social function of primate vocalizations

Only a small proportion of the vocalizations given by monkeys and apes occur in the form of alarm or inter-group calls. Instead, the most common calls given by many non-human primates are low amplitude grunts, coos, or trills that are given at close range and occur in the context of social interactions or group movement. Many of these calls appear to function to initiate and facilitate social interactions. For example, in Japanese macaques (*Macaca fuscata*), grooming interactions are often initiated when one female vocalizes to a potential partner (Masataka 1989; Sakuro 1989). Similarly, in stump-tailed macaques (*Macaca arctoides*), individuals that grunt to mothers before attempting to handle their infants are less likely to receive aggression than are individuals that remain silent (Bauers 1993).

From a functional perspective, these calls are interesting because they are in many ways analogous to human speech. Typically, there is no obvious response to the calls from nearby listeners, and it certainly seems as if these vocalizations, like many human conversations, function simply to mediate social interactions and grease the social wheels. Note, however, that it is difficult to describe the function of these calls without adopting an intentional vocabulary on behalf of the signaller. If a call serves to mollify an opponent or a subordinate mother, it seems almost essential that the signaller be able to recognize her partner's anxiety and to signal her own benign intent.

To examine the function of these close-range vocalizations in more detail, we carried out a detailed study of the grunts given by free-ranging female baboons (*Papio cynocephalus ursinus*) in the Okavango Delta, Botswana. Our work focused on an habituated group that included between 19 and 23 adult females (see Cheney et al. 1995).

Like adult females in many species of Old World Monkeys, female baboons form stable, linear dominance hierarchies (Seyfarth 1976; Hausfater et al. 1982; Smuts & Nicolson 1989). Although most affinitive interactions occur among close kin, adult females also interact with unrelated females, particularly if those females have infants. Normally, if a dominant female approaches a subordinate female, the subordinate is supplanted and moves away. Frequently, however, the dominant female vocalizes to the subordinate, using a low pitched, tonal grunt (Seyfarth, Cheney & Owren, unpublished data). These grunts seem to have an appeasing function, because they increase the probability of a subsequent friendly interaction, such as grooming or infant handling.

We recorded 2,698 incidents in which one female approached another that ranked lower than herself; in 621 (23%) of these cases the dominant female grunted to the subordinate. There were 17 females that could approach at least one lower-ranking, unrelated individual. For 15 of the 17, the mean frequency of approaches to all possible partners that was followed by a friendly interaction was higher if the dominant female first grunted than if she did not (Figure 1a; one-tailed Wilcoxin matched-pairs signed-ranks test, 1 tie, $t = 1$, $P < 0.001$). Similarly, for 14 of 17 individuals the mean frequency with which a female supplanted her lower-ranking partner was higher when she did not call than when she did (Figure 1b; $t = 10$, $P < 0.001$). Results were unaffected by the relative difference in rank

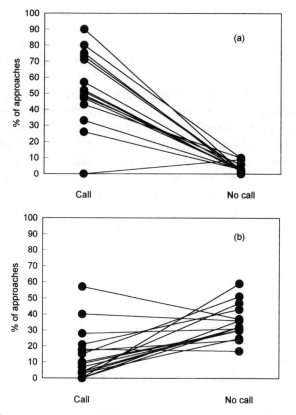

Figure 1. The mean proportion of 17 females' approaches toward subordinate partners that was followed by either (a) friendly behaviour by the dominant or (b) a supplant of the subordinate. Approaches are divided according to whether the dominant female grunted as she approached or whether she remained silent.

between the two females. Grunts, therefore, appeared to mediate and facilitate social interactions among unrelated adult females.

If grunts or other vocalizations do function to facilitate affinitive interactions, they might also be expected to play a role in reconciling opponents following aggression. Non-human primates are frequently aggressive toward one another, yet they live in relatively stable, cohesive social groups. Recent studies have suggested that opponents may mollify the effects of aggressive competition by reconciling soon after fighting or threatening one another (e.g. de Waal & van Roosmalen 1979; de Waal & Yoshihara 1983; York & Rowell 1988; Aureli et al. 1989; Cheney & Seyfarth 1989; Judge 1991; Aureli 1992; Cords 1992, 1993). Two animals are said to have reconciled if, within minutes of behaving aggressively, they interact in a friendly way by touching, hugging, grooming, or approaching one another.

No study, however, has yet considered the role that vocalizations might play in reconciling former opponents.

Baboon females do sometimes grunt to one another after aggression. In an effort to examine the role of grunts in reconciling opponents, we carried out a number of systematic observations of aggressors and their victims. Whenever two females were involved in an aggressive interaction, we followed the aggressor for 10 minutes to determine whether she subsequently interacted with her victim in any way (Silk, Cheney & Seyfarth, in preparation). In 5% of 502 samples, the aggressor subsequently interacted in a friendly manner with her opponent by touching her, grooming her, or interacting with her infant. Eighty-five per cent of these friendly interactions also included a grunt by the aggressor. In 9% of all cases, the aggressor only grunted to her victim and did not interact with her in any other way.

These observations suggested that vocalizations alone, even in the absence of other affinitive interactions, might function to reconcile opponents. Nevertheless, the significance of the grunts themselves was difficult to assess simply from observations, because grunts often occurred in conjunction with other friendly behaviour, such as grooming or infant handling. To determine whether grunts might function to reconcile opponents even in the absence of other affinitive interactions, therefore, we designed a series of playback experiments (for details of the experimental protocol see Cheney *et al.* 1995).

In conducting these experiments, we first waited until a higher-ranking female, A, had threatened or chased an unrelated, lower-ranking female, B. We then followed A for 10 minutes to determine whether she interacted affinitively with her opponent, and, if so, what form this affinitive interaction took. After this period, but within the next 30 minutes, we played a tape-recording of A's distress scream to B and videotaped B's response. Screams were played back to subjects under three conditions: (1) after A had been aggressive to B and did not interact with her again; (2) after A had been aggressive to B and then grunted to B without interacting with her in any other way; and, (3) after a period of at least 90 minutes in which A and B had not interacted.

We chose screams as playback stimuli because they mimicked a context in which subordinate females are sometimes attacked by dominant individuals. When a female baboon receives aggression from a higher-ranking female or male, she typically screams at her opponent. Frequently, she then 'redirects' aggression by threatening a more subordinate individual. We hypothesized that a subordinate female that heard the scream of an unrelated, higher-ranking individual would interpret this call as a potential threat to herself (see discussion in Cheney *et al.* 1995). We predicted that B would react strongly to the sound of A's scream if A had recently threatened

B but had not reconciled (i.e. grunted) with her. B's response in this context should be stronger than it was following a control period when the two females had not interacted. If, however, A had grunted to B after threatening her, B's anxiety should be diminished. We predicted that B's response after vocal 'reconciliation' would be similar to her response following the control period of no interaction.

There were 15 dyads that met all three test conditions. If a dominant female had grunted to her subordinate opponent following a fight, the opponent responded for a significantly shorter period of time to that female's scream than she did following a fight when no further interaction had taken place (Figure 2; one-tailed Wilcoxon matched-pairs signed-ranks test, $n = 15$, 1 tie, $t = 17.5$, $P < 0.025$). Subordinate subjects also responded less strongly to dominant females' screams after a control period of no interaction than after a fight with no reconciliation (Figure 2; $n = 15$, 1 tie, $t = 24$, $P < 0.05$). In contrast, subordinate subjects' responses to dominant females' screams following a fight with a vocal 'reconciliation' were statistically indistinguishable from their responses following a control period of no interaction (Figure 2; $n = 15$, 2 ties, $t = 47.5$, NS).

There were 14 other dyads that met two of the three test conditions described above. For seven dyads, 'fight with no vocal reconciliation' could be compared with the 'no prior interaction' control. For seven other dyads, 'fight with vocal reconciliation' could be compared with the 'no prior interaction' control. Results from these trials further supported the

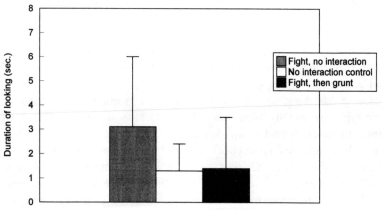

Figure 2. The duration of subjects' responses to the screams of dominant opponents after (1) the dominant threatened the subject and did not interact with her again; (2) the two females had not interacted for at least 90 minutes; and (3) the dominant threatened the subject and then reconciled by grunting to her. Histograms show means and standard deviations for 15 dyads in each of the three conditions. Subjects' responses were scored as looking in the direction of the speaker.

hypothesis that grunts functioned to restore opponents' relationships to baseline levels of tolerance. A significant number of subjects responded more strongly to their opponent's scream after a fight when they had not reconciled than after the control period ($n = 7$, $t = 1$, $P < 0.01$). If, however, the dominant female had grunted to her opponent, the opponent's response was the same as after the control period ($n = 7$, 1 tie, $t = 3.5$, NS).

Some studies of macaques have suggested that proximity alone may serve a reconciliatory function (de Waal 1989; Cords 1993). And, because baboons typically grunt when in relatively close proximity to one another, it might be argued that proximity, rather than the vocalization, was the reconciliatory mechanism.

In 23% of the 'no reconciliation' fights, dominant opponents approached their victims within the next 10 minutes without vocalizing or interacting with them in any other way. Had proximity alone acted to reconcile opponents, subjects that had simply been approached by their opponents following a fight should have responded as weakly to the playbacks as did subjects that received a grunt. This, however, was not true. Subjects that had only been approached responded significantly more strongly than did subjects that had also received a grunt when they were approached (Mann-Whitney U test, $N_1 = 5$, $N_2 = 22$, $U = 22.5$, $P < 0.05$).

The mechanisms underlying monkeys' calls

Both observations and experiments suggest that vocalizations constitute a major component of reconciliatory behaviour in female baboons. Even in the absence of more overt friendly behaviour, baboon grunts act to restore the relationships of opponents to baseline tolerance levels. Grunts serve to mediate and repair social relationships. They also function to initiate and facilitate affinitive contact between individuals of disparate ranks that might not otherwise interact.

What, however, are the mechanisms underlying apparently reconciliatory grunts? Do dominant females give grunts with the intent of appeasing their former victims? One explanation for the prevalence of vocalizations following conflicts is that dominant females grunt in order to alleviate their opponent's anxiety and to reassure them that they are no longer angry. An equally plausible explanation, however, is that dominant females simply grunt to their victims because they are in a friendly mood and wish to interact with their opponents' infants.

Although these two explanations are functionally equivalent, they are based on quite different underlying mental mechanisms. The first explanation focuses on the signaller, and assumes that calling individuals attribute mental states different from their own to their audience. The second focuses on the audience, and assumes that listeners respond to calls on the basis of behavioural contingencies. This latter explanation requires that subordinate females learn, through experience, that grunts signal a low probability of attack; as a result, their anxiety is diminished when a dominant female grunts to them.

Despite their functional equivalence, the distinction between these two explanations is crucially important to any discussion concerned with the evolution of language. If, as Grice and others have argued, true linguistic communication cannot occur unless both speaker and listener take into account each other's states of mind, then monkeys cannot be said to communicate unless they use calls like reconciliatory grunts with the intent of influencing each others' beliefs and emotions. By contrast, if monkeys are incapable of recognizing the relationship between what an individual says and what she thinks, a call that functions to reconcile an opponent will be based on fundamentally different underlying mental mechanisms than reconciliation in the human sense of the term, in which individuals deliberately act to appease or overcome the distrust or animosity of another.

In fact, there is very little evidence that monkeys or other animals ever take into account their audience's mental states when calling to one another. Consider alarm calls, for example. The alarm calls of many birds and mammals are not obligatory, but depend on social context. Individuals often fail to give alarm calls when there is no functional advantage to be gained by alerting others; for instance, when they are alone or in the presence of unrelated individuals (e.g. ground squirrels, Sherman 1977; downy woodpeckers, Sullivan 1985; vervet monkeys, Cheney & Seyfarth 1985; roosters, Gyger *et al.* 1986). However, while this 'audience effect' clearly requires that a signaller monitor the presence and behaviour of group companions, it does not demand that he also distinguish between ignorance and knowledge on the part of his audience. Indeed, in all species studied thus far, signallers call regardless of whether or not their audience is already aware of danger. Vervet monkeys, for example, will continue to give alarm calls long after everyone in their group has seen the predator and retreated to safety.

Experiments with captive rhesus (*Macaca mulatta*) and Japanese macaques have demonstrated that mothers do not alter their alarm calling behaviour depending upon the mental states of their offspring. When given the opportunity to alert ignorant offspring of potential danger, they do not change their alarm calling behaviour (Cheney & Seyfarth 1990a). Similarly, if vervet monkeys attributed mental states different from their own to others, they might be expected to correct or instruct their offspring in the appropriate use of alarm calls. This they never do. Infant vervets give eagle alarm calls to many bird species, like pigeons, that pose no danger to them.

Adults, however, never correct their offspring when they make inappropriate alarm calls, nor do they selectively reinforce them when they give alarm calls to real predators, like martial eagles. Instead, infant vervets seem to learn appropriate usage simply by observing adults (Seyfarth & Cheney 1986).

In summary, there is no doubt that the alarm calls given by monkeys function to inform nearby listeners of quite specific sorts of danger. They seem, however, simply to mirror the intent and state of the signaller, and they fail to take into account their audience's mental states.

A similar disregard for one's audience's mental states seems to characterize the contact and food calls given by many species of animals. Despite numerous attempts to test the hypothesis that foraging animals share information about the location of food or each other's relative positions in the group progression, no study has yet been able to demonstrate that individuals deliberately inform one another. For example, although carrion birds and bats that feed on widely dispersed food sources could potentially share information at common roosting sites, individuals appear to locate food either by following others or simply by finding it themselves (e.g. crows: Richner & Marclay 1991; turkey vultures: Prior & Weatherhead 1991; bats: Wilkinson 1992; red kites: Hiraldo *et al.* 1993).

Even in the case of non-human primates, evidence for intentional information sharing is lacking. Although listeners can potentially use calls to maintain contact with signallers or to locate food resources, the proximate cause of the calls appears to be the current state or status of the signaller. There is no indication that signallers selectively answer the calls of separated individuals, or that they call more upon discovering a new food source than upon returning to a tree that was recently visited by many group members. For example, capuchins (*Cebus capucinus*) and squirrel monkeys (*Saimiri sciureus*) give progression calls primarily when they themselves are moving or about to move (Boinski 1991, 1993). Spider monkeys (*Ateles geoffroyi*) call when they arrive at a fruiting tree, but the calls only function to recruit other subgroups a small proportion of the time (Chapman & Levebre 1990). Similarly, although chimpanzees often give pant hoots upon arrival at large unoccupied fruiting trees, parties that call are not joined more than parties that remain silent (Clark & Wrangham 1994), nor are individuals that remain silent punished for failing to alert others (but see Hauser & Marler 1993 for a possible exception in rhesus macaques). These observations have forced some revision of the hypothesis that calls such as chimpanzees' pant hoots function to alert others to food (Wrangham 1977). Indeed, current evidence suggests that the calls may instead function to signal the caller's status (Mitani & Nishida 1993; Clark & Wrangham 1994).

Baboon contact barks

When moving through wooded areas, female and juvenile baboons often give loud barks that can be heard up to 500 metres away (see also Byrne 1981). These 'contact barks' can potentially function to maintain group cohesion because, upon hearing one or more barks, an individual that has lost contact with others knows immediately where at least some group members are.

Because contact barks are often temporally clumped, with many individuals giving calls at roughly the same time, baboons often appear to be answering one another. What is not clear, however, is whether baboons give such calls with the intent of maintaining contact with each other, or whether the calls simply reflect the signaller's own circumstances (i.e. separated from the group). Hypotheses based on mental state attribution predict that individuals will answer the contact barks of others even when they themselves are in the centre of the group progression and at no risk of becoming separated from others. If, however, baboons are incapable of understanding that other individuals' mental states can be different from their own, they should be unable to recognize when another individual has become separated from the group unless they themselves are also peripheral and at risk of becoming separated. Under these circumstances, contact barks will simply reflect the state and location of the signaller.

To test between these two hypotheses, we gathered data on the social context of the contact barks given by 23 adult females over a three month period. Analysis of almost 2000 individually identifiable barks revealed a highly significant clumping of calls. Indeed, 92% of the calls given by females occurred in the five minutes following a previous call from either another female, the caller herself, or both (see Cheney et al. 1996, for details of the sampling protocol).

If females had given 'answering' calls at random, then 96% (22/23) of each individual's calls should have followed a call by another female, and 4% (1/23) should have occurred following one of her own calls. In fact, the mean proportion of 'answering' calls that followed a call by another female was 74%. Twenty-two of the 23 females gave fewer contact barks in the five minutes following a contact bark by another female than would have been expected by chance (two-tailed binomial test, $P < 0.001$). Even close kin failed to answer each other's contact barks more often than expected by chance.

In contrast, the mean proportion of contact barks given by females that followed one of their own contact barks was 66%. All 23 females 'answered' themselves at least 10 times more than expected by chance ($P < 0.001$).

These data argue against the hypothesis that calls were clumped in time because females were answering one another. Instead, it seems that clumping of calls occurred primarily because each female herself, when she called, was likely to give a number of calls one after the other.

As a further test of the hypothesis that females did not answer the contact barks of other females, but instead gave barks depending primarily on their own position, we carried out a series of 36 playback experiments, in which we played to subjects the contact bark of a close female relative (either a mother, daughter, or sister) (Cheney et al. 1996).

In 19% of trials, subjects did in fact 'answer' their relative's contact bark by giving at least one bark themselves within the next five minutes. (In one additional experiment the subject called in the seventh minute after the playback.) In no case did other, unrelated females in the vicinity respond to the playbacks with a call.

At first inspection, these results might be taken as weak evidence for the selective exchanging of contact barks among close kin. Closer examination, however, reveals that subjects 'answered' playbacks of their relatives' barks primarily when they themselves were peripheral and at risk of becoming separated from the group. Subjects that were in the last third of the group progression were significantly more likely to answer their relatives' contact barks than were subjects that were in the first two thirds (Figure 3; $X^2 = 4.43$, $P < 0.05$). Similarly, they were significantly more likely to give answering barks when there was no other female within 25 metres than if there was at least one other female nearby (Figure 3; $X^2 = 5.86$, $P < 0.05$).

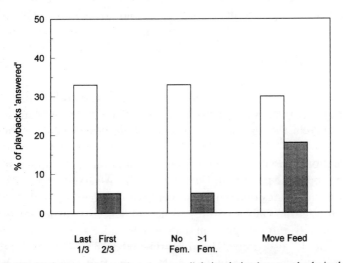

Figure 3. The context in which subjects 'answered' their relatives' contact barks in the 5 minutes following playback. Data are based on 36 trials involving 18 subjects.

Subjects were also more likely to call when the group was moving rather than feeding, though not significantly so.

Both observations and experiments suggest, therefore, that baboons do not give contact barks with the intent of sharing information, even though the calls may ultimately function to allow widely separated individuals to maintain contact with one another. Like the progression, contact, and food calls given by other species of primates, baboon contact barks appear to reflect the signaller's own state and position rather than the state and position of others.

DISCUSSION

The vocalizations of non-human primates share a number of similarities with human speech. Many of the calls given by vervet monkeys, for example, are functionally semantic and serve to denote objects or events in the external world. Vervets seem to compare and classify calls according to their meaning, and not just their acoustic properties. They judge some acoustically different calls to be the same when the calls refer to similar events. The monkeys behave, in other words, as if they recognize the referential relation between calls and the things for which they stand.

The calls given by monkeys during social interactions also appear to serve many of the same purposes as human speech, in the sense that they act to mediate social interactions, to appease, and to reconcile. Other calls function to inform individuals about the caller's location and to maintain group contact and cohesion.

Despite these functional similarities, however, the mental mechanisms underlying non-human primate vocalizations appear to be fundamentally different from the mechanisms underlying adult human speech. When calling to one another, monkeys seem to lack one of the essential requirements of human speech: the ability to take into account their audience's mental states.

Explanations based on mental state attribution make quite specific predictions about the pattern and context of calls. A vervet or macaque that attributes mental states different from her own to others should adjust her alarm or inter-group calls according to her audience's knowledge, and she should selectively inform ignorant individuals more than knowledgeable ones. She should also correct her offspring when it gives alarm calls to inappropriate species. A dominant female baboon that attributes emotions to others should grunt to a subordinate victim in order to alleviate her victim's anxiety even though, being dominant, she feels no anxiety herself. Similarly, a baboon capable of attributing confusion or anxiety to others should answer other individuals' contact barks regardless of her own position in the group progression.

Despite a variety of tests, however, there is no evidence that monkeys attribute mental states to one another. Monkeys appear not to call with the intent of providing information or influencing listeners' beliefs. Instead, listeners appear to respond to calls based on learned behavioural contingencies.

Although vervet monkeys, like many other species of birds and mammals, may vary their rates of alarm calling depending upon the composition of their audience, they do not act deliberately to inform ignorant individuals more than knowledgeable ones (Cheney & Seyfarth 1990a, 1990b). A vervet's alarm call alerts other animals regardless of whether or not they are already aware of the danger. In a like manner, infant vervets are not explicitly instructed to respond to some prey species rather than others. Instead, they learn to recognize their predators by observing the behaviour of adults (Seyfarth & Cheney 1986).

In the case of baboons' reconciliatory grunts, it seems likely that dominant females grunt to their former victims because they wish to interact in a friendly way with them, usually because these individuals have young infants (Cheney *et al.* 1995; Silk, Cheney, & Seyfarth, in preparation). Through past experience, and perhaps also be observing the interactions of others, the victims learn that grunts honestly signal a low probability of aggression. They therefore relax when their former opponents approach.

Similarly, baboons give contact barks when they are at the group's periphery and at risk of becoming separated from others. Through experience, listeners learn that they can maintain contact with at least a subset of the group simply by listening to other individuals' calls.

In all cases, listeners are able to extract relevant information about a call's function based on their own experiences. Their responses need not take into account the signaller's mental states at all. Indeed, in each case, the meaning and function of the cells are to a large part determined by the listener rather than the signaller. Upon hearing a vervet's inter-group wrr, the listener deduces that another group is nearby, and this representation allows her to ignore any subsequent inter-group vocalizations, even those with different acoustic properties. Upon hearing a dominant baboon's grunt, the subordinate listener deduces that she will not be attacked. Upon hearing another baboon's contact bark, the listener deduces the group's location and direction of travel. In each case, the listener extracts rich, semantic information from a signaller who may not, in the human sense, have intended to provide it.

From the listener's perspective, then, non-human primate vocalizations share many similarities with human semantic signals. Not only do calls function to inform others of specific features of the environment and the signaller's emotions and intentions, but they also appear to be judged and classified according to the representations which they instantiate in the listener's mind.

From the signaller's perspective, however, there are striking discontinuities between non-human primate vocalizations and human language, at least as it manifested in adults. These discontinuities are based not so much on the formal properties of the calls themselves than on the mental mechanisms underlying call production. In marked contrast to adult human language, the calls of monkeys do not seem to take into account listeners' mental states. As a result, monkeys cannot communicate with the intent of appeasing those who are anxious or informing those who are ignorant.

There is no doubt that the vocal communication of non-human primates mediates complex social relationships and results in the transfer of quite specific sorts to information. Equally clearly, non-human primate vocalizations affect listeners' mental states, in the sense that they change what other individuals know about the world and affect what they are likely to do. Compared with human language, however, the vocalizations of monkeys achieve this end almost by accident, without individuals being aware of the features of the system in which they are participating. Monkeys, and perhaps also apes, are skilled at monitoring each other's behaviour. There is little evidence, however, that they are equally adept at monitoring each other's states of mind. A challenge for the future will be to identify the selective factors that might have favoured the evolution of mental state attribution in the language and behaviour of our early ancestors.

REFERENCES

Aureli, F. 1992: Post-conflict behaviour among wild long-tailed macaques (*Macaca fascicularis*). *Behavioral Ecology and Sociobiology* 31, 329–337.

Aureli, F., van Schaik, C. & van Hooff, J.A.R.A.M. 1989: Functional aspects of reconciliation among captive long-tailed macaques (*Macaca fascicularis*). *American Journal of Primatology* 19, 39–51.

Bauers, K.A. 1993: A functional analysis of staccato grunt vocalizations in the stumptailed macaque (*Macaca arctoides*). *Ethology* 94, 147–161.

Bennett, J. 1976: *Linguistic Behaviour*. Cambridge: Cambridge University Press.

Boinski, S. 1991: The coordination of spatial position: A field study of the vocal behaviour of adult female squirrel monkeys. *Animal Behaviour* 41, 89–102.

Boinski, S. 1993: Vocal coordination of troop movement among white-faced capuchin monkeys, *Cebus capucinus*. *American Journal of Primatology* 30, 85–100.

Byrne, R.W. 1981: Distance vocalisations of Guinea baboons (*Papio papio*) in Senegal: an analysis of function. *Behaviour* 78, 283–313.
Chapman, C.A. & Levebre, L. 1990: Manipulating foraging group size: spider monkey food calls at fruiting trees. *Animal Behaviour* 39, 891–896.
Cheney, D.L. 1981: Inter-group encounters among free-ranging vervet monkeys. *Folia primatologica* 35, 124–146.
Cheney, D.L. & Seyfarth, R.M. 1985: Vervet monkey alarm calls: Manipulation through shared information? *Behaviour* 93, 150–166.
Cheney, D.L. & Seyfarth, R.M. 1988: Assessment of meaning and the detection of unreliable signals in vervet monkeys. *Animal Behaviour* 36, 477–486.
Cheney, D.L. & Seyfarth, R.M. 1989: Redirected aggression and reconciliation among vervet monkeys, *Cercopithecus aethiops*. *Behaviour*, 110, 258–275.
Cheney, D.L. & Seyfarth, R.M. 1990a. Attending to behaviour versus attending to knowledge: Examining monkeys' attribution of mental states. *Animal Behaviour* 40, 742–753.
Cheney, D.L. & Seyfarth, R.M. 1990b. *How Monkeys See the World: Inside the Mind of Another Species*. Chicago: University of Chicago Press.
Cheney, D.L., Seyfarth, R.M. & Silk, J.B. 1995: The role of grunts in reconciling opponents and facilitating interactions among adult female baboons. *Animal Behaviour* 50, 249–257.
Cheney, D.L., Seyfarth, R.M. & Palombit, R. 1996: Function and mechanisms underlying baboon contact barks. *Animal Behaviour* 52.
Clark, A.P. & Wrangham, R.W. 1994: Chimpanzee arrival pant-hoots: do they signify food or status? *International Journal of Primatology* 15, 185–205.
Cords, M. 1992: Post-conflict reunions and reconciliation in long-tailed macaques. *Animal Behaviour* 44, 57–61.
Cords, M. 1993: On operationally defining reconciliation. *American Journal of Primatology* 29, 255–267.
de Waal, F.B.M. 1989: *Peacemaking Among Primates*. Cambridge, Massachusetts: Harvard University Press.
de Waal, F.B.M. & van Roosmalen, A. 1979: Reconciliation and consolation among chimpanzees. *Behavioral Ecology and Sociobiology* 5, 55–66.
de Waal, F.B.M. & Yoshihara, D. 1983: Reconciliation and redirected affection in rhesus monkeys. *Behaviour* 85, 224–241.
Dennett, D.C. 1987: *The Intentional Stance*. Cambridge, MA: MIT/Bradford Books.
Grice, H.P. 1957: Meaning. *Philosophical Review* 66, 377–388.
Gyger, M., Karakashian, S.J. & Marler, P. 1986: Avian alarm-calling: Is there an audience effect? *Animal Behaviour* 34, 1570–1572.
Hauser, M.D. & Marler, P. 1993: Food-associated calls in rhesus macaques (*Macaca mulatta*). II. Costs and benefits of call production and suppression. *Behavioral Ecology* 4, 206–212.
Hausfater, G., Altmann, J. & Altmann, S. 1982: Long-term consistency of dominance relations in baboons. *Science* 217, 752–755.
Hiraldo, F., Heredia, B., & Alonso, J.C. 1993: Communal roosting of wintering red kites, *Milvus milvus*: Social feeding strategies for the exploitation of food resources. *Ethology* 93, 117–124.
Hockett, C.F. 1960: Logical considerations in the study of animal communication. In *Animal Sounds and Communication* (ed. W.E. Lanyon & W.N. Tavolga), pp. 292–340. Washington, D.C.: American Institute of Biological Sciences.
Jackendoff, R. 1994: *Patterns in the Mind*. New York: Basic Books.
Judge, P.D. 1991: Dyadic and triadic reconciliation in pigtail macaques (*Macaca nemestrina*). *American Journal of Primatology* 23, 225–237.
Masataka, N. 1989: Motivational referents of contact calls in Japanese macaques. *Ethology* 80, 265–273.

Matsuzawa, T. 1985: Color naming and classification in a chimpanzee (*Pan troglodytes*). *Journal of Human Evolution* 14, 283–291.

Mitani, J. & Nishida, T. 1993: Contexts and social correlates of long-distance calling by male chimpanzees. *Animal Behaviour* 45, 735–746.

Morton, E.S. 1977: On the occurrence and significance of motivation-structural rules in some bird and animal sounds. *American Naturalist* 111, 855–869.

Pinker, S. 1994: *The Language Instinct: How the Mind Creates Language*. New York: William Morrow & Co.

Povinelli, D.J. 1993: Reconstructing the evolution of mind. *American Psychologist* 48, 493–509.

Premack, D. 1976: *Intelligence in Ape and Man*. Hillsdale, NJ: Lawrence Erlbaum Assoc.

Prior, K.A. & Weatherhead, P.J. 1991: Turkey vultures foraging at experimental food patches: a test of information transfer at communal roosts. *Behavioral Ecology and Sociobiology* 28, 385–390.

Richner, H. & Marclay, C. 1991: Evolution of avian roosting behaviour: a test of the information centre hypothesis and of a critical assumption. *Animal Behaviour* 41, 433–438.

Robinson, J.G. 1984: Syntactic structures in the vocalizations of wedge-capped capuchin monkeys, *Cebus nigrivittatus*. *Behaviour* 90, 46–79.

Sakuro, O. 1989: Variability in contact calls between troops of Japanese macaques: a possible case of neutral evolution of animal culture. *Animal Behaviour* 38, 900–902.

Savage-Rumbaugh, E.S. 1986: *Ape Language: From Conditioned Response to Symbol*. New York: Columbia University Press.

Seyfarth, R.M. 1976: Social relationships among adult female baboons. *Animal Behaviour* 24, 917–938.

Seyfarth, R.M. & Cheney, D.L. 1986: Vocal development in vervet monkeys. *Animal Behaviour* 34, 1640–1658.

Seyfarth, R.M. & Cheney, D.L. 1992: Meaning and mind in monkeys. *Scientific American* 267, 122–129.

Seyfarth, R.M., Cheney, D.L. & Marler, P. 1980a. Monkey responses to three different alarm calls: Evidence for predator classification and semantic communication. *Science* 210, 801–803.

Seyfarth, R.M., Cheney, D.L. & Marler, P. 1980b. Vervet monkey alarm calls: Semantic communication in a free-ranging primate. *Animal Behaviour* 28, 1070–1094.

Sherman, P.W. 1977: Nepotism and the evolution of alarm calls. *Science* 197, 1246–1253.

Smuts, B. & Nicolson, N. 1989: Reproduction in wild female baboons. *American Journal of Primatology* 19, 229–246.

Struhsaker, T.T. 1967: Auditory communication among vervet monkeys (*Cercopithecus aethiops*). In *Social Communication among Primates* (ed. S.A. Altmann), pp. 281–324. Chicago: University of Chicago Press.

Sullivan, K. 1985: Selective alarm-calling by downy woodpeckers in mixed-species flocks. *Auk* 102, 184–187.

Tiles, J.E. 1987: Meaning. In *The Oxford Companion to the Mind* (ed. R.L. Gregory), pp. 450–454. Oxford; Oxford University Press.

van Hooff, J.A.R.A.M. 1994: Understanding chimpanzee understanding. In *Chimpanzee Cultures* (ed. R.W. Wrangham, W.C. McGrew, F.B.M. de Waal & P.G. Heltne), pp. 267–284. Cambridge, MA: Harvard University Press.

Wilkinson, G.S. 1992: Information transfer at evening bat colonies. *Animal Behaviour* 44, 501–518.

Wrangham, R.W. 1977: Feeding behaviour of chimpanzees in Gombe National Park, Tanzania. In *Primate Ecology* (ed. T.H. Clutton-Brock), pp. 178–212. New York: Academic Press.

York, A.D. & Rowell, T.E. 1988: Reconciliation following aggression in patas monkeys, *Erythrocebus patas*. *Animal Behaviour* 36, 502–509.

Why Culture is Common, but Cultural Evolution is Rare

ROBERT BOYD* & PETER J. RICHERSON†

*Department of Anthropology, University of California,
Los Angeles, CA 90024, USA
†Division of Environmental Studies, University of California,
Davis, CA 95616, USA

Keywords: cultural evolution; social learning; dual inheritance; imitation.

Summary. If culture is defined as variation acquired and maintained by social learning, then culture is common in nature. However, cumulative cultural evolution resulting in behaviours that no individual could invent on their own is limited to humans, song birds, and perhaps chimpanzees. Circumstantial evidence suggests that cumulative cultural evolution requires the capacity for observational learning. Here, we analyse two models of the evolution of psychological capacities that allow cumulative cultural evolution. Both models suggest that the conditions which allow the evolution of such capacities when rare are much more stringent than the conditions which allow the maintenance of the capacities when common. This result follows from the fact that the assumed benefit of the capacities, cumulative cultural adaptation, cannot occur when the capacities are rare. These results suggest why such capacities may be rare in nature.

INTRODUCTION

CULTURAL VARIATION IS COMMON IN NATURE. In creatures as diverse as rats, pigeons, chimpanzees, and octopuses, behaviour is acquired through social learning. As a result, the presence of a particular behaviour in a

© The British Academy 1996.

population makes it more likely that individuals in the next generation will acquire the same behaviour which, in turn, results in persistent differences between populations that are not due to genetic or environmental differences.

In sharp contrast, cumulative cultural evolution is rare. Most culture in non-human animals involves behaviours that individuals can, and do, learn on their own. There are only a few well documented cases in which cultural change accumulates over many generations leading to the evolution of behaviours that no individual could invent—the only well documented examples are song dialects in birds, perhaps some behaviours in chimpanzees, and of course many aspects of human behaviour.

We believe that this situation presents an important evolutionary puzzle. The ability to accumulate socially learned behaviours over many generations has allowed humans to develop subtle, powerful technologies, and to assemble complex institutions that permit us to live in larger, and more complex societies than any other mammal species. These accumulated cultural traditions allow us to exploit a far wider range of habitats than any other animal, so that even with only hunting and gathering technology, humans became the most widespread mammal on earth. The fact that simple forms of cultural variation exist in a wide variety of organisms suggests that intelligence and social life alone are not sufficient to allow cumulative cultural evolution. Cumulative cultural change seems to require some special, derived, probably psychological, capacity. Thus we have the puzzle, if cultural traditions are such a potent means of adaptation, why is this capacity rare?

In this paper we suggest one possible answer to this question. We begin by reviewing the literature on animal social learning. We then analyse two models of the evolution of the psychological capacities that allow cumulative cultural evolution. The results of these models suggest a possible reason why such capacities are rare.

CULTURE IN OTHER ANIMALS

There has been much debate about whether other animals have culture. Some authors define culture in human terms. That is, the investigator essays human cultural behaviour and extracts a number of 'essential' features. For example Tomasello *et al.* (1993) argue that culture is learned by all group members, faithfully transmitted, and subject to cumulative change. Then to be cultural, the behaviour of other animals must exhibit these features. Moreover, a heavy burden of proof is placed on those who would claim culture for other animals—if there is any other plausible interpretation, it is preferable. Others (McGrew 1992; Boesch 1993) argue that a double standard is being applied. If the behavioural variation observed among chimpanzee populations were instead observed among human populations, they argue, anthropologists would regard it as cultural.

Such debates make little sense from an evolutionary perspective. The psychological capacities that underpin human culture must have homologies in the brains of other primates, and perhaps other mammals as well. Moreover, the functional significance of social transmission in humans could well be related to its functional significance in other species. The study of the evolution of human culture must be based on categories that allow human cultural behaviour to be compared to potentially homologous, functionally related behaviour of other organisms. At the same time, such categories should be able to distinguish between human behaviour and the behaviour of other organisms because it is quite plausible that human culture is different in important ways from related behaviour in other species.

Here we define cultural variation as differences among individuals that exist because they have acquired different behaviour as a result of some form of social learning. Cultural variation is contrasted with genetic variation, differences between individuals that exist because they have inherited different genes from their parents, and environmental variation, differences between individuals due to the fact that they have experienced different environments. Cultural variation is often lumped together with environmental variation. However, as we have argued at length elsewhere (Boyd & Richerson 1985), this is an error. Because cultural variation is transmitted from individual to individual it is subject to population dynamic processes analogous to those that effect genetic variation and quite unlike the processes that govern other environmental effects. Combining cultural and environmental effects into a single category conceals these important differences.

There is much evidence that cultural variation, defined this way, is very common in nature. In a review of social transmission of foraging behaviour, Levebre & Palameta (1988) give 97 examples of cultural variation in foraging behaviour in animals as diverse as baboons, sparrows, lizards, and fish. Song dialects are socially transmitted in many species of songbirds. Three decades of study shows that chimpanzees have cultural variation in subsistence techniques, tool use, and social behaviour (Wrangham et al. 1994; McGrew 1992).

There is little evidence, however, of cumulative cultural evolution in other species. With a few exceptions, social learning leads to the spread of behaviours that individuals could have learned on their own. For example, food preferences are socially transmitted in rats. Young rats acquire a preference for a food when they smell the food on the pelage of other rats (Galef 1988). This process can cause the preference for a new food to spread within a population. It can also lead to behavioural differences among populations living in the same environment, because current foraging behaviour depends on a history of social learning. However, it does not lead to the cumulative evolution of new, complex behaviours that no individual rat could learn on its own.

In contrast, human cultures do accumulate changes over many generations, resulting in culturally transmitted behaviours that no single human individual could invent on their own. Even in the simplest hunting and gathering societies people depend on such complex, evolved knowledge and technology. To live in the arid Kalahari, the !Kung San need to know what plants are edible, how to find them during different seasons, how to find water, how to track and find game, how to make bows and arrow poison, and many other skills. The fact that the !Kung can acquire the knowledge, tools, and skills necessary to survive the rigors of the Kalahari is not so surprising—many other species can do the same. What is amazing is that the same brain that allows the !Kung to survive in the Kalahari, also permits the Inuit to acquire the very different knowledge, tools, and skills necessary to live on the tundra and ice north of the Arctic circle, and the Ache the knowledge, tools, and skills necessary to live in the tropical forests of Paraguay. There is no other animal that occupies a comparable range of habitats or utilizes a comparable range of subsistence techniques and social structures. Two kinds of evidence indicate that such differences result from cumulative cultural evolution of complex traditions. First, such gradual change is documented in both the historical and archaeological records. Second, cumulative change leads to a branching pattern of descent with modification in which more closely related populations share more derived characters than distantly related populations. Although the possibility of horizontal transmission among cultural lineages makes reconstructing such cultural phylogenies difficult for 'cultures' (Boyd *et al.* in press), patterns of cultural descent can be reconstructed for particular cultural components, such as language or technologies.

Circumstantial evidence suggests that the ability to acquire novel behaviours by observation is an essential for cumulative cultural change. Students of animal social learning distinguish *observational learning or true imitation*, which occurs when younger animals observe the behaviour of older animals and learn how to perform a novel behaviour by watching them, from a number of other mechanisms of social transmission which also lead to behavioural continuity without observational learning (Galef 1988; Visalberghi & Fragazy, 1990; Whiten & Ham 1992). One such mechanism, *local enhancement*, occurs when the activity of older animals increases the chance that younger animals will learn the behaviour on their own. If younger, naive individuals are attracted to the locations in the environment where older, experienced individuals are active they will tend to learn the same behaviours as the older individuals. Young individuals do not acquire the information necessary to perform the behaviour by observing older individuals. Instead, the activity of others causes them to be more likely to acquire this information through interaction with the environment. Imagine a young monkey acquiring its food preferences as it follows its mother around. Even if the young monkey never pays any attention to what its mother eats, she will lead it to locations where some foods are common and others rare, and the young monkey may learn to eat much the same foods as mom.

Local enhancement and observational learning are similar in that they both can lead to persistent behavioural differences among populations, but only observational learning allows cumulative cultural change (Tomasello et al. 1993). To see why, consider the cultural transmission of stone tool use. Suppose that on their own in especially favourable circumstances, an occasional early hominid learned to strike rocks together to make useful flakes. Their companions, who spent time near them, would be exposed to the same kinds of conditions and some of them might learn to make flakes too, entirely on their own. This behaviour could be preserved by local enhancement because groups in which tools were used would spend more time in proximity to the appropriate stones. However, that would be as far as it would go. Even if an especially talented individual found a way to improve the flakes, this innovation would not spread to other members of the group because each individual learned the behaviour anew. Local enhancement is limited by the learning capabilities of individuals, and the fact that each new learner must start from scratch. With observational learning, on the other hand, innovations can persist as long as younger individuals are able to acquire the modified behaviour by observational learning. To the extent that observers can use the behaviour of models as a starting point, observational learning can lead to the cumulative evolution of behaviours that no single individual could invent on its own.

Most students of animal social learning believe that observational learning is limited to humans, and perhaps, chimpanzees and some bird species. Several lines of evidence suggest that observational learning is not responsible for cultural traditions in other animals. First, many of the behaviours, like potato washing in Japanese macaques, are relatively simple, and could be learned independently by individuals in each generation. Second, new behaviours like potato washing often take a long time to spread through the group, a pace more consistent with the idea that each individual had to learn the behaviour on its own. Finally, extensive laboratory experiments capable of distinguishing observational learning from other forms of social transmission like local enhancement have usually failed to demonstrated observational learning (Galef 1988; Whiten & Ham 1992; Tomasello *et al.* 1993; Visalberghi 1993), except in humans and song birds. (In many song birds, song traditions are transmitted by imitation, but little or nothing else is.) The fact that observational learning appears limited to humans seems to confirm that observational learning is necessary for cumulative cultural change. However, one must be cautious here because most students of animal social learning refuse to invoke observational learning unless all other possible explanations have been excluded. Thus, there actually may be many cases of observational learning that are interpreted as social enhancement or some putatively simpler mechanism. A few well controlled laboratory studies do apparently show some true imitation in non-human animals (Heyes 1993; Dawson & Foss 1965), and striking anecdotes suggest that observational learning may occur in organisms as diverse as parrots (Pepperberg 1988) and orangutans (Russon & Galdikas 1993).

Adaptation by cumulative cultural evolution is apparently not a by-product of intelligence and social life. Cebus monkeys are among the world's cleverest creatures. In nature, they use tools and perform many complex behaviours, and in captivity, they can be taught extremely demanding tasks. Cebus monkeys live in social groups and have ample opportunity to observe the behaviour of other individuals of their own species. Yet good laboratory evidence suggests that cebus monkeys make no use of observational learning. This suggests that observational learning is not simply a by-product of intelligence and the opportunity to observe conspecifics. Rather, observational learning seems to require special psychological mechanisms (Bandura 1986). This conclusion suggests, in turn, that the psychological mechanisms that enable humans to learn by observation are adaptations have been shaped by natural selection because culture is beneficial. Of course, this need not be the case. Observational learning could be a by-product of some other adaptation that is unique to humans, such as bipedalism, dependence on complex vocal communication, or the capacity for deception. However, given the great importance of culture in human affairs, it is reasonable to think about the possible adaptive advantages of culture. In what follows we consider the two mathematical models of the evolution of the capacity for observational learning based on this assumption.

MODELS OF THE EVOLUTION OF SOCIAL LEARNING

The maintenance of cultural variation involves two quite different processes (Figure 1). First, there must be some kind of *transmission* of information

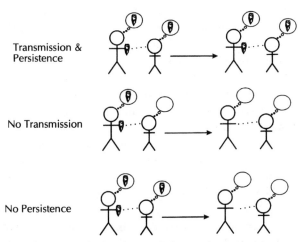

Figure 1. The maintenance of cultural transmission requires both the accurate transmission of mental representations from experienced to inexperienced individuals, and the persistence of those representations through the lives of individuals until such time that they act as models for others.

from one brain to another. Consider, for example, the maintenance of the use of a particular kind of tool. Individuals have information stored in their brain that allows them to manufacture and use the tool. For use of the tool to persist through time, observing tool use and manufacture must cause individuals in the next 'generation' to acquire information that allows them to manufacture and use the same tool. (We put generation in quotes because the same model can be used to represent culture change occuring on much shorter time scales. See Boyd & Richerson 1985: 68–69.) As we have seen, this transmission may occur because individuals can learn how to make and use tools by observation, or because observation stimulates them to learn on their own how to make and use the tool, for example by local enhancement. Second, individuals must preserve the information that allows them to make and use the tool until such time that they serve as models for the next generation of individuals. Such *persistence* may fail to occur for two different reasons: individuals may forget how to make or use the tool, or they may, as a result of interacting with the environment, modify the information stored in their brains so that they make or use the tool in a significantly different way. Without both transmission and persistence there cannot be culturally transmitted variation.

Our previous work on the evolution of culture (Boyd & Richerson 1985, 1988, 1989, in press) has focused on the evolution of persistence. All of the models analysed in these studies assume that transmission occurs, and consider the evolution of genes that affect the extent to which behaviour acquired by imitation is modified by individual learning. They differ in how the trait is modelled (discrete vs. continuous), how environmental variation is modelled, whether individuals are sensitive to the number of models who exhibit a particular cultural variant, and a number of other features. This work leads to the robust conclusion that natural selection will favour individuals who do not modify culturally acquired behaviour when individual learning is costly or error prone, and environments are variable, but not too variable. Thus, natural selection can favour persistence. (See Rogers 1989 for a related model.)

In several papers, Feldman and his co-workers (Cavalli-Sforza & Feldman 1983a, 1983b; Aoki & Feldman 1987) have considered the evolution of genes that affect transmission. In these models it is assumed that there is a beneficial trait that can only be acquired by cultural transmission, not by individual learning. They further allow for the possibility that successful transmission requires new behaviour both on the part of the individual acquiring the behaviour and in the individual modelling the behaviour. Thus there are two different genetic loci, one affecting the behaviour of the transmitter and a second affecting the behaviour of the receiver. For transmission to evolve, there must be substitutions at both loci. These models are very relevant to the evolution of communication systems. However, they cannot address the questions posed here because the culturally transmitted trait cannot be acquired or modified by individual learning.

Here we consider two models of the evolution of psychological capacities that allow the transmission of behaviour that can be acquired or modified through individual learning. Each model is designed to answer the same basic question: What are the conditions under which selection can favour a costly psychological capacity that allows individuals to acquire behaviour by imitation? The primary difference between the models is the nature of the culturally transmitted behaviour. In the first model, the behaviour is discrete—individuals are either skilled or unskilled, and the skill can be acquired either by social or individual learning. In the second model, there is a continuum of behaviours subject to stabilizing selection. Only the continuous trait model allows true cumulative cultural change leading to behaviours that individuals cannot learn on their own. However, the discrete model allows us to investigate the effects of several factors that are difficult to include in the continuous character model. As we will see, both models tell a similar story about why there is a selective barrier to the evolution of the capacity for observational learning, and why capacities that allow local enhancement and related mechanisms do not face a similar barrier.

Discrete character model

Consider an organism that lives in a temporally variable environment that can be in an infinite number of states. In each state, individuals can acquire a skill which increases fitness, so that unskilled individuals have fitness W_0, and skilled individuals have fitness $W_0 + D$. Each generation there is a probability γ that the environment switches from its current state to a different state. When this occurs, the old skill is no longer useful in the new environment.

There are two genotypes with different learning rules. *Individual learners* acquire the skill appropriate to the current environment with probability δ at a cost C_I. *Social learners* observe n randomly selected members of the previous generation. If there is a skilled individual among the n, an imitator acquires the skill at cost C_S. Otherwise they acquire the skill with probability δ at a cost C_I. The ability to acquire the skill by social learning reduces the fitness of an individual an amount K. Thus, parameters C_I and C_S give the variable costs of individual and social learning respectively, and K gives the fixed cost associated with the capacity for social learning.

It is shown in the appendix that social learning can increase when rare, and is the only ESS when the following condition holds

$$(1 - (1 - \delta)^n)(1 - \gamma)[D(1 - \delta) + C_I - C_S] > K \tag{1}$$

When (1) is true, social learning has higher fitness than individual learning no matter what the mix of the two types in the population. The term in square brackets gives the fitness benefit of acquiring the skill through social rather than individual learning—$C_I - C_S$ is the advantage that results from the fact that social learning may reduce the cost of acquiring the trait, and $D(1 - \delta)$ is the advantage that results from being more likely to acquire the skill. Sensibly, the latter term implies that the fitness advantage of social learning increases as the likelihood that individuals will learn the trait on their own, δ, decreases. The less likely it is that individual learners will acquire the skill, the bigger the relative advantage that accrues to social learning. The fitness benefit is discounted by the two factors on the left hand side of expression (1). The term $1 - \gamma$ expresses the fact that social learning is only beneficial if the environment has not changed, and term $1 - (1 - \delta)^n$ gives the probability that at least one of the n individuals from the previous generation will have acquired the behaviour when social learning is rare. Notice that this latter term decreases as the probability of learning the trait decreases. Thus the net advantage of social learning is highest at intermediate values of δ, when there is a good chance that individuals will learn the skill on their own, but also a good chance that they won't.

When (1) is not satisfied, there is a range of conditions in which social learning cannot increase when rare, but is an ESS once it becomes common. In this analysis we are limited to the case $n = 1$ because when $n > 1$ the dynamics of the cultural traits are nonlinear, and such systems are difficult to analyse in autocorrelated random environments. With this assumption, social learning is an ESS when:

$$\frac{\delta(1-\gamma)(D(1-\delta)+C_I-C_S)}{\gamma+(1-\gamma)\delta} > K \tag{2}$$

To compare this expression with (1), notice that when $n = 1$, $1-(1-\delta)^n = \delta$, and thus, the benefit of social learning when it is common is the benefit when rare divided by the term $\gamma+(1-\gamma)\delta$. When individual learners are likely to acquire the skill (so that δ is large), the conditions for social learning to increase when rare (1) and to persist when common (2) will be similar. However, when individual learners are unlikely to acquire the skill ($\delta \ll 1$) and the rate of environmental change is slow ($\gamma \ll 1$), social learning will be able to persist when common under a much wider range of conditions than it can increase when it is rare. When social learning is rare, most of the population will be individual learners who have little chance of acquiring the skill. As a consequence, social learning will provide little benefit because there will be few skilled individuals to observe. When social learning is common, the population will slowly accumulate the skill over many generations. If the environment does not change too often, the social learning population will spend most of the time with the skill at high frequency, and thus the cost of the capacity for social learning need only be less than the net benefit of acquiring the skill by individual learning.

Continuous character model

Consider an organism that is characterized by a single quantitative character that is subject to stabilizing selection. During generation t the optimum value of the quantitative character is θ_t. Each generation there is a probability γ that the environment changes. If the environment does not change then $\theta_{t+1} = \theta_t$. If it does change, then θ_{t+1} is a normal random variable with mean Θ, and variance H. Notice that this assumption implies that Θ is the long run optimum trait value.

Each individual acquires its trait value through a combination of genetic transmission, imitation, and individual learning. The adult trait value, x, is given by:

$$x = (1-a)[(1-i)\Theta + iy] + a\theta_t \tag{3}$$

The term $(1 - i)\Theta + iy$ represents a 'norm of reaction' which forms the basis for subsequent individual learning. It is acquired as the result of a combination of a genetically acquired norm of reaction at the long run optimum, Θ, and the observed trait value, y, of a randomly selected member of the previous generation. The parameter i governs the relative importance of genetic inheritance and imitation in determining the norm of reaction. When $i = 0$, the norm of reaction is completely determined by an innate, genetically inherited value. As i increases, the observed trait value of another individual has greater influence on the trait until, when $i = 1$, the norm of reaction is completely determined by observational learning. Because observational learning is assumed to require special purpose cognitive machinery, individuals incur a fitness cost proportional to the importance of observational learning in determining their norm of reaction, iC. Thus, C measures the incremental cost of the capacity for observational learning. Individuals adjust their adult behaviour from the norm of reaction toward the current optimum a fraction a. To capture the idea that cumulative change is possible we assume that a is small, so that the repeated action of learning and social transmission can lead to fitness increases that could not be attained by individual learning.

With these assumptions it is shown in the appendix that a population in which most individuals do not imitate can be invaded by rare individuals who imitate a little bit only if

$$(1 - \gamma)aH > C \tag{4}$$

The parameter H is a measure of how far the population is from the optimum in fitness units, on average, immediately after an environmental change. Since a population without imitation always starts from the same norm of reaction, Θ, the term aH is a measure of the average fitness improvement due to individual learning in a single generation. Thus, (4) says that imitation can evolve only when the benefit of imitating what individuals can learn on their own is sufficient to compensate for the costs of the capacity to imitate.

In contrast, the condition for social learning to be maintained once it is common is much more easily satisfied. It is shown in the appendix that a population in which $i = 1$ can resist invasion by rare alleles that reduce the reliance on imitation whenever:

$$\frac{(1 - \gamma)aH}{\gamma + (1 - \gamma)a} > C \tag{5}$$

If the rate at which the population adapts by individual learning, a, is greater than the rate at which the environment changes, γ, then a population in which social learning is common spends most of its time with the mean behaviour near the optimum. Thus, (5) says that imitation is evolutionarily stable as long as the cost of the capacity is less than a substantial fraction of the total improvement in fitness due to many generations of social learning.

DISCUSSION

Both of these models tell a similar story about the evolution of capacities that allow social learning. When social learning is rare, the only useful behaviour that is present in the population, and thus the only behaviour that can be acquired by social learning, is behaviour that individuals can learn on their own. In contrast, when social learning is common the population accumulates adaptive behaviour over many generations, and, as long as the environment does not change faster than adaptive behaviour accumulates, social learning allows individuals to acquire behaviours that are much *more* adaptive than they could acquire on their own.

This result provides a potential explanation for why cultural variation is so common in nature, but cumulative cultural evolution so rare. Capacities that increase the chance that individuals will learn behaviours that they could learn on their own will be favoured as long as they are relatively cheap. On the other hand, even though the benefits of cumulative cultural evolution are potentially substantial, selection cannot favour a capacity for observational learning when rare. Thus unless observational learning substantially reduces the cost of individual learning, it will not increase because there is an 'adaptive valley' that must be crossed before benefits of cumulative cultural change are realized. This argument suggests, in turn, that it is likely that the capacities that allow the initial evolution of observational learning must evolve as a side effect of some other adaptive change. For example, it has been argued that observational learning requires that individuals have what psychologists and philosophers call a 'theory of mind (Cheney & Seyfarth 1990; Tomasello et al. 1993).' That is, imitators must be able to understand that others have different beliefs and goals than they. Lacking such a theory, typical animals cannot make a connection between the acts of other animals and their own goal states, and thus can't interpret the acts of other animals as acts they might usefully perform. A theory of mind may have initially evolved to allow individuals to better predict the behaviour of other members of their social group. Once it had evolved for that reason it could be elaborated because it allowed observational learning and cumulative cultural evolution.

REFERENCES

Aoki, K. & Feldman, M.W. 1987: Toward a theory for the evolution of cultural communication: Coevolution of signal transmission and reception *Proceedings of the National Academy of Sciences, U.S.A.* 84, 7164–8.

Bandura, A. 1986: *Social Foundations of Thought and Action: A Social Cognitive Theory.* Englewood Cliffs, NJ: Prentice Hall.

Boesch, C. 1993: Aspects of transmission of tool-use in wild chimpanzees. In *Tools, Language, and Cognition in Human Evolution* (ed. K. R. Gibson & T. Ingold) pp. 171–189. Cambridge University Press, Cambridge

Boyd, R. & Richerson, P.J. 1985: *Culture and the Evolutionary Process.* University of Chicago Press, Chicago.

Boyd, R. & Richerson, P.J. 1988: An evolutionary model of social learning: the effects of spatial and temporal variation. In *Social Learning, Psychological and Biological Perspectives* (ed. T. Zentall & B.G. Galef), pp. 29–48. Lawrence Erlbaum Associates, Inc., Hillsdale, NJ.

Boyd, R. & Richerson, P.J. 1989: Social Learning as an Adaptation. *Lectures on Mathematics in the Life Sciences* 20, 1–26.

Boyd, R. & Richerson, P.J. 1995: Why does culture increase human adaptability? *Ethology and Sociobiology* 16, 125–141.

Boyd, R., Richerson, P.J., Borgerhoff Mulder, M. & Durham, W.H. (In press) Are cultural phylogenics possible? In *Human Nature, Between Biology and the Social Sciences* (ed. P. Weingart, P.J. Richerson, S. Mitchell & S. Maasen).

Cavalli-Sforza, L.L. & Feldman, M.W. 1983a: Paradox of the evolution of communication and of social interactivity. *Proceedings of the National Academy of Sciences, U.S.A.* 80, 2017–2021.

Cavalli-Sforza, L.L. & Feldman, M.W. 1983b: Cultural versus genetic adaptation. *Proceedings of the National Academy of Sciences, U.S.A.* 80, 4993–4996.

Cheney, D. & Seyfarth, R. 1990: *How Monkeys See the World.* University of Chicago Press, Chicago.

Dawson, B.V. & Foss, B.M. 1965: Observational learning in budgerigars. *Animal Behaviour* 13, 470–474.

Galef, B.G. 1988: Imitation in animals: History, definitions, and interpretation of data from the psychological laboratory. In *Social Learning, Psychological and Biological Perspectives* (ed. T. Zentall & B.G. Galef, Jr.), pp. 3–29. Lawrence Erlbaum Associates, Inc., Hillsdale, New Jersey.

Heyes, C.M. 1993: Imitation, culture and cognition. *Animal Behaviour* 46, 999–1010.

Levebre, L. & Palameta, B. 1988: Mechanisms, ecology, and population diffusion of socially-learned, food-finding behaviour in feral pigeons. In *Social Learning, Psychological and Biological Perspectives* (ed. T. Zentall & B.G. Galef, Jr.), pp. 141–165. Lawrence Erlbaum Associates, Inc., Hillsdale, New Jersey.

McGrew, W. 1992: *Chimpanzee Material Culture.* Cambridge University Press, Cambridge.

Pepperberg, I. 1988: The importance of social interaction and observation in the acquisition of communicative competence: Possible parallels between avian and human learning. In *Social Learning, Psychological and Biological Perspectives* (ed. T. Zentall & B.G. Galef, Jr.), pp. 3–29. Lawrence Erlbaum Associates, Inc., Hillsdale, New Jersey.

Rogers, A.R. 1989: Does biology constrain culture? *American Anthropologist* 90, 819–831.

Russon, A.E. & Galdikas, B. 1993: *Journal of Comparative Psychology.*

Tomasello, M., Kruger, A.C. & Ratner, H.H. 1993: Cultural learning. *Behavioral and Brain Sciences* 16, 495–552.

Visaberghi, E. 1993: Capuchin monkeys: A window into tool use in apes and humans. In *Tools, Language, and Cognition in Human Evolution* (ed. K.R. Gibson & T. Ingold), pp. 138–150. Cambridge University Press, Cambridge.

Visalberghi, E. & Fragazy, D.M. 1990: Do monkeys ape? In *Language and Intelligence in Monkeys and Apes* (ed. S. Parker & K. Gibson), pp. 247–273. Cambridge University Press, Cambridge.

Whiten, A. & Ham, R. 1992: On the nature and evolution of imitation in the animal kingdom: A reappraisal of a century of research. In *Advances in the Study of Behavior, Vol. 21* (ed. P.J.B. Slater, J.S. Rosenblatt, C. Beer & M. Milkinski), pp. 239–283. Academic Press, New York.

Wrangham, R.W., McGrew, W.C., DeWaal, F.B.M. & Heltne, P.G. 1994: *Chimpanzee Cultures*. Havard University Press, Cambridge.

APPENDIX

Analysis of discrete character model

Individual learners always have the same fitness:

$$W_I = W_0 + \delta D - C_I. \tag{A1.1}$$

The expected fitness of social learners depends on the frequency of social learners in the previous generation, q, the frequency of skilled individuals among social learners, p, and whether the environment has changed during the previous generation.

$$W_S = \gamma(W_0 + \delta D - C_l)$$
$$+ (1 - \gamma)(W_0 + \pi(D - C_s) + (1 - \pi)(\delta D - C_l)) \tag{A1.2}$$

where π is the probability that at least one of the n individuals in the sample of models has acquired the skill favoured in the previous environment, and can be calculated as below:

$$\pi = \sum_{i=0}^{n} \binom{n}{i} q^i (1-q)^{n-i} [1 - (1-p)^i (1-\delta)^{n-i}]. \tag{A1.3}$$

To understand this expression assume that there are i social learners among the n models observed by a given, naive social learner. The probability that all i of the social learners are not skilled is $(1-p)^i$, and the probability that the remaining $n - i$ individual learners are not skilled is $(1-\delta)^{n-i}$, and therefore, the probability that there is at least one skilled individual among the n given that there are i social learners is $1 - (1-p)^i(1-\delta)^{n-i}$. Then to calculate π take the expectation over all values of i.

Thus, social learners will have higher fitness in a particular generation if

$$W_S - W_I = \pi(1 - \gamma)(D(1 - \delta) + C_I - C_S) - K > 0. \tag{A1.4}$$

We consider two special cases. Case 1: $q \approx 0$, $\pi \approx 1 - (1 - \delta)^n$. When social learners are rare, they will observe only individual learners, and thus the probability of observing at least one skilled individual does not depend on q or p. Thus, social learning will increase when rare if

$$(1 - (1 - \delta)^n)(1 - \gamma)(D(1 - \delta) + C_I - C_S) - K > 0. \tag{A1.5}$$

Immediately after an environmental change, the frequency of skilled individuals among social learners is δ, and then increases monotonically until the next environmental change. Thus the expected value of π is greater than $(1 - (1 - \delta)^n)$, and if social learning can increase when rare it will continue to increase until it reaches fixation.

Case 2: $n = 1$, $\pi = 1 - q(1 - p) - (1 - q)(1 - \delta)$. Assume that selection is sufficiently weak so that the effect of selection on cultural evolution can be ignored (i.e., on dynamics of p), and genetic evolution (the dynamics of q) responds to the stationary distribution of p.

Then the frequency of the currently favoured behaviour after learning and imitation is

$$p' = \begin{cases} \delta & \text{if environment changes} \\ (qp + (1-q)\delta)(1 - \delta) + \delta & \text{if environment does not change.} \end{cases} \tag{A1.6}$$

Suppose at some time t the probability density for p is $f_t(p)$ with mean P_t. Then the mean of $f_{t+1}(p)$ given by

$$P_{t+1} = \int [(1 - \gamma)((qp + (1-q)\delta)(1 - \delta) + \delta) + \gamma\delta] f_t(p) dp. \tag{A1.7}$$

Integrating yields the following recursion for P_t

$$P_{t+1} = \gamma\delta + (1 - \gamma)[(qP_t + (1-q)\delta)(1 - \delta) + \delta]. \tag{A1.8}$$

Thus the equilibrium value of mean frequency of the favoured behaviour is:

$$P = \frac{\delta + (1 - \gamma)(1 - q)\delta(1 - \delta)}{1 - (1 - \gamma)(1 - \delta)q}. \tag{A1.9}$$

Assume that selection is weak enough that the dynamics of q respond to the stationary distribution of p. Then, since the expression for W_s is linear in p when $n = 1$, we can substitute P for p. With this assumption

$$\pi = \frac{\delta}{1 - (1 - \gamma)(1 - \delta)q}. \tag{A1.10}$$

Notice that $\pi > \delta$, which implies that social learners are more likely on average to acquire the skill. Substituting A1.10 into A1.4 yields the following condition for social learning to increase in frequency

$$\frac{(1 - \gamma)(D(1 - \delta) + C_I - C_S)\delta}{1 - (1 - \gamma)(1 - \delta)q} > K. \tag{A1.11}$$

Analysis of continuous character model

Since we are free to determine the scale of measurement of trait values, we can, without loss of generality set $\Theta = 0$. Then the mean value of x in the population during generation t, X_t, is:

$$X_t = (1-a)iX_{t-1} + a\theta_t. \tag{A2.1}$$

The logarithm of the fitness of an individual with adult trait value x is proportional to:

$$\ln(W) \propto -(x - \theta_t)^2 - C(i). \tag{A2.2}$$

Thus the expected fitness of an individual whose behavioural acquisition is governed by the parameter i is

$$E\{\ln(W)\} \propto -(1-a)^2 E\{(iX_{t-1} - \theta_t)^2\} - C(i). \tag{A2.3}$$

Consider the competition between two genotypes. The common type has development characterized by parameter i and the rare type by $i + \delta$, where δ is very small. If one assumes that changes in i have no effect on the variance of the trait among the invading type individuals, the expected fitness of the invading type is approximately proportional to

$$E\{\ln(W)\} \propto -(1-a)^2$$

$$\times [(i^2 + 2i\delta)E\{X_{t-1}^2\} - 2(i+\delta)E\{X_{t-1}\theta_t\} + \theta_t^2] - C(i) - \frac{\partial C}{\partial i}\delta. \tag{A2.4}$$

Combining expression A2.3 and A2.4 shows that the invading type will increase in frequency if

$$-(1-a)^2[2i\delta E\{X_{t-1}^2\} - 2\delta E\{X_{t-1}\theta_t\}] - \frac{\partial C}{\partial i}\delta > 0. \tag{A2.5}$$

To calculate $E\{X_{t-1}\theta_t\}$ first notice that

$$\theta_t = \begin{cases} \theta_{t-1} & \text{with probability } 1-\gamma \\ \varepsilon & \text{with probability } \gamma \end{cases} \tag{A2.6}$$

where ε is an independent normal random variable with mean zero and variance H. Thus it follows that

$$E\{\theta_t X_{t-1}\} = (1-\gamma)E\{\theta_{t-1} X_{t-1}\} + \gamma E\{X_{t-1}\varepsilon\}. \tag{A2.7}$$

Multiply both sides of A2.1 by θ_t and taking the expectation with respect to the joint stationary distributions yields:

$$E\{\theta_t X_t\} = (1-a)iE\{\theta_t X_{t-1}\} + aH. \tag{A2.8}$$

Combing A2.8 and A2.9 yields the following expression for $E\{X_{t-1}\theta_t\}$:

$$E\{X_{t-1}\theta_t\} = \frac{(1-\gamma)aH}{1-i(1-\gamma)(1-a)}. \tag{A2.9}$$

To calculate $E\{X_{t-1}^2\}$ square both sides of A2.1, take the expectation, and using A2.9 solve:

$$E\{X_{t-1}^2\} = \frac{a^2 - 2i(1-a)E\{X_{t-1}\theta_t\}}{1-i^2(1-a)^2}. \tag{A2.10}$$

Substituting A2.9 and A2.10 into A2.5 and simplifying yields (4) and (5) in the text.

An Evolutionary and Chronological Framework for Human Social Behaviour

ROBERT A. FOLEY

*Department of Biological Anthropology, University of Cambridge,
Downing Street, Cambridge, CB2 3DZ*

Keywords: human evolution; social evolution; environment of evolutionary adaptedness; human behavioural ecology.

Summary. Human social behaviour is the product of millions of years of evolution. The details of the chronological and phylogenetic context in which human behaviour evolved can provide information about both the historical depth of specific behaviours and the reasons underlying their evolution. The chronological framework is described, and the ecological basis for human social evolution discussed. Eight key 'events' and time periods are identified: 35 million years ago (35 Myr), 25 Myr, 15 Myr, 5 Myr, 2 Myr, 300,000 years ago (300 Kyr), 100 Kyr and 30 Kyr. Critical developments occur in these periods when such attributes as compulsive sociality, male kin-bonding and changes in life history strategy and parenting behaviour occur. It is argued that a key factor in hominid social evolution is the conjunction of male kin-bonding and selection for energetically expensive offspring; that the shift to modern human behaviour occurs over a prolonged period in excess of 200 Kyr; and that the human evolutionary heritage (the EEA) is not unitary.

INTRODUCTION

TWO APPROACHES HAVE DOMINATED the research into the evolution of human social behaviour. One is primatology, and the extrapolation of the behaviour of extant non-human primates, and the principles underlying animal behaviour, to both humans and the ancestral hominids. The other is anthropology, and the inference of evolutionary history from either human universals or the specific behaviours of hunter-gatherers. In this paper I want to explore a third approach—the chronological and phylogenetic context for hominid evolution. The access that palaeobiology can provide to the timing and evolutionary context of changes in social behaviour can potentially fill the gap between living human and living ape. Palaeobiology can provide direct (albeit patchy) evidence for the path by which the baseline of primate social behaviour has been extended to the full modern human repertoire. In particular, I shall attempt to show that human social behaviour was not an inevitable evolutionary product, nor just a chance event, but the outcome of specific interactions between populations and their environments occurring cumulatively over millions of years. The aim will be to show that while the generalities of behavioural and ecological theory provide powerful models for social evolution, it is how these operate at particular times and in particular places that is paramount. The chronological pattern that will emerge will hopefully throw light on the nature of our evolutionary inheritance and the adaptive basis for human social behaviour. In the first part of the paper I shall present a phylogenetic and chronological context for human social evolution, while in the second I shall discuss the processes by which this occurred.

EVIDENCE FOR A TIMESCALE FOR HUMAN SOCIAL EVOLUTION

35 million years (35 Myr): the anthropoids and the origins of society

The 'origins of society' are often considered to be a classic problem in anthropology. Advances in the study of animal behaviour have, however, greatly modified this perception. Sociality, as distinct from a tendency to aggregate or gregariousness (Charles-Dominique 1977; Lee 1994), requires both the formation and the maintenance of relationships between members of a stable unit (Hinde 1976; Dunbar 1988). As such, a number of mammalian and avian groups can be considered to have maintained sociality, while the social insects fall outside this definition. Amongst the mammals, however, the primates exhibit the most flexible sociality. Out of the over 175 species of anthropoid primates, all but the orang-utan (*Pongo pygmaeus*) are social in the sense that individuals associate with each other for extended periods, interact in patterned ways and form relationships that can be defined by their qualities and intensity.

The orang-utan provides an interesting exception, which in itself demonstrates the ubiquity of primate sociality. While prolonged associations between members of the opposite sex are rare, there are affinities between males and specific females within discrete home ranges (Galdikas 1985). Associations between females have also been observed, especially when several mother-infant pairs congregate. It is not yet known if the females aggregate with the same other individuals over time. Orang-utans thus appear to be sociable within the context of solitary life, and their sociality is based on knowledge of individuals within a larger 'neighbourhood' where group size is limited to one for ecological reasons. In effect, orang-utan sociality could be considered as highly sophisticated, occurring as it does in the absence of visual and tactile reinforcement.

Anthropoid primates can thus be considered as compulsively social, and to have exploited sociality as their core adaptation rather than morphological specialization (Jolly 1984; Dunbar 1988). If sociality occurs on all branches of the anthropoid clade, it is an ancestral or plesiomorphic trait for the group as a whole, or at least one which appeared early in anthropoid evolution. While the origins of this group are still far from clear, both the genetic and palaeontological evidence agree on the monophyly of the platyrrhines and catarrhines and places their last common ancestor at least 35 Myr ago (Kay & Fleagle 1994). Far from being a uniquely human phenomenon, sociality based on interactions, relationships and individual knowledge has a much deeper evolutionary heritage. This capacity for creating and maintaining social systems is likely to be an anciently embedded biological trait.

The anatomical and physiological correlates of social behaviour—the tangible evolutionary evidence—lies in the large brains and prolonged life history traits of the anthropoids. It can be argued that the hormonal and biochemical mechanisms mediating behaviour such as aggression, arousal, conciliation and so on, are similar across anthropoids, including humans (Keverne 1995). We can thus propose that humans are compulsively social as *anthropoid primates* rather than as humans. The primary inference to be drawn is that the origins of society, that hoary chestnut of anthropology, lie in the Eocene or Miocene, and not in recent prehistory or history.

25 Myr: finite social space and the kinship as the basis for social organization

While the capacity for social living may extend back over 30 Myr, the structure of those societies and associated behaviours have evolved over the subsequent periods. Foley & Lee (1989) have argued that social variation is based on two quantifiable elements: spatial association and kinship, and as a result only a finite number of social systems can occur (Figure 1). The Finite Social System model allows us to explore the range of variation that has occurred within the primates, relative to all potential states that can exist. Furthermore it allows for the specification of particular evolutionary

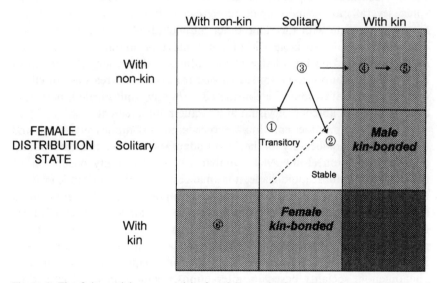

Figure 1. The finite social space model of social organization (Foley & Lee 1989). Social structure arises from the way in which males and females associate with members of their own sex—no association, with non-kin or with kin. Associations between males and females may be either transitory or stable (indicated by diagonal line, shown in central cell only). Where females associate with their own kin then female kin-bonding will occur. Where males associate with kin the group will be male kin-bonded. Associations between males and females may be either transitory or stable. The proposed basal hominoid social system is solitary males with stable associations with females who are not related to each other (3), as is found among gorillas. From this cell hominoids have evolved a diverse array of social systems—solitary (1) in orangs, monogamous (2) in gibbons, and male kin-bonded in chimpanzees (4). Humans have extended this with the presence of inter-generational lineages (5). It should be noted that all the hominoid social systems are evolutionarily adjacent to each other, in contrast to the female kin-bonded systems found in the cercopithecoids.

pathways in transitions between social states. When anthropoid sociality is considered, the non-random distribution of states is striking. Firstly, stable associations between males and females are the norm, in contrast to social states among ungulates and carnivores (Lee 1994). With the evolution of menstrual cycling among anthropoids, maintaining continual access to females becomes a male priority leading to a continual male presence irrespective of the female-female associations. Secondly, there is considerable congruence between social state and phylogenetic relatedness (Foley & Lee 1989). This can be interpreted in a number of ways. There may be phylogenetic inertia within social evolution with ancestral states being important determinants of subsequent evolutionary pathways (see below). Alternatively, since the model places states in relation to adjacency, shifts between states may follow pathways constrained by the plausibility or stability of intermediate states.

In this model, kinship arises through sex-specific dispersal, while the costs and benefits of co-residing with kin are ecological or reproductive in origin (Wrangham 1979, 1980). Female kin-bonding is the most common independently evolved state, due to the ecological advantages of female kin co-operative control of resources among primates exploiting clumped, patchy, relatively large food resources. Male kin-bonding is infrequent and associated with control of females as a dispersed and patchy resource, when those females can be localized in time and space. Monogamy (a phylogenetic rarity among the anthropoids) appears to be associated with the inability of females to co-reside in relation to the resource base, as well as some significant and essential component of male contribution to infant survival. Monogamy may well be an unstable state, for if males can acquire more females then they will do so, mapping polygyny onto either female kin-bonded states or resulting in male kin states. The determination of core social states and the probability of different transitions between states have yet to be determined in a phylogenetically controlled analysis.

However, some interesting general patterns are apparent. The platyrrhines show considerable diversity in social state and indeed occupy the greatest number of different states. This diversity of social states may reflect both their early origin and their monophyletic radiation with subsequent niche separation within diverse New World habitats. Amongst the catarrhines, the number of states observed is lower. However, within these states are strong phylogenetic patterns which can throw light on the evolution of catarrhine social behaviour.

15 Myr: catarrhine social phylogeny and the evolution of male kin-bonding

That there may be a significant phylogenetic effect in patterns of social evolution provides an important avenue for investigating the timing and nature of human behavioural evolution. The phylogeny of the hominoids is now well established by molecular and anatomical evidence. Among living taxa, the hominoids are the sister clade of the Cercopithecoidea. Among the Hominoidea the hylobatids are the sister clade of all other hominoids, while among humans and the great apes, chimpanzees and humans appear to be the most closely related. Gorillas are the sister clade of the human/chimpanzee clade, and the orang is the sister group of all African apes and humans (Figure 2) (Williams & Goodman 1989).

The hominoids are remarkably diverse socially, as seen in Figure 2. The gibbons are primarily monogamous, the orang-utan solitary, the gorilla has single male groups, and the chimpanzee has a fission-fusion community

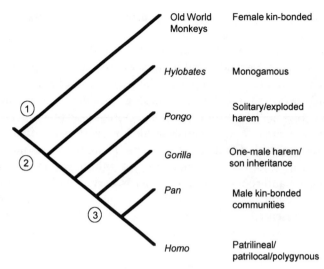

Figure 2. Evolutionary relationships of the catarrhines (Old World monkeys and apes), showing social systems. Female kin-bonding evolved only in the Old World monkey clade (1). Male residence appears to have become established among basal hominoids (2), while male kin-bonding evolved in the chimpanzee/human clade (3).

made up of related males and unrelated females. The bonobo or pygmy chimpanzee is less well understood, but also appears to be male kin-bonded with the addition of strong relationships between males and individual females (Smuts et al. 1987). Humans are socially variable, but a dominant pattern is patrilocality and unilineal descent groups, usually based on males. Despite this diversity, however, there are significant phylogenetic patterns; although the hominoids all occupy different cells within the finite social space model, they are all adjacent to each other, and as such it is possible to reconstruct the evolution of their social behaviour using phylogenetic techniques (Foley 1989; and see Rendall & Difiore 1995 for a recent analysis).

Figure 2 also shows the phylogeny of the hominoids with the inferred points of key elements of social evolution superimposed. The key observation is that while the cercopithecoids show extensive female kin-bonding, this is absent from the hominoid clade. The inference would be that among stem hominoids sociality was not characterized by large groups with a kin-based organization. As is evidenced by the gibbons, orang and gorilla, the fundamental social niche of the hominoids is likely to have been small social units made up of one male, one or more females, and young. The period of early diversification or the Hominoidea, from 25 to 10 Myr, perhaps established the small core units of hominoid social life. Gibbons, orangs and gorillas all represent variations on this theme.

The main shift in this pattern occurs on the stem hominid/chimpanzee clade. Kin-based social organizations develop with the establishment of larger communities. These, though, in contrast to those found in the cercopithecoids, are male based. Both common chimpanzees and humans, while maintaining the characteristic sub-units of small groups, have organized above them a larger community. The cores of these communities are the residential males. Females leave these communities at maturity and join other ones.

If this reconstruction is the case, then the phylogenetic context for human social evolution consists of small 'family' units as part of a general hominoid ancestry, and male kin-bonding, male residence, and female dispersal as an African ape ancestry, established in the late Miocene, between 8 and 5 Myr.

5 Myr: savanna socioecology

The divergence of the ancestors of humans and chimpanzees is thought to have occurred between 7 and 5 Myr, and there is sound fossil evidence for the existence of bipedal hominids by around 4 Myr. In all probability any differences between these early hominids and other apes in terms of behaviour were likely to have been at the level of variation displayed between extant hominoid groups, rather than being significant differences in 'grade'.

General socioecological principles might allow some predictions to be made about the social behaviour of the australopithecines. The primary characteristics of the early hominids, and the difference between them and other apes, lies in their bipedal locomotion. This has generally been associated with the occupation of more open environments, environments that became more widespread in eastern Africa during the later Miocene and Pliocene. Although it is likely that the early hominids retained considerable levels of arboreal activity, especially in feeding (Susman et al. 1985), the ecological implications of foraging in more savanna conditions are profound. Resources are more spatially patchy, seasonally variable, more dispersed and overall less abundant than those in forests (Foley 1987a). Time for foraging, day range length, and home range area are all likely to increase. The implication would be that while communities may remain large and a significant element of social organization (due at least in part to increased predator pressure, but also as a consequence of inter-group competition and the advantage of male coalitions), the actual foraging and day-to-day functional groupings may also have been smaller (Foley 1993). This may have led to a strengthening of the ancestral hominoid sub-units. It is interesting to note that among humans and bonobos there is an intensification of the strengths of attachment within the smaller social units, essentially the formation of something approaching a family structure. From an evolutionary point of view it is unclear whether this is an independent evolution in the bonobo and hominid lineage or whether it reflects the ancestral condition from which the common chimpanzee has departed. A strong case can be made for the former, and that stronger association in sub-units occurs in response to food occurring in larger patches (White & Wrangham 1988; Foley 1989). If this interpretation is correct, the social organization of the bonobo may be considered of interest less because it represents that of the last common ancestor, as some have claimed, but because it is convergently derived in two lineages in response to similar selective pressures.

There is an increasing consensus that the australopithecines are closer in behaviour to the African apes than to modern humans; in 'grade' terms, their level of organization is likely to have been similar to that of the apes, and the array of australopithecines are likely to have exhibited a range of variations on social behaviour, but variations that would have been little more than a variant on that seen in chimpanzees today.

2 Myr: expensive offspring and the socioecological basis for encephalization

It has been suggested that one response to the seasonality of the more open environments of eastern Africa would have been a greater reliance upon meat-eating, particularly as a dry season response to the movements of large, herbivorous mammals (Foley 1987a). For many populations such a response may not have been possible, due to the absence of herds of ungulates as they underwent seasonal migrations. These surviving, or even thriving, populations of australopithecines, between 3 and 1 Myr, would have maintained the ancestral and conservative patterns of social behaviour. In contrast, those able to invest more foraging effort into meat or other high quality resources, would have entered a new resource structure, and some life history and social changes would have occurred.

The next major event in the evolution of human social behaviour is the increasing brain size that occurs within genus *Homo*. Among the australopithecines there is very little increase in brain size, when body size is taken into account. Maximum Encephalization Quotient (EQ) for extant apes is little over 2.0, and australopithecine EQ ranges from 2.4 to 3.1. An increase to 3.3 can be observed for *Homo ergaster* at 1.6 Myr (McHenry 1992). Thereafter brain size increases steadily, if not spectacularly, over the subsequent million years.

It has been widely argued that increased social complexity underlies this increase in encephalization (Humphrey 1976; Dunbar 1992). However, while sociality might be a major selective force, it does not explain the conditions under which this selection can lead to evolutionary change (Foley 1995a). The key question with this particular part of hominid evolution is what enables early *Homo* to have larger brains, given that these are highly expensive tissue that impose great energetic costs on mothers and infants alike. Foley & Lee (1991) have calculated these additional costs as up to 9% of an infant's nutritional requirements, while Leonard & Robertson (1992) have estimated that the additional size of the human brain means that brain metabolism for humans accounts for 22% of BMR, whereas for the chimpanzee it would only be 8%. In view of the additional costs involved in encephalization, there must not only be positive selective pressure in favour of larger brains, but also a more secure ecological basis. It can be inferred that the social evolution implied by encephalization is dependent upon energetic changes.

Meat-eating may be proposed as a significant part of the change in hominid energetics (Foley & Lee 1991). Meat is a high quality resource, providing both ample energy and protein. As such it can be an important contribution to the additional costs of larger brains. As various authors have shown, there is a link between large brains and high quality food supplies, and in the case of early *Homo* meat rather than plant foods may well be the critical resource added to the early hominid diet breadth. There is certainly evidence for *Homo ergaster* of an increase in hominid involvement in animal butchery, although whether this is through hunting or scavenging is a matter of debate. However, from the point of view of evolutionary ecology the fact that meat is acquired is of greater significance than the means by which it is acquired.

If meat is a means by which higher levels of encephalization can be sustained, then a number of questions about the means by which this occurred can be considered. One such means, as pointed out by Martin *et al.* (1985) and Aiello & Wheeler (1995), is that less gut tissue is required with greater levels of carnivory. As gut tissue is also very metabolically costly, then higher levels of meat-eating can also lead to reduced overall growth and maintenance costs as the gut becomes smaller. Smaller gut size is a corollary of larger brains and higher levels of meat-eating. To this could also be added two further observations: first, that data show that with *Homo* at 2.0 Myr, and *Homo ergaster* between 1.5 and 1.2 Myr, stone technology both develops and is greatly enhanced (Schick & Toth 1993); and second, that with *Homo ergaster* thorax shape, and by implication gut size, shifts to the pattern found in humans rather than apes (Ruff & Walker 1993). Together with the dietary and anatomical evidence, this all points to a change in the way energy was acquired and metabolized by early hominids, and in turn an implied change in the way the hominids were organized socially.

The argument here is that a shift to more open and seasonal environments under local ecological conditions where animal resources were abundant, especially in the dry season, led to greater levels of meat-eating. This was the essential cause, through different populations living in different environmental conditions, of the divergence of the genus *Homo* from the trends found among the australopithecines (Foley 1987a). Greater meat-eating provided more energy, allowed for reduced energy expenditure, and acted as a selective pressure leading to greater levels of sociality. As a corollary of this, during the period 2.0 to 1.0 Myr the expected shift in social organization might well have been towards more intense and extensive male alliances. The phylogenetic heritage of male kin-bonding, evolved for reasons related to longevity and male access to females, provided a premium in terms of foraging behaviour under these new ecological conditions. Females associated with male groups that were numerically larger and effective at acquiring, and probably protecting, resources.

300,000 years (300 Kyr): the 1000 gram brain and evolution of human life history strategies

The period between 1.6 and 0.3 Myr has often been viewed as one of evolutionary stasis (Rightmire 1981). The stability of both morphology and technology lends considerable credence to this view in terms of overall grade of biological organization and behavioural adaptation. However, it should be noted that during this time the hominid range expanded very markedly, and there was considerable evolutionary divergence between populations. Such divergence is likely to have incorporated behavioural and social diversity, at least at the level known for chimpanzee populations today (McGrew 1992), and more likely on a greater scale.

Nonetheless around 300 Kyr, in addition to various biogeographical and behavioural (archaeological) changes, there is an important shift in the rate of encephalization (Aiello & Dunbar 1993). Over the following 250 Kyr brain size increases from a nominal '*Homo erectus*' baseline of around 1000 grams, to one that overlaps with the range found in modern humans. Archaic *sapiens* and Neanderthals both have high levels of encephalization, comparable with those found in anatomically modern humans.

The ecological and social conditions that gave rise to the selection for this acceleration in trend are hard to pinpoint. Certainly there are some significant archaeological changes, principally associated in Africa and Europe with the development of prepared flake technologies (the Middle Palaeolithic / Middle Stone Age) (Schick & Toth 1993), and there is also the possibility that fire may be systematically used at this point. According to Aiello & Dunbar (1993), at this stage group size reaches a critical threshold where grooming is insufficient for the maintenance of social relationships within a group, and language supplants grooming as the primary means of social lubrication, resulting in rapid evolutionary and social changes. Alternatively, Foley & Lee (1991) have argued that 1000 grams represents, in terms of brain energy expenditure, the point at which modern human growth rates and developmental strategies would be necessary to sustain the very high metabolic costs of brain growth. Essentially it can be argued that the energy needed by both mother and infant during development would be so high that the principal means for solving this problem lie in slower growth rates. This change in life history strategy—the well-known shift from more rapid ape growth patterns to the extended and delayed pattern of maturation found in modern humans—would have profound social and demographic consequences: a longer period of infant altriciality and dependence, longer inter-birth intervals, delayed onset of first reproduction. A corollary of this might also be an extension of longevity, with profound consequences for competition between males and mating strategies.

It can perhaps be argued that the period between 300 Kyr and 200 Kyr was of critical importance in the evolution of human social behaviour. The energetic costs of reproduction associated with larger brains resulted in a change in life history strategies, leading to new patterns of social behaviour and organization. Underpinning both social and life history changes might be changes in foraging pattern associated with technology or some other extractive strategy. The consequences of this change could well have been either directly, or indirectly through group size, selection for much greater levels of communication, and hence the evolution of language. Such a timing for what might broadly be considered the origins of language functionally equivalent to that found in modern humans would be consistent with the morphological data, with the acceleration of brain size evolution that occurs at this time, and the apparently 'modern' behaviours associated with Neanderthals (Mellars 1996) who would be as much descendants of these archaic groups as *Homo sapiens* itself.

100 Kyr: dispersal, group size and territoriality

Anatomically modern humans—*Homo sapiens*—are present in Africa from around 140 Kyr. Both morphological and genetic evidence support the view that the origins of modern humans lie within Africa, and that they disperse from their area of endemism to other parts of the world over the subsequent 60 Kyr (Lahr & Foley 1994). By and large both the evolution of anatomically modern humans and their dispersal are not associated with any markedly visible evidence for a change in behaviour, although within Africa temporary changes in technology do occur at various points (Foley

1987b; Klein 1992; Brookes *et al.* 1995). These, however, are transient, and the first 80 Kyr of the existence of anatomically modern humans do not appear to be characterized by any behaviour that is significantly different from that of contemporary archaic populations such as the Neanderthals. Significant changes of behaviour occur from around 40 Kyr, but these are regionally variable. The most dramatic shift occurs in Europe and the Mediterranean, where Upper Palaeolithic blade industries occur; other regions, such as Australia and eastern Asia do not undergo any apparent change during the later Pleistocene (Mellars 1991; Klein 1992). Furthermore, the European evidence, which is the most abundant, shows that there is as much change between early and late Upper Palaeolithic as there is between the Middle and Upper Palaeolithic (Clark & Lindly 1989). This is indicated morphologically with the loss of skeletal robusticity that occurs (Lahr 1996), or the relatively sudden flowering of art during the Magdalenian.

Technologically, anatomically modern humans appear to be highly variable, with very distinct regional and temporal patterns occurring (Foley 1987b). It is hard to sustain the view that on behavioural grounds the appearance of anatomically modern humans was a rapid or dramatic revolution in the hominid world. Events occurred cumulatively and multiply over a period of 100 Kyr, culminating in the full colonization of the world and the shift to agriculture during the period 15 to 5 Kyr. Indeed, it is the high potential for dispersal itself, rather than any specific behaviour, that seems to characterize modern humans (Gamble 1993).

The question this raises is—what is the basis for these dispersals? In the light of what has already been established in hominid social evolution, it can be argued that the ancestral social condition for modern humans consisted of moderately large communities, with coalitions of males linked by kinship, and unrelated females attached to specific males or possibly several males. Given the evidence for changes in life history parameters, particularly the slower growth, delayed maturation and increased longevity, such groups are likely to have at least an element of inter-generational lineage structuring (patrilineal, given the pattern of male residence). Assuming conditions of net local population growth, two significant characteristics would arise from this ancestral social organization. The first is that with male kin-bonded groups, communities would be at least partially closed to each other and hostile, resulting in some form of territorial or agonistic behaviour between communities. A consequence of this would be that overall group or community size would be an advantage, particularly in terms of numbers of males within a coalition. The competitive advantages that would arise from this would, however, also lead to both social and ecological pressures. As group size goes up, competition for both resources and reproduction would increase. This would lead to the second of the two ancestral tendencies occurring—demographic fission of communities. Primate social groups tend to split into two when they reach group sizes that are greater than can be socially or ecologically maintained (Dunbar 1992, this volume). Fission of groups would in turn be a factor promoting geographical dispersal and leading to the colonization of new regions and localities, whether or not they might already be occupied by hominids.

The overall effect would be to produce kin-based communities, benefiting from a tendency towards larger size, but ultimately with limits on that size. Fission of male kin-bonded groups would in effect lead to a segmented lineage structure dispersed and dispersing across the region as a whole, and thus producing larger-scale networks of cultural groups, and a regional pattern of ethnic differentiation. An important element in this model is that it explains why there should be such regional and chronological variation in the appearance of what has been referred to as modern behaviour, or at least its archaeological manifestation. Art, 'symbolic behaviour', blade technology, and so on (such as those found in the European Upper Palaeolithic), are not so much manifestations of a radically different form of behaviour but of specific demographic and ecological conditions arising from the successful dispersals of the descendants of the middle and early Upper Pleistocene African hominid populations.

30 Kyr: demography and the agricultural revolution

A case may be made that the appearance of anatomically modern humans heralds the end of genetically-based evolution, and the Upper Palaeolithic the end of long term behavioural evolution. As it is clearly not the case that social evolution was terminated at these points, it may be that at this time models based on Darwinian evolution should be abandoned in favour of alternative ones drawing on theories derived from the social sciences. However, some suggestive observations can be made that might indicate that neo-Darwinian principles may still be of use in understanding more recent social evolution.

The first of these might be that the new chronology for the evolution of 'modern' human anatomy and behaviour brings the development of agriculture into a new relationship with these major evolutionary events. The orthodox chronology denotes hunting and gathering as a stable adaptation over periods of hundreds of thousands of years, if not millions, with agriculture as a short and recent aberration over the last 10 Kyr. However, if modern human behaviour is more recent in origin, then so too is what is generally referred to as the hunter-gatherer adaptive lifestyle—the suite of traits associated with small flexible bands of bilaterally related individuals, with a sexual division of labour and food-sharing (Foley 1988). More significantly, there will be considerable regional variation, with, for example, modern hunter-gatherers being present in Africa over a period of 100 Kyr, in Australia over perhaps 60 Kyr, and in Europe only 35 Kyr, while in the New World the figure may be less than 20 Kyr. These timescales undermine the major evolutionary contrast between hunting and gathering and agriculture.

In view of the fact that modern humans dispersed into most parts of the world during the late Pleistocene, and that these dispersals were repeated many times (including during the process of agriculture itself), it would appear that human demography over this period would have been far more dynamic than traditionally perceived. The shift from small mobile hunter-gatherer groups to larger and more sedentary farming communities should be seen as an interaction between the basic social structures described above as the adaptations and heritage of the last 100 Kyr, and the new and more packed demographic conditions.

This perspective has implications for considering the process of ecological intensification that occurs at the end of the Pleistocene in many parts of the world. It has traditionally been thought that hunter-gatherer adaptations are essentially homeostatic, with either dispersal or reduced fertility acting to keep populations within the level of carrying capacity. The apparent stability of hunter-gatherers over very long periods of time would be cited as evidence of this (Lee & DeVore 1968). However, under this new chronology for hunter-gatherers, in many parts of the world change seems to have been continuous, and intensification of foraging behaviour, sometimes leading to food production, occurs relatively rapidly after colonization. Elsewhere it has been argued (Foley 1988) agriculture can be considered as an evolutionary response to demographic constraint and declining resources. The economic innovations involved, with smaller territories, larger group size, and more control over resources by both males and females, would have acted to maintain the proposed ancestral social organization consisting of polygynous family groups linked by alliances of male kin organized patrilinealy. Ironically, hunter-gatherers, as we are able to observe them today, are very different socially from these earlier manifestations, and it is probable that the eclectic and flexible foraging systems they are able to employ in the resource depleted post-Pleistocene are only possible under novel social organization, albeit one modified from the ancestral conditions.

The implication is that the human phenomena of the late Pleistocene, including such elements as the development of larger social networks, intense use of art and symbols, and the development of agriculture itself,

EVOLUTIONARY FRAMEWORK FOR HUMAN BEHAVIOUR 109

should be seen as an integral part of the dispersal of modern humans, and that the contrast in evolutionary terms between hunter-gatherers and agriculturalists should not be overestimated. In this light, modern hunter-gatherers may represent considerable economic conservatism, but more novel social structures, while agriculturalists have radically new economic systems but more conservative social organizations. That these also give rise to new reproductive and behavioural strategies—early weaning, larger group sizes—should be seen as a significant element of the human evolutionary heritage.

EVOLUTIONARY CONDITIONS AND THE EVOLUTION OF HUMAN SOCIAL BEHAVIOUR

The EEA and the chronology of human social evolution

The previous section outlined a chronology for the evolution of human social behaviour. It should be stressed that this chronology, while based on empirical evidence, must remain tentative given the nature of the data involved. It does, however, serve a number of purposes. First, it emphasizes that neither human social behaviour, nor the capacity for 'modern' behaviour, evolved as a single package, but was the result of combining new elements with ancestral conditions over extensive periods of time. The timescale for this is over the last 35 Myr, but even within the last 1 Myr and last 100 Kyr there is considerable chronological variation. Second, the nature of social evolution is, like any evolutionary process, additive; novel components are derived from existing ones, and also become combined with more ancient ones. In the case of human social evolution, for example, the social complexity observable today has arisen from building the catarrhine patterns of kin-bonding on the already existing anthropoid propensity for sociality. This accumulation of new elements, with interactions between ancient and modern elements giving rise to new behaviours, has continued throughout human evolution. A third implication is that while there is an evolutionary heritage to human behaviour, it is not unitary. Some elements are much older than others. In the chronology described here, for example, relationships between members of the same sex have remained more stable than relationships between the different sexes. Another suggestion might be that alliance and coalition structures are more conservative than mating tactics. Figure 3 summarizes the principal phylogenetic and chronological contexts for the evolution of human social behaviour.

A fourth aspect that should be emphasized is that a closer attention paid to chronology demonstrates that human evolution is made up of multiple events, and that there is no key trigger that forces the hominid lineage in an inevitable

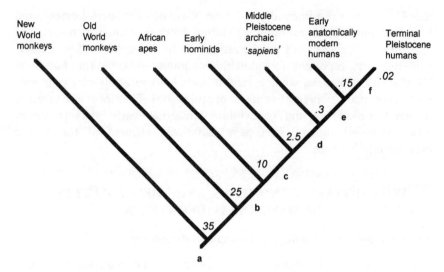

Figure 3. Cladogram showing the proposed points of evolution for key characteristics in the evolution of modern human social behaviour. Numbers indicate approximate age of branching events in millions of years. Letters indicate the evolution of the following characteristics: (a) compulsive sociality; (b) male residence patterns, small social units, and male kin-bonding; (c) increased parental effort and stronger male-female bonding; (d) modern human life history parameters and associated changes in maternal behaviour and longevity; language and more flexible use of technology may also have its origins at this point; (e) fission-fusion dispersal behaviour and inter-group competition; (f) adaptive responses to demographic crowding and resource depletion. See text for a full discussion.

direction leading to modern humans. Each event is the product of particular evolutionary, ecological and demographic interactions, the time and place of which are both significant and contingent. There is a strong and comprehensible adaptive basis for the evolution of human social behaviour, but this can only be analysed by paying attention to the context in which it has evolved.

There are a number of implications of this perspective; one of these relates to evolutionary psychology and the concept of the environment of evolutionary adaptiveness (EEA) (Symonds 1979). The EEA refers to the optimal environment to which humans are adapted. As such it is primarily taken to be the conditions that shaped human evolution. In practice the hunter-gatherer lifestyle has been assumed to be the human EEA, and by implication deviations from that lifestyle take humans into less optimal environments, and hence can lead to maladaptations, both physical and psychological. Reference to the chronological and phylogenetic context for the evolution of human behaviour shows that the EEA is by no means unitary. Different elements have different time depths, and by implication different mechanisms. In addition the social environment has evolved additively, and the various characteristics may also be treated partially at least as independent variables. Furthermore some traits may be very deeply embedded in human evolutionary history, while others are more ancient. Indeed, the phylogenetic context described here implies that many of these are shared with other primates. Finally, hunting and gathering is only one element of the human evolutionary environment, and other parts of our species' history, including denser and more settled agricultural communities, may be significant rather than just being aberrant developments.

The hominid paradox: male kin-bonding, expensive offspring and variation in life history strategy

One of the most important observations derived from the chronological and phylogenetic patterns described here is that the hominoid clade shows no tendency towards female kin-bonding. In particular, the African ape clade displays a trend towards female dispersal from their natal groups and the formation of kin-based male coalitions. On this evidence it has been argued that male kin-bonding may have been the basis for larger community structure during the course of hominoid evolution. The other major trend identified was towards the greater energetic costs of reproduction due to the larger brains. A shift towards higher quality food, especially meat, was identified as a mechanism underlying this, leading to associated reduced costs due to smaller guts. In addition, slower growth rates and delayed maturation appear to be related to the increase in brain size.

One response to high energetic costs on mothers is for females to recruit helpers. Allomothering is extensive among mammals, but Lee (1989) has shown that this occurs either under conditions of strong female dominance hierarchies, involving coercion and suppression of reproduction in some individuals, or else where females can recruit kin, especially female kin. The paradox of hominid evolution is that the context in which very expensive offspring have evolved is a social environment where females do not live in female kin-bonded groups, and therefore that their options for recruiting allomothers and helpers is limited. The major question that arises is how females have been able to cope?

Two strategies may be suggested. One is provisioning by males, and this has traditionally been seen as an important element in human evolution, the development of the pair-bond, close attachment between individual males and females, and food sharing. Leaving aside the question of the timing of such developments, the ethnographic evidence is ambiguous about the extent to which food is shared, especially between putative fathers and mothers. An alternative explanation is the evolution of more extended life history parameters.

One of the most striking things about human biology is the change in life history variables. Humans have slower growth rates, mature later, have long inter-birth intervals, and extended lifespans, although as Lee has pointed out, the difference from African apes has often been exaggerated. There is evidence that life history parameters are correlated with brain size (Harvey et al. 1987), with larger brains being associated with an extension of lifespan and slower and delayed rates of maturation. Smith (1989) has shown that this can be seen among primates in the context of dental development, and it also possible to see during the course of hominid evolution that as brain size increases, growth rates slow down (Smith 1993; Bromage & Dean 1985; Foley & Lee 1991; Foley 1995b).

This very marked departure from an ape life history baseline is perhaps the key to solving the paradox of human social evolution. With pressure on reproductive costs for females determined by selection for large brains, and in the absence of easily available female kin as allomothers, changes in life history in turn became the means by which the high costs of reproduction could be borne. The subsequent consequences in terms of parental behaviour, infant survivorship and demography, and social relationships in turn would have been significant, and acted as further selection for larger brains and more expensive offspring.

The interesting aspect of this link between social behaviour, in terms of the evolutionary heritage of male kin-bonded groups, and expensive offspring and the adaptive consequences for life history strategy, is that there is a further element of evolutionary heritage involved. When compared with cercopithecoids, hominoids display far more variation in body size and life history parameters. Cercopithcoids are far more conservative and homogeneous as a group in this context. Thus it could be argued that in undergoing evolutionary change in life history strategies in response to the selective pressures associated with large brains and complex sociality, the hominids were extending an evolutionary potential that had already occurred extensively within the Hominoidea.

A multiple event model for the evolution of modern humans

The central issue in palaeoanthropology over the last decade has been the origins of modern humans (Mellars & Stringer 1989; see Lewin 1993 for a review). This debate has been particularly useful in focusing research into the differences between archaic and modern humans, and avoiding simply treating other species of hominids as incipient modern humans. One element that has entered the literature as a result of this research effort has been the idea of a marked contrast between modern humans and archaic hominids, not just in terms of anatomy, but also behaviour (Binford 1989; Mellars 1996).

EVOLUTIONARY FRAMEWORK FOR HUMAN BEHAVIOUR

The success of the Recent African Origin model for *Homo sapiens*, which is linked to the notion of modern humans being something of a radical evolutionary departure, has brought with it a number of associated ideas. These ideas include the following: that there was a significant shift in biological and behavioural capacity and potential that occurred in a small population in Africa in the late middle Pleistocene; that there is a link between biological and behavioural change, and broad synchrony in these changes; that this resulted in a major adaptive shift, with descendants of the African populations (modern humans) being biologically, behaviourally, technologically and cognitively superior to archaic hominids; that this superiority led to rapid dispersals across the globe, and the displacement of non-modern populations such as the Neanderthals (Stringer & Gamble 1993); and that this dispersal is broadly associated with an Upper Palaeolithic technology and fully modern cognition and behaviour (Klein 1992); that this transformation from the archaic to modern is essentially a revolution ('the human revolution'), in which the differences between the two are discontinuous and more marked than differences between other hominid species (Mellars 1991); and that in contrast to the fully cultural behaviour of modern humans, archaic hominids lacked a number of key traits and the potential for modern behaviour. A number of different mechanisms have been proposed as the basis for the human revolution, of which language and symbolic thought (the symbolic explosion) are predominant (Mellars 1991; Knight *et al.* 1995).

The chronology presented here raises a number of serious doubts on the association of these components, principally due to the temporal spread of events (Lahr & Foley 1994). The genetically (mtDNA and Y chromosome) identified founder population in Africa probably existed in excess of 250 Kyr ago (Stoneking 1993; Dorit *et al.* 1995). The evolution of anatomically modern humans occurred between 150 and 100 Kyr, and indeed there is no clear anatomical boundary between archaic and modern; Lahr (1994) has shown that there is continual and regionally variable evolution in human cranial form over the period 100 to 10 Kyr. Behavioural change associated with the Upper Palaeo-lithic occurs no earlier than 45 Kyr, and then only in Europe and the Mediterranean. In Africa, Australia and East Asia there is no such transition, although it can be argued that the African Later Stone Age is functionally equivalent. However, by the time that the European Upper Palaeolithic was in place, the divergence of the main trajectories in human populations was already established. Events less than 50 Kyr, such as the so-called symbolic explosion in Europe, must be regionally specific, not universal or global human traits. Within Africa there are technological changes in the direction associated with the Upper Palaeolithic extending back almost to 100 Kyr, but these are ephemeral and firmly embedded in a

Middle Stone Age technology. Furthermore, dispersals occur throughout this period, such that the resulting geographical pattern is not a simple one of supplanting, but repeated demographic palimpsests with greater or lesser patterns of replacement. The result is that there is considerable genetic diversification occurring not just in early Upper Pleistocene, but in the terminal Pleistocene and Holocene.

The fact that the events associated with modern human origins are spread over more than 200 Kyr suggests that this is not a single revolutionary change. The chronology set out here implies three linked 'events'. The first of these is essentially biological and behavioural, occurring approximately around 300 Kyr, and is linked to the expansion of the brain and changes in life history strategy. It may also be associated with the development of Middle Stone Age technologies. This event probably also involves the successful dispersal of populations, most likely from an African source. Language may well have been present at this stage. The second is the evolution of anatomically modern humans, again in Africa, over a period of more than 50 Kyr. This evolution does not seem to be associated with any behavioural change, and indeed early modern humans are associated with the same Middle Stone Age technologies as Neanderthals and other archaics. However, by 100 Kyr these populations were also undergoing dispersals, not as a single expansion but as a series of multiple dispersals. The third is the regionally variable process of intensification and demographic flux that occurs after 30 Kyr, and continues through into the Holocene with the development of agriculture. It is this last phase that is associated with the rapid development of technology and the 'explosion' of symbolic life in some parts of the world.

In terms of the patterns of social evolution of concern here, this multiple event model does not imply any major transformation in the structure of human social organization in the late Pleistocene, nor necessarily in the cognitive abilities of the late Pleistocene humans. Rather any social evolution that occurred would be the transformation of the existing male kin-bonded systems in response to local demographic and ecological conditions.

CONCLUSIONS

This paper has set out to establish the evolutionary pathway to human social behaviour. I have argued that the best means of approaching this problem is to use the phylogenetic and palaeobiological context to establish a chronology of events, and that this chronology can then serve as the basis for exploring the functional and adaptive links between the various elements. In considering the available evidence it is clear that human social evolution has been the result of the addition of new traits successively. Many elements are shared with anthropoids, with other hominoids and with the African apes. The key elements identified here include the establishment of compulsive sociality among anthropoids (>35 Myr), the development of male kin-bonding in African apes (10–5 Myr), and the life history strategies evolved in response to expensive offspring in the course of the evolution of *Homo* (2–0.3 Myr).

This perspective undermines the notion that human social evolution is the result of a single trigger change, or that it is the product of highly improbable chance events. Instead I have stressed the fact that the pattern of hominid social evolution has been the consequence of specific contexts— demographic, biological, environmental and social—occurring at particular times and in particular places. The conjunction of specific ecological conditions with the phylogenetic heritage of apes and hominids has led to the evolution of the complex social behaviour, and its underlying cognitive skills, found in humans today.

Note. I thank P. C. Lee and M. M. Lahr for comments on an earlier draft of this paper.

REFERENCES

Aiello, L.C. & Dunbar, R.I.M. 1993: Neocortex size, group size and the origin of language in the hominids. *Current Anthropology* 34, 184–93.

Aiello, L.C. & Wheeler, P. 1995: The expensive tissue hypothesis. *Current Anthropology* 36, 199–222.

Binford, L.R. 1989: Isolating the transition to cultural adaptations: an organizational approach. In *The Emergence of Modern Humans: Biocultural Adaptations in the Later Pleistocene* (ed. E. Trinkaus), pp. 18–41. Cambridge: Cambridge University Press.

Bromage, T.G. & Dean, M.C. 1985: Re-evaluation of the age at death of Plio-Pleistocene fossil hominids. *Nature* 317, 525–528.

Brookes, A.S., Helgren, D.M. & Cramer, J.S. *et al.* 1995: Dating and context of three Middle Stone Age sites with bone points in the Upper Semliki Valley, Zaire. *Science* 268, 548–552.

Charles-Domique, P. 1977: *Ecology and Behaviour of Nocturnal Primates.* New York: Columbia University Press.

Clark, G.A. & Lindly, J.M. 1989: The case of continuity: observations on the biocultural transition in Europe and Western Asia. In *The Human Revolution* (ed. P. Mellars & C. Stringer), pp. 626–676. Edinburgh: Edinburgh University Press.

Dorit, R.L., Akashi, H. & Gilbert, W. 1995: Absence of polymorphism at the ZFY locus on the human Y chromosome. *Science* 268, 1183–5.

Dunbar, R.I.M. 1988: *Primate Social Systems.* London: Croom Helm.

Dunbar, R.I.M. 1992: Neocortex size as a constraint on group size in primates. *Journal of Human Evolution* 22, 469–493.

Foley, R.A. 1987a: *Another Unique Species: Patterns in Human Evolutionary Ecology.* London: Longman.

Foley, R.A. 1987b: Hominid species & stone tools assemblages: how are they related? *Antiquity* 61, 380–392.

Foley, R.A. 1988 Hominids, humans & hunter-gatherers: an evolutionary perspective. In *Hunters and Gatherers 1: History, Evolution and Social Change* (ed. T. Ingold, D. Riches & J. Woodburn), pp. 207–221. Oxford: Berg.

Foley, R.A. 1989: The evolution of hominid social behaviour. In *Comparative Socioecology: the Behavioural Ecology of Humans and Other Mammals* (ed. V. Standen & R. Foley), pp. 474–493. Oxford: Blackwell Scientific Publications.

Foley, R.A. 1993: The influence of seasonality on hominid evolution. In *Seasonality and Human Ecology* (ed. S.J. Ulijaszek & S. Strickland), pp. 17–37. Cambridge: Cambridge University Press

Foley, R.A. 1995a: The causes and consequences of human evolution. *Journal of the Royal Anthropological Institute* 1, 67–86.

Foley, R.A. 1995b: The evolution and adaptive significance of human maternal behaviour. In *Human and Non-Human Primate Mothers: an integrated approach* (ed. C. Pryce, R.D. Martin, & D. Skuse,), pp. 27–36. Zurich: Karger.

Foley, R.A. and Lee, P.C. 1989: Finite social space, evolutionary pathways and reconstructing hominid behaviour. *Science* 243, 901–6.

Foley, R.A. & Lee, P.C. 1991: Ecology and energetics of encephalization in hominid evolution. *Philosophical Transactions of the Royal Society, London* Series B 334, 223–232.

Gamble, C. 1993: *Timewalkers*. London: Allen Lane.

Galdikas, B. 1985: Orangutan sociality at Tunjung Puting A. *Journal of Primatology* 9, 101–19.

Harvey, P.H., Martin, R.D. & Clutton-Brock, T.H. 1987: Life histories in comparative perspective. In *Primate Societies* (ed. B.B. Smuts, D.L. Cheney, R.M. Seyfarth, R.W. Wrangham & T.T. Struhsaker), pp. 181–196. Chicago: University of Chicago Press.

Hinde, R.A. 1976: Interactions, relationships and social structure. *Man* 11, 1–17.

Humphrey, N.K. 1976: The social function of intellect. In *Growing Points in Ethology* (ed. P.P.G. Bateson & R.A. Hinde), pp. 303–317. Cambridge: Cambridge University Press.

Jolly, A. 1984: *The Evolution of Primate Behaviour* (2nd Ed.). New York: MacMillan.

Kay, R. & Fleagle, J. 1994: *Anthropoid Origins*. New York: Plenum.

Keverne, E.B. 1995: Neurochemical changes accompanying the reproductive process: their significance for maternal care in primates and other mammals. In *Human and Non-Human Primate Mothers: an integrated approach* (ed. C. Pryce, R.D. Martin, & D. Skuse) pp. 69–79. Zurich: Karger.

Klein, R.G. 1992: The archaeology of modern human origins. *Evolutionary Anthropology* 1, 5–15.

Knight, C., Power, C. & Watts, I. 1995: The human symbolic revolution: a Darwinian account. *Cambridge Archaeological Journal* 5, 75–114.

Lahr, M.M. 1994: The multiregional model of modern human origins: a reassessment of its morphological basis. *Journal of Human Evolution* 26, 23–56.

Lahr, M.M. 1996: *The Evolution of Human Diversity*. Cambridge: Cambridge University Press.

Lahr, M.M. & Foley, R.A. 1994: Multiple dispersals and the origins of modern humans. *Evolutionary Anthropology* 3, 48–60.

Lee, P.C. 1989: Family structure, communal care and female reproductive effort. In *Comparative Socioecology: the Behavioural Ecology of Humans and Other Mammals* (ed. V. Standen & R. Foley), pp. 323–340. Oxford: Blackwell Scientific Publications.

Lee, P.C. 1994: Social structure and evolution. In *Behaviour and Evolution* (ed. P.J.B. Slater & T.R. Halliday), pp. 266–303. Cambridge: Cambridge University Press.

Lee, R.B. & DeVore, I. 1968: Problems in the studies of hunters and gatherers. In *Man the Hunter* (ed. R.B. Lee & I. DeVore), pp. 3–12. Chicago: Aldine.

Leonard, W.R. & Robertson, M.L. 1992: Nutritional requirements and human evolution: a bioenergetics model. *American Journal of Human Biology* 4, 179–195.

Lewin, R. 1993: *The Origin of Modern Humans*. Washinton, DC: Smithsonian Institution.
Martin, R.D., Chivres, D.J. & MacLarnon, A. 1985: Gastrointestinal allometry in primates and other mammals. In *Size and Scaling in Primate Biology*, pp. 61–89. New York: Plenum.
McGrew, W.C. 1992: *Chimpanzee Material Culture*. Cambridge: Cambridge University Press.
McHenry, H. 1992: How big were early hominids? *Evolutionary Anthropology* 1, 15–20.
Mellars, P. 1991: Cognitive changes and the emergence of modern humans in Europe. *Cambridge Archaeological Journal* 1, 63–76.
Mellars, P.C. 1996: *The Neanderthal Legacy*. Princeton NJ: Princeton University Press.
Mellars, P. & Stringer, C. (eds) 1989: *The Human Revolution*. Edinburgh: Edinburgh University Press.
Rendall, D. & Difiore, A. 1995: The road less travelled: phylogenetic perspectives in primatology. *Evolutionary Anthropology* 4, 43–52.
Rightmire, G.P. 1981: Patterns in the evolution of *Homo erectus*. *Paleobiology* 7, 241–246.
Ruff, C. & Walker, A.C. 1993: Body size and shape. In *The Nariokotome Skeleton* (ed. A.C. Walker & R.E. Leakey), pp. 234–265. Cambridge: Harvard University Press.
Schick, K. & Toth, N. 1993: *Making Silent Stones Speak*. New York: Simon and Schuster.
Smith, B.H. 1989: Dental development as a measure of life history in primates. *Evolution* 43, 683–688.
Smith, B.H. 1993: The physiological age of WT15000. In *The Nariokotome Skeleton* (ed. A.C. Walker & R.E. Leakey), pp. 195–220. Cambridge: Harvard University Press.
Smuts,B, Seyfarth, R. & Cheney, D. *et al.* 1987: *Primate Societies*. Chicago: Chicago University Press.
Stoneking, M. 1993: DNA and recent human evolution. *Evolutionary Anthropology* 2, 60–73.
Stringer, C. & Gamble, C. 1993: *In Search of the Neanderthals*. London: Thames and Hudson.
Susman, R.L., Stern, J.T. & Jungers, W.L. 1985: Locomotor adaptations in the Hadar hominids. In *Ancestors: the Hard Evidence* (ed. E. Delson), pp. 184–192. New York: Alan Liss.
Symonds, D. 1979: *The Evolution of Human Sexuality*. Oxford: Oxford University Press.
White, F. and Wrangham, R.W. 1988: Feeding competition and patch size in the chimpanzee species *Pan paniscus* and *Pan troglodytes*. *Behaviour* 105, 148–163.
Williams, S.A. & Goodman, M. 1989: A statistical test that supports a human/chimpanzee clade based on non-coding DNA sequence data. *Molecular Biology and Evolution* 2, 338–346.
Wrangham, R.W. 1979: On the evolution of ape social systems. *Social Science Information* 18, 335–68.
Wrangham, R.W. 1980: An ecological model of female bonded primate groups. *Behaviour* 75, 262–300.

Friendship and the Banker's Paradox: Other Pathways to the Evolution of Adaptations for Altruism

JOHN TOOBY & LEDA COSMIDES

Center for Evolutionary Psychology,
University of California, Santa Barbara, CA 93106, USA

Keywords: reciprocity; altruism; co-operation; social exchange; reciprocal altruism; evolutionary psychology.

Summary. The classical definition of altruism in evolutionary biology requires that an organism incur a fitness cost in the course of providing others with a fitness benefit. New insights are gained, however, by exploring the implications of an adaptationist version of the 'problem of altruism', as the existence of machinery designed to deliver benefits to others. Alternative pathways for the evolution of altruism are discussed, which avoid barriers thought to limit the emergence of reciprocation across species. We define the Banker's Paradox, and show how its solution can select for cognitive machinery designed to deliver benefits to others, even in the absence of traditional reciprocation. These models allow one to understand aspects of the design and social dynamics of human friendship that are otherwise mysterious.

FROM A SELECTIONIST TO AN ADAPTATIONIST ANALYSIS OF ALTRUISM

THE ANALYSIS OF THE EVOLUTION OF ALTRUISM has been a central focus of modern evolutionary biology for almost four decades, ever since Williams, Hamilton, and Maynard Smith caused researchers to appreciate its significance (Williams & Williams 1957; Hamilton 1963, 1964; and Maynard

Smith 1964). The related concepts of conflict and co-operation have since developed into standard tools of evolutionary thought, and their use has transformed our understanding of everything from inter-organism interactions and kinship (Hamilton 1964) to inter-gene and within organism interactions and structures. For example, when applied to the genome these concepts lead straightforwardly to the derivation of the set of principles of intragenomic conflict that govern much about how genetic systems and intra-individual structures evolve (e.g., Cosmides & Tooby 1981). Indeed, pursuing the logic of conflict and co-operation has even led to a transformation in how biologists think of fitness itself—not just in the addition of kin effects to individual reproduction (Hamilton 1964), but also in the reconsideration of what entities it is proper to assign fitness to. It is clear now that sexually reproducing individuals cannot properly be assigned fitnesses, nor can they be correctly characterized as inclusive fitness maximizers, because the genome contains multiple sets of genes whose fitnesses cannot all be maximized by the same set of outcomes (Cosmides & Tooby 1981; Dawkins 1982; Haig 1993). For this reason, fitnesses can only coherently be assigned to genes or sets of co-replicated genes rather than to individual organisms or groups. By this and other routes, the careful analysis of co-operation and conflict has led inexorably to the recognition that genic selection is the fundamental level driving the evolutionary process, with individual selection analyses as often inexact and frequently problematic oversimplifications. In this new world of biological analysis, folk concepts like 'self-interest' and 'individual' have no exact counterparts, and their uncritical use can lead away from the proper understanding of biological phenomena.

There are two evolutionary pathways to altruism that have been proposed so far, kin selection, and reciprocal altruism. We think there are other pathways in addition to these two, and after revisiting the logic of reciprocal altruism we would like to explore several of them. Williams (1966) introduced the core of the reciprocal altruism argument, which was greatly expanded upon by Trivers (1971), and fitted into the Prisoner's Dilemma formalism by Axelrod & Hamilton (1981; Boyd 1988). The argument is that altruistic acts can be favoured if they cause the target of the altruism to subsequently reciprocate the act. A population of reciprocating designs is stable against invasion by nonreciprocators if part of the design is the detection of nonreciprocation and the subsequent exclusion of nonreciprocators. This argument is, in fact, a transplantation into biology of the fundamental economic insight that self-interested agents can increase their own welfare through contingently benefiting others through acts of exchange, i.e., by exploiting the potential for realizing gains in trade, to use terminology from economics. The reciprocal altruism argument involves the exploration of only one branch of the more inclusive set of logically possible exchange relationships—the branch in which there is a delay between the time at which the agent takes the altruistic action and her discovery of whether the act is contingently compensated. The natural category of exchange relationships and their timing and contingency is larger than this one line of analysis, and for this reason, we tend to term the more inclusive set of relationships *social exchange*.

Classically, the analysis of the problem of altruism follows logically from its standard definition: An altruistic act is one that lowers the direct individual reproduction of the organism committing the act while simultaneously raising the direct individual reproduction of another organism (Williams & Williams 1957; Hamilton 1964; Maynard Smith 1964). Viewed in this way, an essential part of the definition of altruism is that the individual committing the altruistic act be incurring a diminution in its direct reproduction—that is, a cost. Altruism is not considered to have taken place unless such a cost is suffered, and the existence of this cost must be demonstrated before there is considered to be a phenomenon to be explained. With cost to direct fitness defining and limiting the class of instances of altruism, the explanatory task becomes one of finding a corresponding and greater consequent benefit to fitness, as when there is a sufficiently offsetting benefit to kin (Williams & Williams 1957; Hamilton 1963, 1964; Maynard Smith 1964). Although the definition of altruism is sometimes widened to include acts that are costly in terms of inclusive fitness, the definition remains cost-centered. As useful as this framework has been, we think that a modification in the classical definition of altruism may open the way to additional insights about biologically interesting social phenomena, particularly in humans. Before discussing this modification, however, it is necessary to review briefly the logic of adaptationism, because the two issues are tied together.

To begin with, we think that some measure of confusion has been generated in evolutionary biology by failing to clearly distinguish the first level of evolutionary functional analysis, selectionist analysis, from the second level of functional analysis, adaptationist analysis (Williams 1966; Symons 1990, 1992; Thornhill 1991). The first is the widespread and often productive practice of analysing behaviour or morphology in terms of its current or even implicitly prospective fitness consequences. If used carefully, this can be a key heuristic tool, and its widespread adoption has contributed to the avalanche of functional insights achieved in the last forty years. However, just as individual selection analyses need to be reformulated into genic selection analyses to sidestep errors and accurately explain the full landscape of biological phenomena, so also selectionist models need to be reformulated into adaptationist analyses to capture more precisely the relationship between selection and phenotypic design (Tooby & Cosmides 1990a, 1992).

Within an adaptationist framework, an organism can be described as a self-reproducing machine. The presence in these organic machines of organization that causes reproduction inevitably brings into existence natural selection, a system of negative and positive feedback, that decreases the frequency of inheritable features that impede or preclude their own reproduction, and that increases the frequency of features that promote their own reproduction (directly, or in other organisms). Over the long run, down chains of descent, this feedback cycle pushes a species' design stepwise 'uphill' towards arrangements of elements that are increasingly improbably well-organized to cause their own reproduction into subsequent generations, within the envelope of ancestral conditions the species evolved in. Because the reproductive fates of the inherited traits that coexist in the same organism are to some significant extent linked together, traits will be selected to enhance each other's functionality (with some important exceptions, see Cosmides & Tooby 1981; Tooby & Cosmides 1990b for the relevant genetic analysis and qualifications). Consequently, accumulating design features will often tend to sequentially fit themselves together into increasingly functionally elaborated machines for trait propagation, composed of constituent mechanisms—adaptations—that solve problems that are either necessary for trait reproduction or increase its likelihood within environments sufficiently similar to ancestral conditions (Dawkins 1986; Symons 1992; Thornhill 1991; Tooby & Cosmides 1990a, 1992; Williams 1966, 1985).

From an adaptationist as opposed to a selectionist perspective, the central object of investigation is identifying and mapping the functional organization of the organism's machinery, and discovering exactly how this ordered arrangement produced propagation within the environment within which the machinery evolved. For the purpose of this engineering analysis, one can define the *environment of evolutionary adaptedness* (EEA) for an adaptation with precision. The EEA is the set of selection pressures (i.e., properties of the ancestral world) that endured long enough to push each allele underlying the adaptation from its initial appearance to effective fixation (or to frequency-dependent equilibrium), and to maintain them at that relative frequency while other necessary alleles at related loci were similarly brought to near fixation. Because moving mutations from low initial frequencies to fixation takes substantial time, and sequential fixations must usually have been necessary to construct complex adaptations, complex functional design in organisms owes its detailed organization to the structure of long-enduring regularities of each species' past. Each functional design feature present in a modern organism is there in response to the repeating elements of past environments, and these regularities must be correctly characterized if the design features are to be understood.

Adaptations are thus recognizable by 'evidence of special design' (Williams 1966)—that is, by whether there is a highly non-random co-ordination between recurring properties of the phenotype and the recurring structure of the ancestral environment, so that when they interacted together they meshed to reliably promote fitness (genetic propagation). The demonstration that features of an organism constitute an adaptation is always, at core, a probability argument concerning how non-randomly functional this co-ordination is. The standards for recognizing special design include such factors as economy, efficiency, complexity, precision, specialization, and reliability (Williams 1966), which are valid in that they index how unlikely a configuration is to have emerged randomly, that is, in the absence of selection. As Pinker and Bloom eloquently put it with respect to the eye, '[t]he eye has a transparent refracting outer cover, a variable-focus lens, a diaphragm whose diameter changes with illumination level, muscles that move it in precise conjunction and convergence with those of the other eye, and elaborate neural circuits that respond to patterns defining edges, colors, motion, and stereoscopic disparity. It is impossible to make sense of the structure of the eye without noting that it appears as if it was designed for the purpose of seeing... Structures that can do what the eye does are extremely low-probability arrangements of matter. By an unimaginably large margin, most objects defined by the space of biologically possible arrangements of matter cannot bring an image into focus, modulate the amount of incoming light, respond to the presence of edges and depth boundaries, and so on' (Pinker & Bloom, 1990).

So, what would an adaptationist view of the problem of altruism be, as opposed to a selectionist view? An adaptationist definition of altruism would focus on whether there was a highly nonrandom phenotypic complexity that is organized in such a way that it reliably causes an organism to deliver benefits to others, rather than on whether the delivery was costly. The existence of such a design is the adaptationist problem of altruism—an evolutionary 'problem' requiring explanation whether that delivery is costly, cost-free, or even secondarily beneficial to the deliverer. Indeed, the greater the cost component, the more this will militate against the emergence or elaboration of machinery designed to deliver benefits, and the less widespread such adaptations for altruism are expected to be. The less costly or more secondarily beneficial the machinery is, the more widespread such adaptations should be, and the more functionally elaborated and improbably functionally organized they will be. Moreover, once altruistic adaptations are in place, selection will act to minimize or neutralize their cost, or even make them secondarily beneficial, to the extent possible.

One reason why cost has been emphasized is, we suspect, because researchers have been attempting to distinguish altruistic acts that are incidental by-products of adaptations designed for other functions from altruistic acts produced by adaptations designed to deliver them. The presence of a cost component does not, however, distinguish these cases. The world is full of costly altruistic acts—every time a gazelle walks toward a hidden lion, altruism (classically defined) is taking place. The important distinction is whether the analysis of cross-generationally recurrent phenotypic structures can support the claim that there is machinery that is well-designed to deliver benefits to other organisms under ancestral conditions. Finding that this machinery produces collateral benefits for the organism not connected with the delivery of altruistic acts to others is irrelevant if these are side-effects of its design: if they do not explain the features of organization that are well-designed for delivering benefits to others, then the adaptationist 'problem of altruism' is still present. To mutate a phrase from George Williams (1966), the issue is not altruism *per se*, but design for altruism, that is, design for benefit delivery.

Of course, part of the adaptationist task involves explaining how the designed delivery of benefits to other organisms is ultimately tributary to the fitness of the genes underlying the altruistic adaptation, and in this task it is necessary to show that the fitness benefits are greater than the costs. However, this explanatory burden exists for the explanation of all adaptations, and not just for altruistic ones. We suggest that, in order to make more progress in understanding altruism, it will be necessary to shift from the selectionist practice of categorizing individual current behaviours as selfish or altruistic to the adaptationist project of investigating the logic of the organization of altruistic machinery, and analysing what problem each element is solving.

Finally, we think that an adaptationist perspective on altruism and aggression makes it clear why, in the biological world, aggression is so much more common a form of social instrumentality than altruism. Because organisms are improbably well-organized collections of matter, entropy ensures that these intricate machines, with so many interdependent parts, will be easy to disrupt. There are only a minuscule number of ways that an organism's parts will fit together so that they function correctly, while there are a vast number of pathways that will 'break' a complexly organized system. Introducing even minute changes into the organization of a single component can result in death (consider the effects of a drop of curare or a tiny puncture to the heart). Unfortunately, the corollary to being organized is that the set of acts that are capable of enhancing the functioning of a complex system is an infinitesimally small subset of the set of all possible acts. Because there are many more ways to damage an organism than to enhance its functioning, evolving designs for delivering damage is easy and hence common, while evolving designs that can deliver narrowly targetted benefits is hard and hence rare. Because the task of correctly identifying and successfully enacting beneficial operations will often be very difficult, we think that such adaptations will frequently require complex computations, and suspect that at least some adaptations for altruism may turn out to rival the complexity of the eye. From this engineering perspective, the existence of cognitive machinery that is functionally organized to deliver benefits to others is a highly improbable state of affairs.

ADAPTATIONISM AND NON-COSTLY ROUTES TO ALTRUISM

So, what new insights might an adaptationist approach to altruism provide? First, it makes clear that there potentially may be, in a species, many distinct and separable sets of adaptations for altruism, designed to deliver benefits to different targets for quite independent reasons. We believe that reciprocal altruism and kin-selected altruism are only two pathways out of a larger set (Tooby & Cosmides 1984, 1989a). If there are a number of independent pathways that cause the evolution of adaptations for altruism, then each type of selection pressure can shape its own distinct set of adaptive devices to serve different ends according to its own independent functional logic.

Second, it allows researchers to consider a far broader variety of definitions of *benefit* and hence of *altruism* than they would under the classical definition, which requires an increase in the target's direct (or even inclusive) fitness. Delivery of some alternative kinds of 'benefits', such as increasing the target's longevity, or increasing the target's ability to act, deserve independent treatment, regardless of whether they would have increased the target's inclusive fitness as a by-product (see below). Since the phenomenon to be explained is functional organization in whatever form it appears, then organization designed to increase the survival of targeted individuals, for example, requires as much explanation as organization designed to increase the target's reproduction. Acting to insure someone's survival or to increase their energy budget fits naturally into the more encompassing common-sense definition of altruism as the conferral of benefits, even if there is no impact on the recipient's reproduction. People want to understand altruism in this broader sense, and its role in social life—not just altruism in the narrow sense.

Third, the abandonment of a cost-centered definition of altruism allows one to see how the evolution of non-kin based altruism might be easier than it is usually considered to be. Many researchers, such as Boorman & Levitt

(1980) and Axelrod & Hamilton (1981), have pointed out that while reciprocation or tit for tat are evolutionarily stable against invasion by defectors, they are selected against when they appear at low frequencies, creating a barrier to the evolution of co-operation (see also Boyd & Lorberbaum 1987). The lone mutant is initially altruistic to each new potential partner, but because its acts are never reciprocated by the surrounding population of defectors, its fitness is lower than theirs. For the mutation to take off, it must appear initially in sufficiently high concentrations that it meets its design-replicas often enough to compensate for its encounter with and exploitation by defectors.

If one ceases to model altruistic acts as necessarily and definitionally costly, however, another pathway to the evolution of machinery designed to provide benefits becomes straightforward (Tooby & Cosmides 1984, 1989a). If one imagines, as a thought experiment, a world in which organisms act without regard to their consequences on others, each organism will be selected to engage in behaviours because of their probable favourable fitness consequences on relevant gene sets it carries. Furthermore, each of its actions can be naturally partitioned into one of three categories, on the basis of its consequences for other organisms: (1) actions that have a beneficial effect on another organism, (2) actions that have an injurious effect on another organism, and (3) actions that have no net effect on another organism. As the animal goes about its affairs, it will continuously, and at no cost to itself, be dispensing collateral benefits and injuries on others. Given this initial state, other organisms will certainly be selected to deploy themselves so as to avoid harm and capture benefits. But they will also be selected to engage in actions that have the net effect of increasing the probability that the actor will 'emit' benefits and of decreasing the probability that it will produce harm.

How might this influence take place? Leaving aside the important topic of manipulation, X could increase the frequency with which Y emits zero-cost behaviours that incidentally benefit X by providing contingent rewards: i.e., by providing benefits to X whenever it engages in a behaviour with side-effects that happen to benefit oneself. Under natural conditions, X may commonly have available many courses of action that benefit itself about equally, but whose collateral consequences on Y might be sharply different. The same is true for Y. For example, if X knows the way back to the camp, but Y is lost, X experiences little cost by allowing Y to follow her home. In such a case, Y needs to create only minor changes in payoffs to change the course of action that X will take. By attaching new payoff contingencies to alternative courses of action, and successfully making these contingencies detectable to the actor, one individual may influence the behaviour of another to its benefit. What one would see emerging in such a world would be the mutual provisioning of benefits between social interactors. In such a scenario, the low initial frequency of the mutant type constitutes no barrier to the evolution of altruistic behaviour, nor is cost intrinsically a barrier either.

What, then, is the mutation and what is the background of pre-existing adaptations that this model requires? The new mutant design is one that contingently responds to the actions of decision-making agents when they are beneficial in nature, by conditionally providing that agent with a detectable corresponding benefit. The model assumes that the mutant is born into a world in which the members of the population have the computational ability to (1) compute and compare the rates of return for alternative courses of action, and (2) use this information in deciding what course of action to pursue on subsequent occasions. Many species have evolved such competences for other purposes (such as foraging). For example, Gallistel (1990) has shown that classical and operant conditioning are produced by computational processes that are formally equivalent to multivariate time series analysis: by analysing correlations, the animal computes the rate of delivery of an unconditioned stimulus when a conditioning event is present and absent. Of course, Garcia & Koelling's work (1966) on learned food aversions in rats was the first in a long line of studies showing that conditioning will not occur unless the animal has 'prior hypotheses' about what causes what (e.g., that food can cause nausea, but not electric shocks), so it is far from inevitable that animals will be able to connect a conditionally delivered social reward to an action they took for other reasons. Nonetheless, it is only necessary that there be a rudimentary, slightly better than random ability to detect social contingency to get the system started. Once started, one would predict that such forces would increasingly shape specialized computational devices so that they could effectively track social agency and social contingency.

This general line of reasoning has motivated our own experimental investigations of how humans interpret and reason about conditional social actions. Human cognitive machinery does, as expected, sharply distinguish inanimate causal conditionals from social conditionals such as social exchanges and threats. More importantly, humans appear to have an independent specialized computational system that is well-designed for reasoning adaptively about the conditional relationships involved in social exchange (Cosmides 1989; Cosmides & Tooby 1992; Gigerenzer & Hug 1992), and another one for conditional social threats (Tooby & Cosmides 1989b, forthcoming). Of these, the experimental investigation of adaptations for reasoning about social exchange has proceeded the farthest, and we have been able to find evidence that the machinery involved has many design features that are specialized for this function (see Table 1).

Table 1. Computational machinery that governs reasoning about social contracts (based on evidence reviewed in Cosmides & Tooby 1992)

Design features:
1. It includes inference procedures specialized for detecting cheaters.
2. The cheater detection procedures cannot detect violations that do not correspond to cheating (e.g., mistakes where no one profits from the violation).
3. The machinery operates even in situations that are unfamiliar and culturally alien.
4. The definition of cheating it embodies varies lawfully as a function of one's perspective.
5. The machinery is just as good at computing the cost-benefit representation of a social contract from the perspective of one party as from the perspective of another.
6. It cannot detect cheaters unless the rule has been assigned the cost-benefit representation of a social contract.
7. It translates the surface content of situations involving the contingent provision of benefits into representational primitives such as 'benefit', 'cost', 'obligation', 'entitlement', 'intentional' and 'agent'.
8. It imports these conceptual primitives, even when they are absent from the surface content.
9. It derives the implications specified by the computational theory, even when these are not valid inferences of the propositional calculus (e.g., 'If you take the benefit, then you are obligated to pay the cost' implies 'If you paid the cost, then you are entitled to take the benefit').
10. It does not include procedures specialized for detecting altruists (individuals who have paid costs but refused to accept the benefits to which they are therefore entitled).
11. It cannot solve problems drawn from other domains; e.g., it will not allow one to detect bluffs and double crosses in situations of threat.
12. It appears to be neurologically isolable from more general reasoning abilities (e.g., it is unimpaired in schizophrenic patients who show other reasoning deficits; Maljkovic 1987).
13. It appears to operate across a wide variety of cultures (including an indigenous population of hunter-horticulturalists in the Ecuadorian Amazon; Sugiyama, Tooby & Cosmides 1995).

Alternative hypotheses eliminated:
1. That familiarity can explain the social contract effect.
2. That social contract content merely activates the rules of inference of the propositional calculus.
3. That social contract content merely promotes (for whatever reason) 'clear thinking'.
4. That permission schema theory can explain the social contract effect.
5. That any problem involving payoffs will elicit the detection of violations.
6. That a content-independent deontic logic can explain the effect.

A parallel and growing body of evidence from cognitive development is showing that human infants have cognitive machinery that makes sharp distinctions between animate and inanimate causation (Leslie 1988, 1994; Gelman 1990; Premack & Premack 1994), and that toddlers have a well-developed 'mind-reading' system, which uses eye direction and movement to infer what other people want, know, and believe (Baron-Cohen 1995; Leslie & Thaiss 1992). These inference systems provide 'privileged hypotheses' about social causation that vastly expand the time frames across which humans can that compute socially contingent changes in rates of return.

In any case, what is critical to this evolutionary pathway is that the organism whose actions are to be influenced be capable of categorizing its actions in terms of their consequences for others, rather than just in terms of their consequences for itself. If the animal cannot do this, then it cannot reliably be induced to repeat, out of the sets of actions it considers equivalent, the specific type of action that delivered the collateral benefit to the animal prepared to reward it. In such cases, mutant individuals equipped with the adaptation to respond to benefits by providing contingent rewards will be selected against, because these rewards will be ineffectual: they will not increase the probability that the target individual will repeat the beneficial action in the future. For such species, this pathway to the evolution of social exchange is closed.

The ability to compute the effects of actions on others, and to categorize such acts in terms of their value to others, is a nontrivial requirement. It may be the rarity of this set of prerequisite adaptations, and not the cost problem, that is a real impediment to the frequent evolution of social exchange (e.g., the 'mind-reading' abilities of other primates appear to be far more limited than our own; Cheney & Seyfarth 1990; Whiten 1991). However, kin-selected machinery for altruism would select for these same prerequisite adaptations, and so the evolution of social exchange may be commonly facilitated by the prior evolution of kin-selected altruistic adaptations. In any case, once adaptations for social exchange have begun to emerge, they will select for increasingly sophisticated computational abilities to model other organisms' values, intentions, principles of categorization and social representation, and responsiveness to social contingency (Cosmides 1985; Cosmides & Tooby 1989; Humphrey 1984; Whiten 1991). For example, one would expect that humans would have a specialized computational device—an implicit 'theory of human nature'—that models what motivations and mental representations others would develop when placed in various evolutionarily recurrent situations. This would function in tandem with the increasingly well-documented 'theory of mind' module (Baron-Cohen *et al.* 1985; Baron-Cohen 1995; Leslie & Thaiss 1992), and other widely discussed mechanisms such as empathy and emotion recognition.

The ability to understand the nature of actions in terms of their meaning and impact on others is a two-edged sword, however. Not only does it facilitate the growth of co-operation, but it also lengthens the reach of extortive threat and makes revenge possible. This is because the argument about collateral benefits applies symmetrically to collateral injury (Tooby & Cosmides 1984, 1989). Organisms can be expected to evolve systems of contingent injury that force other animals to take their interests into account when choosing their courses of action. The evolution of threats and revenge similarly depends on the nature of the interpretive machinery a species has. If another animal lacks the capacity to categorize acts based on the injury they cause, then punishing it is ineffective, and vindictive designs will not evolve. This may be why most species are limited to proximate deterrence and immediate threat, rather than to more complex intercontingent strategies such as revenge.

In any case, once adaptations for delivering contingent rewards and adaptations for detecting contingent rewards become present in the same population, the population can evolve without impediment towards full social exchange. The increasing ability of the members of a species to detect and produce social contingency and to represent what is valuable and injurious to others frees the altruistic dynamics from an initial context in which actions with beneficial side-effects for others are undertaken for other purposes. Once contingency can be detected, contingent reward can become the sole reason an action is taken. As the evolutionary process continues, the adaptations involved can be increasingly accurately described as serving the function of delivering benefits to others.

The costs of actions may not be relevant to an adaptationist *definition* of altruism, but they are relevant to understanding some of the *design features* of adaptations for delivering benefits. To influence each other in a well-calibrated way, animals must be able to accurately estimate the costs and benefits of an action to self and others, and to predict what actions others will take in the absence of a contingently provided benefit (see, e.g., Cosmides & Tooby 1989, on baselines). The size of a contingently delivered benefit will change the landscape of payoffs: X may engage in actions that it formerly avoided because the costs outweighed the benefits, because a contingently delivered benefit now makes them worthwhile. Y should be designed to deliver an optimal reward level: one that yields the greatest average net benefit to itself in terms of prospectively altered dispositions to act in the other animal. If inducements are too weak, the benefit may not be delivered. If inducements are 'too strong'—that is, if X would have delivered the same benefits in response to smaller inducements, then the reward might be wasteful. A key computational component is the ability to map the world of costs and benefits according to the psychology of a potential exchange partner (or antagonist), and to judge whether its beneficial (or harmful) acts were 'intentional'—i.e., generated because of the impact they could be expected to have on one's own behaviour. The latter would allow one to determine when a social contingency has appeared or been withdrawn; to distinguish exchanges explicitly arrived at from noisier, more probabilistic sequences; to monitor others for cues of valuation, and so on.

CRISIS MANAGEMENT

The Banker's paradox

> If thou wouldst get a friend, prove him first, and be not hasty to credit him. For some man is a friend for his own occasion, and will not abide in the day of thy trouble... Again, some friend is a companion at the table, and will not continue in the day of thy affliction... If thou be brought low, he will be against thee, and will hide himself from thy face... A faithful friend is a strong defence: and he that hath found such a one hath found himself a treasure. Nothing doth countervail a faithful friend... *From Ecclesiastes 6*

Many people become angry when they first hear the evolutionary claim that the phenomenon of friendship is solely based on the reciprocal exchange of favours, and deny that their friendships are founded on such a basis. Similarly, many people report experiencing a spontaneous pleasure when they can help others without any expectation or anticipation of reward. Their memory of the pleasure is not diminished by not ever having received a reward in return. Indeed, explicit linkage between favours or insistence by a recipient that she be allowed to immediately 'repay' are generally taken as signs of a lack of friendship. What is going on? One widely accepted interpretation is that these denials are simply the deceptive surface of human social manipulation. We think, however, that narrow exchange contingency does not capture the phenomenology or indeed the phenomenon of friendship. We propose that the altruistic adaptations that underlie friendship do not map onto the structure of tit for tat or any other standard model of reciprocal altruism based on alternating sequences of contingent favours.

One dimension of difference is illustrated by what we will call the *Banker's Paradox*. Bankers have a limited amount of money, and must choose who to invest it in. Each choice is a gamble: taken together, they must ultimately yield a net profit, or the banker will go out of business. This set of incentives leads to a common complaint about the banking system: that bankers will only loan money to individuals who do not need it. The harsh irony of the Banker's Paradox is this: just when individuals need money most desperately, they are also the poorest credit risks and, therefore, the least likely to be selected to receive a loan.

This situation is analogous to a serious adaptive problem faced by our hominid ancestors: exactly when an ancestral hunter-gatherer is in most dire need of assistance, she becomes a bad 'credit risk' and, for this reason, is less attractive as a potential recipient of assistance. If we conceptualize contingent benefit-benefit interactions as social exchange (rather than more narrowly as reciprocation), then individuals rendering assistance can be seen as facing a series of choices about when to extend credit and to whom.

Assisting one individual may take time, resources, or be dangerous to oneself—it therefore precludes other worthwhile activities, including assisting others. From this perspective, exchange relationships are analogous to economic investments. Individuals need to decide who they will invest in, and how much they will invest. Just as some economic investments are more attractive than others, some people should be more attractive as objects of investment than others.

Computational adaptations designed to regulate such decisions should certainly take into account whether an individual will be willing to repay in the future (i.e., are they a cheater?). But they should also assess whether the person will be in a position to repay (i.e., are they a good credit risk?), and whether the terms of exchange will be favourable (will this exchange partnership ultimately prove more profitable than the alternatives it will preclude?). If the object of investment dies, becomes permanently disabled, leaves the social group, or experiences a permanent and debilitating social reversal, then the investment will be lost. If the trouble an individual is in increases the probability of such outcomes when compared to the prospective fortunes of other potential exchange partners, then selection might be expected to lead to the hardhearted abandonment of those in certain types of need. In contrast, if a person's trouble is temporary or they can easily be returned to a position of full benefit-dispensing competence by feasible amounts of assistance (e.g., extending a branch to a drowning person), then personal troubles should not make someone a less attractive object of assistance. Indeed, a person who is in this kind of trouble might be a more attractive object of investment than one who is currently safe, because the same delivered investment will be valued more by the person in dire need. The attractiveness of extending the branch can be compared to nursing someone with a life-threatening disease for months: the cost is high, and the outcome is uncertain.

For hunter-gatherers, illness, injury, bad luck in foraging, or the inability to resist an attack by social antagonists would all have been frequent reversals of fortune with a major selective impact. The ability to attract assistance during such threatening reversals in welfare, where the absence of help might be deadly, may well have had far more significant selective consequences than the ability to cultivate social exchange relationships that promote marginal increases in returns during times when one is healthy, safe, and well-fed. Yet selection would seem to favour decision rules that caused others to desert you exactly when your need for help was greatest. This recurrent predicament constituted a grave adaptive problem for our ancestors—a problem whose solution would be strongly favoured if one could be found. What design features might contribute to the solution of this problem?

Becoming irreplaceable: The appetite for individuality

One key factor is replaceability or substitutability. Consider X's choice between two potential objects of investment, Y and Z. Each helps X in different ways; the magnitude of the benefits Z delivers are higher than the magnitude of the benefits that Y delivers, but the types of benefits that Y supplies can be supplied by no one else locally. Consider the alternative payoffs when one or the other enters a crisis and requires help. Extending 'credit' to a person in crisis may easily have a negative payoff if the *kind* of benefits that she customarily delivers could be easily supplied by others. To the extent an individual is in social relationships in which the assistance she delivers to her partners could easily be supplied in her absence by others, then there would be no necessary selection for her partners to help her out of difficulty. A 'replaceable' person would have been extremely vulnerable to desertion. In contrast, extending credit has a higher payoff if the person who is currently in trouble customarily delivers types of benefits (or has some other value) that would be difficult to obtain in her absence. Selection should favour decision rules that cause X to exhibit loyalty to Y to the extent that Y is irreplaceably valuable to X. In other words, Y's associates will invest far more in rescuing her than they would if she lacked these unique distinguishing properties (Tooby & Cosmides 1984, 1989a). Y may be helped, and Z abandoned even though the benefits Z delivers are greater.

If Banker's Paradox dilemmas had been a selection pressure, then one would expect to see adaptations that caused humans to:

1 have an appetite to be recognized and valued for their individuality or exceptional attributes;

2 be motivated to notice what attributes they have that others value but cannot obtain as easily elsewhere;

3 be motivated to cultivate specialized skills, attributes, and habitual activities that increase their relative irreplaceability;

4 be motivated to lead others to believe that they have such attributes;

5 preferentially seek, cultivate, or maintain social associations and participate in social groups where their package of valued attributes is most indispensable, because what they can differentially offer is what others differentially lack;

6 preferentially avoid social circles in which what they can offer is not valued or is easily supplied by others; and,

7 be jealous or rivalrous when someone within their social circle develops abilities to confer similar types of benefits, or when someone with similarly valued attributes enters their social circle. Such jealousy would motivate and organize actions that drive off attribute-rivals and that inhibit

individuals who value the actor from developing potential relationships with others who could supply the same type of assistance.

Although we are unaware of any experimental studies specifically of these traits, we think many aspects of human social and mental life show clear evidence of them. Much of social life seems to consist of a continual movement to find and occupy individualized niches that are unusually other-benefitting but hard to imitate, accompanied by a shuffling of social associations in search of configurations where the parties are most highly mutually valued. Indeed, the cross-culturally general motivation for status (as opposed to dominance) is arguably a product, in some measure, of this kind of selection pressure. Calling someone irreplaceable, or stressing how they will be (or have been) missed is a ubiquitous form of praise. Many other phenomena seem to be obvious expressions of a psychology organized to deal with the threat of social replaceability. These include everything from complaints about feeling anonymous in modern mass societies to the incessant fissioning off of smaller social groups whose members cultivate a mutual sense of belonging and discourage transactions with outgroup members. More significantly, the growth of irreplaceability as a feature of hominid life would have had powerful secondary impacts on hominid evolution. For example, individuals could pursue more productive, but more injury producing subsistence practices, such as large game hunting.

The motivation to discover and occupy unique niches of valued individuality is facilitated by the many forces that act to spontaneously locate individuals in unique 'starting positions' (Tooby & Cosmides 1988). These include, obviously, the fact that each individual's talents and shortcomings will be somewhat different due to random genetic variation, the accidents of ontogeny, and the different kinship, demographic and social circumstances they are born into. One might expect selection for adaptations that guide an individual not only to hone those skills that she can do well in an absolute sense, but to put special effort into those skills that she does relatively well, so that she 'product-differentiates' herself. Indeed, the most common and basic meaning humans apply to the issue of ability-acquisition is a social meaning—ability relative to others—rather than an absolute standard. Competences that everyone shares are not even noticed. In any case, Plomin & Daniels' work (1987) on the effects of nonshared environment provides strong evidence that individuals do product-differentiate themselves, even among their siblings, as does Sulloway's pioneering work on birth order (forthcoming).

Fair weather friends and deep engagement

The archetypal concept of the fair weather friend implies that there is also another kind of friend, a close or true friend—someone who is deeply engaged in your continued survival and in your physical and social welfare (but not necessarily in promoting the propagation of the genes you carry). It is this kind of friend that the fair weather friend is the counterfeit of. If you are a hunter-gatherer with few or no individuals who are deeply engaged in your welfare, then you are extremely vulnerable to the volatility of events—a hostage to fortune. Indeed, the higher the variance or volatility of the environment inhabited, the more individuals ought to care about friendships.

But if you wait until you are in trouble to determine whether anyone cares, it may be too late, if the answer is 'no'. When times are good, close friends who are deeply committed to you and casual exchange partners for whom you are replaceable may behave very similarly to each other. Moreover, since it is advantageous for anyone to be categorized as a close friend by someone who is not in difficulty, humans face the adaptive problem of friendship mimicry. The adaptive problem of discriminating true friends from fair weather friends would have been a formidable signal detection problem for our ancestors. One would expect the human psychological architecture to contain subsystems designed to sift social events for cues that would reduce uncertainty about the relative engagement different individuals have in one's welfare, i.e., assess the genuineness of friendship. Of course, the most ecologically valid evidence is what people actually do when you are genuinely in trouble. One would expect that assistance received in such times would be far more computationally meaningful, and cause a far greater change in attitude toward the giver than assistance rendered at other times. Phenomenologically, individuals seem to be deeply moved at such times, find such acts deeply memorable, and often subsequently feel compelled to communicate that they will never forget who helped them.

Given these facts and hypotheses, modern life creates a paradox. For the purposes of friendship assessment, different events and time periods will vary substantially in their informativeness, and certain types of events such as a period of personal trouble will be particularly clarifying. Yet, the human psychological architecture will obviously have been selected to avoid genuine and unnecessary personal difficulties. Safer, more stable modern environments may, therefore, be leaving people in genuine and uncharacteristically protracted doubt as to the nature of their relationships, and whether anyone is deeply engaged in their welfare. Because of the lack of clarifying events, an individual may have many apparently warm social contacts, and yet feel lonely, uneasy, and hungry for the confident sensation of deep social connectedness that people who live in environments that force deep mutual dependence routinely enjoy.

Although there are other kinds of cues, the basic structure of the clarifying event our minds are designed to monitor is one in which a particular individual has the opportunity to help, and that help would be of great value to the recipient. If they fail to help you when such help would be a deliverance, and the cost to them would not have been prohibitive, then it is a mistake to waste one of your scarce friendship niches on them (see below). Their level of commitment is revealed by the magnitude of the cost they are willing to incur per unit of benefit they are willing to deliver. Although there are many other variables that are important—such as how alert they are for opportunities to help, and how effective they can be at helping—the presence of deep engagement is a key variable.

NICHE LIMITATION MODELS OF FRIENDSHIP

Human hunter-gatherers, along with all other prisoners of space and time, have finite time and energy budgets, and cannot be in more than one place at a time. The decision to spend time with some individuals is, therefore, the decision not to spend time with others. Close spatial association is the prime factor that produces opportunities to help and be helped. For a hunter-gatherer, who one chooses to associate with will facilitate or preclude, over time, the development of computational states in others that are beneficial over the long run. From this perspective, each individual can be thought of as having a restricted number of *friendship* or *association niches*, and faces the computational problem of filling these slots with individuals from whom they will reap the best long-term outcomes. If an individual has a limited number of association niches, then the logic of the adaptations underlying friendship may be considerably different than that suggested by the standard model of reciprocation.

What factors would a well-designed computational device take into account in deciding how these niches should be filled?

1 *Number of slots already filled.* Adaptations should be designed to compute how many individuals in one's social world are deeply engaged in one's welfare, and how much uncertainty there is in this computation. If the number is high, then other factors, such as efficiency in exchange relationships or short run return to investment, might be weighted more heavily. If the number is low, or the individual is uncertain about the commitment of her friends, then adaptations should motivate counter-measures: activities that increase the likelihood of friend recruitment or consolidation should become more appealing.

2 *Who emits positive externalities?* The ongoing rewards of interacting with a person can take many forms other than specific acts of altruism.

Behaviours that are not undertaken as intentional acts of altruism often have side-effects that are beneficial to others—what economists call positive externalities. Some potential associates exude more positive externalities than others. For a knowledge-generating and knowledge intensive species such as ours, such situations abound. Someone who is a better wayfinder, game locator, tool-maker, or who speaks neighbouring dialects is a better associate, independent of the intentional altruistic acts she might direct toward you. Similarly, there are an entire array of joint returns that come about through co-ordinated action, such as group hunting or joint problem-solving. Individuals may vary in their value as friends and associates because they contribute to the general success, or because their attributes mesh especially well with yours or with other members of your cooperative unit.

3 *Who is good at reading your mind?* Dyads who are able to communicate well with each other, and who intuitively can understand each other's thoughts and intentions will derive considerably more from co-operative relationships than those who lack such rapport.

4 *Who considers you irreplaceable?* All else equal, it is better to fill a friendship niche with a person who considers you difficult to replace. This person has a bigger stake in your continued health and well-being than an individual who can acquire the kind of benefits you provide elsewhere.

5 *Who wants the same things you want?* A person who values the same things you do will continually be acting to transform the local world into a form that benefits you, as a by-product of their acting to make the world suitable for themselves. Trivial modern cases are easy to see: e.g., a roommate who likes the same music or who doesn't keep setting the thermostat to a temperature you dislike. Ancestrally, associates who shared affinities would have manifested many important mutual positive externalities, such as those who share enemies; those who have the same stake in the status of a coalition; spouses or affines who share a joint stake in the welfare of a set of children, and so on. There are likely to have been recurrent disputes and stable social divisions, and an individual is automatically benefitted by the existence of others who shared the same interest in the outcome. A person who your enemies fear, or a person who attracts more suitors than she can handle, may be a more valuable associate than a reliable reciprocator whose tastes differ widely from your own.

These and many other factors should be processed by the computational machinery that generates what we phenomenally experience as spontaneous liking. Many of them are attributes rather than act-histories, which offers an explanation for why we often experience a spontaneous and deep liking for someone on first exposure.

In other words, not only do individual humans have different reproductive values that can be estimated based on various cues they manifest, but they also have different association values. One dimension of this value is the partner-independent component, while the other component will vary specifically with respect to the individual attributes of each other potential partner. Adaptations that evolved to regulate association should be designed to fill niches with partners whose association delivers the most net rewards, and who value the individual highly and specifically. The tendency to dispense benefits contingent upon specific reciprocation is not the logic that defines association-value. Although the disposition to make alternating exchanges may not be completely irrelevant to an individual's value as an associate, it is neither a necessary nor a sufficient attribute. It can be trumped by other factors.

Of course, who you can associate with depends not only on who you like, but on who likes you, as well as larger scale structures of friend and family clustering. The computational architecture should be designed to deploy one's choices, acts, and attributes so as to make one's own association value high, and to attract the best distribution of friends into one's limited set of association niches. When this deployment is not effective enough to recruit a worthwhile set of friends, then the architecture should initiate other measures. Increasing the delivery of beneficial acts to others is one possibility, but the analysis above suggests other operations that might be effective: moving into new social worlds, initiating mateships (which have the potential to be a specialized kind of deep engagement association), conceiving children, increasing one's aggressive skills, searching for new positive externalities to exploit, moderating one's negative externalities, ending unfavourable relationships, chasing off association rivals, cultivating irreplaceability, resorting to extortion, and so on—each of which could lead to favourable reconfigurations of one's social world.

The dynamics of this kind of world are considerably different from what the co-operator-defector models, in isolation, suggest. In a world of limited friendship niches, the issue is not necessarily cheating *per se*, but the relative returns of different, mutually exclusive associations. Losing a valued friend, being able to spend less time with the friend, becoming less valued by that friend, or at the extreme, social isolation, may be more costly than being cheated. (This is not to say, however, that one cannot be cheated by a friend.) One way of modelling such a situation is as a Hobbesian bidding war of all against all, waged with the benefits of association, gated by the effectively limited number of friendship niches an individual has. The possibility that a friend will switch between friendships (or rather between mutually exclusive time-association budgets) on the basis of the relative rewards generated by each is the force that keeps the stream of benefits flowing and calibrated. In such a world, the adaptations will be designed to monitor all returns from a relationship, not just those from concrete acts of material assistance, reciprocally exchanged. It will be advantageous to be a high quality associate, and so individuals should feel a spontaneous pleasure in discovering effective ways of helping their friends, without looking for any contingent return. Instead of being cheated, the primary risk is experiencing a world increasingly devoid of deeply engaged social partners, or sufficiently beneficial social partners, or both. Adaptations should be designed to respond to signs of waning affection by increasing the desire to be liked, and mobilizing changes that will bring it about.

Friendship versus exchange

Accordingly, the phenomenology of friendship unsurprisingly reflects the pleasure you experience in someone's company, the pleasure you feel knowing they enjoy your company, the affection generated by an ease of mutual understanding, the desire to be thoughtful and considerate, the satisfaction in shared interests and tastes, how deeply you were moved by those who helped you when you were in deep trouble, how much pleasure it gave you to be able to help friends when they were in trouble, the trust you have in your friends, and so on. Explicit contingent exchange and turn-taking reciprocation are the forms of altruism that exist when trust is low and friendship is weak or absent, and treating others in such a fashion is commonly interpreted as a communication to that effect. The injection of explicit contingent exchange into existing friendships (e.g., buying a friend's car) is experienced as awkward. It seems to be a pervasive expression of human psychology that people in repeated contact feel the need to rapidly transform relationships that began in commercial transactions into something 'more'—with signs that indicate the relationship is no longer one simply of contingent exchange, but of friendship. Those of us who live in modern market economies engage in explicit contingent exchanges—often with strangers—at an evolutionarily unprecedented rate. We would argue that the widespread alienation many feel with modern commercial society is the result of an evolved psychological architecture that experiences this level of explicit contingent exchange in our lives as a message about how deeply (or rather, how shallowly) we are engaged with others.

Runaway friendship

The issues of irreplaceability and association value have a variety of implications about the functional organization of human social psychology. One of the most interesting implications of this model is how the detection of strong valuation should select for design features that construct a strong reflected valuation: a mirroring effect. By the argument of the Banker's

Paradox, if you are unusually or uniquely valuable to someone else—for whatever reason—then that person has an uncommonly strong interest in your survival during times of difficulty. The interest they have in your survival makes them, therefore, highly valuable to you. The fact that they have a stake in you means (to the extent their support is not redundant to you) that you have a stake in them. Moreover, to the extent they recognize this, the initial stake they have in you may be augmented. Our psychological adaptations should have evolved in response to these dynamics. For example, because you may be the only route through which your maiden aunt can propagate the genes she bears, her psychological architecture may recognize you as being uniquely valuable to her. Because she would sacrifice everything for you (let us assume), that makes her in turn an unusual or perhaps uniquely valuable person in your social universe. Because she values you, you have a corresponding stake in her survival and in the maintenance of her ability to act on your behalf. A risky action to save her life would not be a case of reciprocal altruism, but of altruism through cyclic valuation.

In the same way that the initial impetus in Fisherian runaway sexual selection may have been minor, the initial stake that one person has in the welfare of another might be minor. But the fact that this gives you a stake in them, which gives them a greater stake in you, and so on, can under the right conditions set up a runaway process that produces deep engagements. The recursive nature of these cyclic valuations can reinforce and magnify each person's association value to the other, far beyond the initial valuations. Friendships may become extremely powerful, despite weak initial conditions. Of course, this requires mutual communication and the ability to detect when someone truly values you (in which deception is certainly possible). But against a background of impoverished social options, it might not take much of an initial asymmetric valuation to get such a mirror relationship running and mutually reinforcing. Indeed, under the right conditions, a simple arbitrary decision may be enough (as in oaths of friendship that are found in many cultures), provided it is in the form of an emotional 'commitment' in the sense meant by Hirschleifer (1987) or Frank (1988). When applied to mate choice, these and many of the other arguments made above may help to illuminate the functional design of the adaptations that regulate romantic love (see also Nozick 1989: Ch. 8).

Finally, we want to emphasize that the benefits that certain of the adaptations for altruism described above are designed to deliver are not necessarily benefits at all in the classical sense of increases in direct reproduction or inclusive fitness. The benefits delivered may sometimes have such effects on the recipient's fitness, but this will be as an incidental by-product of the design of the adaptation. It is not the functional product of the adaptation—that is, what the adaptation was designed to do. For example, in some of these cases, the function of the altruistic act was to extend the recipient's lifespan or otherwise preserve whatever properties make the recipient willing and able to continue supplying benefits to you. If the recipient's fitness increases as a result, this is a side-effect of the computational design and, therefore, irrelevant to the selection pressure that shaped it. Meaningful alternative models of the evolution of altruism might be developed by looking at the delivery of energy, or survival through high-risk episodes, or what might be called agency altruism—increasing the ability of other agents to take effective action. By moving beyond the classical definition of altruism, which requires a fitness cost to the deliverer and a fitness benefit to the recipient, evolutionarily oriented researchers can construct a much richer family of models of altruism which may better account for the diverse array of altruistic adaptations in humans and other species.

Note. We would like to warmly acknowledge Irven DeVore, at whose Simian Seminar the first version of this paper was presented, for his input. He is particularly thanked for the intellectually stimulating effect that his frequent remark 'Tooby, you can be replaced!' had on the central ideas of this paper. We would also like to thank Daphne Bugental, David Buss, Steve Pinker and Don Symons who, as always, provided enlightening insights, and Betsy Jackson, Rob Kurzban, and Melissa Rutherford for their help on manuscript preparation. We are also deeply indebted to the many members of the Center for Evolutionary Psychology's regular seminar, and the seminar on the evolutionary psychology of coalitions, with whom many of these ideas were discussed. The preparation of this paper was supported, in part, by generous grants from the Harry Frank Guggenheim Foundation, the James S. McDonnell Foundation, and the National Science Foundation (#BNS9157-449 to John Tooby).

REFERENCES

Axelrod, R. & Hamilton, W.D. 1981: The evolution of cooperation. *Science* 211, 1390-1396.
Baron-Cohen, S. 1995: *Mindblindness: An essay on autism and theory of mind.* Cambridge, MA: MIT Press.
Baron-Cohen, S., Leslie, A. & Frith, U. 1985: Does the autistic child have a 'theory of mind'? *Cognition* 21, 37-46.
Boorman, S. & Levitt, P. 1980: *The Genetics of Altruism.* NY: Academic Press.
Boyd, R. 1988: Is the repeated prisoner's dilemma a good model of reciprocal altruism? *Ethology and Sociobiology* 9, 211-222.
Boyd, R. & Lorberbaum, J. 1987: No pure strategy is evolutionarily stable in the repeated Prisoner's Dilemma game. *Nature* 327, 58-59.
Cheney, D. & Seyfarth, R. 1990: *How Monkeys See the World.* Chicago: University of Chicago Press.
Cosmides, L. 1985: *Deduction or Darwinian algorithms? An explanation of the 'elusive' content effect on the Wason selection task.* Doctoral dissertation, Department of Psychology, Harvard University: University Microfilms, #86-02206.

Cosmides, L. 1989: The logic of social exchange: Has natural selection shaped how humans reason? Studies with the Wason selection task. *Cognition* 31, 187–276.

Cosmides, L. & Tooby, J. 1981: Cytoplasmic inheritance and intragenomic conflict. *Journal of Theoretical Biology*, 89, 83–129.

Cosmides, L. & Tooby, J. 1989: Evolutionary psychology and the generation of culture, Part II. A computational theory of social exchange. *Ethology and Sociobiology* 10, 51–97.

Cosmides, L. & Tooby, J. 1992: Cognitive adaptations for social exchange. In *The Adapted Mind: Evolutionary psychology and the generation of culture* (ed. J. Barkow, L. Cosmides & J. Tooby), pp. 163–228. NY: Oxford University Press.

Dawkins, R. 1982: *The Extended Phenotype*. NY: Oxford.

Dawkins, R. 1986: *The Blind Watchmaker*. NY: Norton.

Frank, R. 1988: *Passions within Reason: The strategic role of the emotions*. NY: Norton.

Gallistel, C.R. 1990: *The Organizations of Learning*. Cambridge, MA: MIT Press.

Garcia, J. & Koelling, R. 1966: Relations of cue to consequence in avoidance learning. *Psychonomic Science* 4, 123–124.

Gelman, R. 1990: First principles organize attention to and learning about relevant data: Number and the animate-inanimate distinction as examples. *Cognitive Science* 14, 79–106.

Gigerenzer, G., & Hug, K. 1992: Domain-specific reasoning: Social contracts, cheating and perspective change. *Cognition* 43, 127–171.

Haig, D. 1993: Genetic conflicts in human pregnancy. *Quarterly Review of Biology* 68, 495–532.

Hamilton, W. D. 1963: The evolution of altruistic behavior. *American Naturalist* 97, 31–33.

Hamilton, W.D. 1964: The genetical theory of social behavior. *Journal of Theoretical Biology* 7, 1–52.

Hirschleiffer, J. 1987: On emotions as guarantors of threats and promises. In *The Latest on the Best: Essays on evolution and optimality* (ed. J. Dupre). Cambridge, MA: MIT Press.

Humphrey, N. 1984: *Consciousness regained*. Oxford: Oxford University Press.

Leslie. A. 1988: Some implications of pretense for the development of theories of mind. In *Developing Theories of Mind* (ed. J.W. Astington, P.L. Harris, & D.R. Olson), pp. 19–46. NY: Cambridge University Press.

Leslie, A 1994: ToMM, ToBY, and Agency: Core architecture and domain specificity. In *Mapping the Mind: Domain specificity in cognition and culture* (ed. L.A. Hirschfeld & S.A. Gelman), pp. 119–148. NY: Cambridge University Press.

Leslie, A. & Thaiss, L. 1992 Domain specificity in conceptual development: Neuropsychological evidence from autism. *Cognition* 43, 225–251.

Maljkovic, V. 1987: *Reasoning in evolutionarily important domains and schizophrenia: Dissociation between content-dependent and content-independent reasoning*. Undergraduate honors thesis, Dept. of Psychology, Harvard University.

Maynard Smith, J. 1964: Group selection and kin selection. *Nature*, 201, 1145–1147.

Nozick, R. 1989: *The Examined Life*. NY: Simon & Schuster.

Pinker, S. & Bloom, P. 1990: Natural language and natural selection. *Behavioral and Brain Sciences*. 13, 707–727.

Plomin, R. & Daniels, D. 1987: Why are children in the same family so different from one another? *Behavioral and Brain Sciences* 10, 1–16.

Premack, D. & Premack, A. 1994: Origins of human social competence. In *The Cognitive Neurosciences* (ed. M Gazzaniga). Cambridge, MA: MIT Press.

Sugiyama, L., Tooby, J. & Cosmides, L. 1995: Testing for universality: Reasoning adaptations among the Achuar of Amazonia. *Meetings of the Human Behavior and Evolution Society*, Santa Barbara, CA.

Sulloway, F. (Forthcoming) *Born to Rebel*.

Symons, D. 1990: A critique of Darwinian anthropology. *Ethology and Sociobiology* 10, 131–144.

Symons, D. 1992: On the use and misuse of Darwinism in the study of human behavior. In *The Adapted Mind: Evolutionary psychology and the generation of culture* (ed. J. Barkow, L. Cosmides & J . Tooby), pp. 37–159.
Thornhill, R. 1991: The study of adaptation. In *Interpretation and Explanation in the Study of Behavior* (ed. M. Bekoff & D. Jamieson). Boulder, CO: Westview Press.
Tooby, J. & Cosmides, L. 1984: Friendship, reciprocity, and the Banker's Paradox. *Institute for Evolutionary Studies Technical Report 84-1.*
Tooby, J. & Cosmides, L. 1988: Can non-universal mental organs evolve? Constraints from genetics, adaptation, and the evolution of sex. *Institute for Evolutionary Studies Technical Report 88-2.*
Tooby, J. & Cosmides, L. 1989a: Are there different kinds of cooperation and separate Darwinian algorithms for each? *Meetings of the Human Behavior and Evolution Society*, Evanston, IL.
Tooby, J. & Cosmides, L. 1989b: The logic of threat. *Meetings of the Human Behavior and Evolution Society*, Evanston, IL.
Tooby, J. & Cosmides, L. 1990a. The past explains the present: Emotional adaptations and the structure of ancestral environments. *Ethology and Sociobiology* 11, 375–424.
Tooby, J. & Cosmides, L. l 990b. On the universality of human nature and the uniqueness of the individual: The role of genetics and adaptation. *Journal of Personality* 58, 17–67.
Tooby, J. & Cosmides, L. 1992: The psychological foundations of culture. In *The Adapted Mind: Evolutionary psychology and the generation of culture* (ed. J. Barkow, L. Cosmides & J. Tooby), pp. 19–136. NY: Oxford University Press.
Tooby, J. & Cosmides, L. (forthcoming) Cognitive adaptations for threat.
Trivers, R.L. 1971: The evolution of reciprocal altruism. *Quarterly Review of Biology* 46, 35–57.
Whiten, A. 1991: *Natural Theories of Mind.* Oxford: Blackwell.
Williams, G.C. 1966: *Adaptation and Natural Selection.* Princeton: Princeton University Press.
Williams, G. C. 1985: A defense of reductionism in evolutionary biology. *Oxford Surveys in Evolutionary Biology* 2, 1–27.
Williams, G.C. & Williams, D.C. 1957: Natural selection of individually harmful social adaptations among sibs with special reference to social insects. *Evolution* 17, 249–253.

The Early Prehistory of Human Social Behaviour: Issues of Archaeological Inference and Cognitive Evolution

STEVEN MITHEN

Department of Archaeology, University of Reading, Reading, RG6 6AA

Keywords: prehistoric archaeology; social behaviour; evolution of the mind.

Summary. Unlike the social behaviour of non-human primates, that of human foragers pervades all domains of behaviour. The natural world, material culture and spatial positioning all play an active role in the social interactions of humans. This pervasiveness of social behaviour is readily apparent in the ethnographic record and can be traced in the archaeological records of the Upper Palaeolithic and Mesolithic periods, for which archaeologists can reconstruct complex patterns of social behaviour. For the Early Palaeolithic, however, the archaeological evidence for social behaviour implies that groups were uniformly small and lacking in social structure. This conflicts with evidence from the fossil and palaeoenvironmental records which suggest social complexity and variability. A resolution of these contradictory lines of evidence is offered in terms of a high degree of domain specific mentality for the Early Humans of the Lower and Middle Palaeolithic. The Early Human mind appears to be one in which social behaviour was relatively isolated from interaction with the natural world and material culture. This is in marked contrast to the pervasiveness of social behaviour among behaviourally modern humans arising from a dramatic increase in cognitive fluidity that becomes apparent in the archaeological record at c. 50,000 years ago.

© The British Academy 1996.

INTRODUCTION

ARCHAEOLOGISTS WHO STUDY EARLY PREHISTORY have the task of reconstructing the social behaviour of hunter-gatherers from the first appearance of stone tools, c. 2.5 million years ago, to the appearance of agricultural communities, originating c. 10,000 years ago in the Near East but not arriving in N.W. Europe until a mere c.5,500 years ago. As such, archaeologists are not simply concerned with reconstructing social behaviour at any one specific time and place in prehistory, but with exploring the process of long term change in human social behaviour. To do this, bridges must be built between behavioural ecology, the theories and models of which dominate research at the beginning of our chronological range, and social anthropology which deals with the uniquely human type of social behaviour that arises towards the end of our period of study. The continuing failure to find some integration between these two disciplines remains a major hindrance to our understanding of the evolution of social behaviour.

In this paper I will suggest that an integration can indeed be developed by drawing on the insights provided by evolutionary psychology and the character of the archaeological record. First I will consider the differences between the social behaviour of human foragers and that of non-human primates. I will then show how the distinctive features of human social behaviour can be traced in the archaeological record to at least 40,000 years ago, the timing of the transition from the Middle to the Upper Palaeolithic in Europe. I will then consider the evidence for social behaviour prior to this time, especially that of the Neanderthals in western Europe. By doing this I will identify a curious paradox between the evidence in the archaeological record, which implies a simple and uniform pattern of social behaviour, and that from the fossil and palaeoenvironmental records which imply social complexity and variability. I will argue that the resolution of this paradox, and indeed an understanding of early prehistory in general, can only be gained by addressing the evolution of the mind, an argument that I have made at greater length elsewhere (Mithen 1996).

THE SOCIAL BEHAVIOUR OF HUMAN FORAGERS AND NON-HUMAN PRIMATES

There is a yawning gulf between the character of social behaviour of human foragers and that of non-human primates. To admit this is not to deny evolutionary continuity but simply to recognize that there is at least 6 million years of evolution separating modern humans from our closest relatives, the chimpanzees. In this light, it would be surprising if the methods used by behavioural ecologists to explain the social interactions of baboons and chimpanzees could work equally well to explain the social behaviour of modern humans. We should expect an evolutionary approach to human social behaviour to involve a fundamentally different set of concepts and models to those which are sufficient for explaining the behaviour of non-human primates. In the same way, the study of human language requires a different set of tools to those used for studying the 'language'/vocalizations of non-human primates, although there must be an evolutionary continuity between these.

For the purposes of this paper there are two distinctive features of human social behaviour that I wish to highlight. The first is the degree of social complexity. Research during the last two decades has made it clear that there is nothing simple about the social behaviour of chimpanzees and other non-human primates. Primate social life is about building friendships and alliances, about manipulation and deception, about acquiring and exploiting social knowledge (Byrne & Whiten 1988; Cheney & Seyfarth 1990; Byrne 1995). Chimpanzees, and many other types of primates, live within a truly tangled social web.

Yet there can also be little doubt that the social behaviour of modern humans in hunter-gatherer societies is several orders of magnitude greater in its complexity and variability. As can be appreciated from many ethnographies, the amount of social knowledge a human individual possesses, the number and spatial extent of their social ties—often extending over many thousands of square miles—the time depth to social relationships, and the extraordinary variability in human social behaviour is in a dramatic contrast to the social behaviour of non-human primates (e.g. Lee 1979; Silberbauer 1981; Leacock & Lee 1982).

A second distinctive feature of human social behaviour is the limited extent to which it constitutes a discrete behavioural domain. When we consider a chimpanzee it appears relatively easy to identify whether it is, or is not, engaging in social behaviour at any one moment in time. Fishing for termites, for instance, does not appear to be have social significance. The skill may have been acquired in a social context (McGrew 1992) but in the actual process of termite fishing, or indeed ant dipping or cracking nuts, there is little of social significance. But when we see chimpanzees grooming, or displaying, or fighting, this is behaviour clearly within a social domain.

Moreover, for many primates the complexity of behaviour within this domain often appears greater than that in the non-social world. This has given rise to the idea of a relatively discrete domain of social intelligence (Humphrey 1976; Byrne & Whiten 1988, Cheney & Seyfarth 1988, 1990). We might indeed characterize the mind of a chimpanzee as in Figure 1, in which tool making and using, foraging and communication are largely controlled by domain-general cognitive processes, such as associative and trial-and-error learning, whereas there is a specialized and largely discrete domain of social intelligence for coping with the particularly challenging problems of the social world (Mithen 1996).

When we turn to human foragers the boundaries between social and non-social behaviour are much more fuzzy, if indeed they exist at all. It becomes impossible to designate behaviour as either social or non-social. This can be illustrated by briefly considering the interaction of human foragers with the natural world and the role of material culture in human social behaviour.

Continuity between the social and natural worlds

The most obvious manner in which the natural world is used by human foragers for social ends is by food sharing through which reciprocal obligations are constructed. The provisioning of women with meat by male hunters is a common feature among modern foragers and the appearance of this type of behaviour has been invoked as a critical feature for the evolution of human social behaviour (e.g. Isaac 1978; Soffer 1994). It has been shown that among the Ache of Paraguay the most efficient hunters, in terms of providing the greatest amount of food to the group, also have the highest reproductive success (Kaplan & Hill 1985). In contrast to modern humans, chimpanzees do not engage in provisioning and their 'food sharing' (e.g. Boesch & Boesch 1989) should be predominantly (if not totally) described as tolerated theft.

The social use of the natural world by human foragers, however, is more profound than simply using the provisioning, sharing and exchange of food as a medium for social interaction. It appears to be ubiquitous among human foragers that the concepts and thought processes which are used for thinking about the natural world, include those used for social interaction. Bird-David (1990), for instance, describes how forest dwelling foragers, such as the Mbuti of Zaire, conceive of the 'forest as parent', it is in effect a social being that gives. Similarly the Inuit living in the Canadian Arctic view their environment as 'imbued with human qualities of will and purpose' (Ridington quoted in Ingold 1993, 440). With regard to modern foragers in general, Tim Ingold argues that 'there are not two worlds of persons (society) and things (nature), but just one world—one environment—saturated with personal powers and embracing both human beings, the animals and plants on which they depend, and the landscape in which they live and move' (1992, 42).

This cognitive fluidity between the social and natural world is exemplified in the phenomena of anthropomorphism and totemism, both being pervasive among hunter-gatherers (Willis 1990), as is particularly evident from their art (Morphy 1989). While chimpanzees may be proficient at attributing thoughts and desires to other individuals (Byrne 1995), it is highly doubtful that they attribute thoughts, desires and intentions to members of other species, as humans do when they anthropomorphize. And it is also highly unlikely that chimpanzees think that other members of their own species may share a common ancestor with animals such as snakes or leopards, as in totemic thought.

An important point to emphasize is that this social understanding of the natural world is not simply epiphenomenal among hunter-gatherers—something that can be safely ignored by those who wish to take an evolutionary approach to understanding human behaviour. For these attitudes to the natural world play a fundamental role in creating and manipulating social relationships. Consider, for instance, the attitude to the polar bear by the Inuit as described by Saladin D'Angular (1990). This animal is 'killed with passion, butchered with care and eaten with delight'. But at the same time it is thought of as another human being, or at least another adult male. When a bear is killed the same constraints apply as to what activities can be undertaken in a camp as when a hunter dies. The bear is thought of as a human ancestor and as a kinsman. Indeed, the Inuit believe that there was a time when polar bears and people could easily change from one kind to another. The important point is that by associating themselves with the polar bear, the Inuit males use the bear as a potent ideological tool to consolidate their domination of women. Indeed the use of nature as a means of establishing and maintaining power relationships is pervasive among hunter-gatherer groups.

In summary, while interaction with the social and natural worlds appear to be discrete domains of behaviour for chimpanzees and other non-human primates, no such distinction can be drawn for human foragers.

Material culture and social interaction

We find the same contrast when we consider the relationship between the domains of technical and social behaviour, which effectively do not exist as separate entities for modern humans (Ingold 1993). As the work of McGrew (1992) and the Boesch's (1983, 1990, 1993; Boesch 1993) have shown, chimpanzees make and use a diverse array of tools, many of which seem well designed for the tasks for which they are used. But the designs and manner of use appear to have no social significance, other than passively reflecting the cultural traditions of the group in some cases.

The tools of modern humans also display very effective designs for their functional tasks (e.g. Oswalt 1976; Torrence 1983; Bleed 1986; Churchill 1993). But at the very same time these tools are used in conducting social relationships. Polly Wiessner (1983) has documented this for the arrows of the Kalahari San. While these are very effective hunting weapons, the shape of the arrow heads also carries information about group affiliation. Their use in hunting the eland, an animal central to San mythology, results in the arrows also having considerable symbolic significance.

The use of material culture for social interaction by modern humans is most evident in body adornment, ubiquitous among modern humans. Randall White (1992, 1993) has stressed that body adornment is not simply a passive reflection of social categories or status, let alone mere decoration, but an active means of engaging in social behaviour. He quotes Strathern: 'what people wear, and what they do to and with their bodies in general, forms an important part of the flow of information—establishing, modifying, and commenting on major social categories, such as age, sex and status' (quoted in White 1992: 539–40). Similarly, Turner stated that 'the surface of the body ... becomes the symbolic stage upon which the drama of socialisation is enacted, and bodily adornment ... becomes the language through which it is expressed' (quoted in White 1992: 539).

Projectile points and body adornment are the most obvious candidates for the social use of material culture, but even the most mundane domestic items are used actively in social strategies (Hodder 1985). As Wobst (1977) argued, one cannot have a half-way house with some items but not others imbued with social information. For modern humans all material culture is actively used in social interaction; in contrast the material culture of chimpanzees plays no role in their social strategies and tactics. For instance while chimpanzees appear to be very concerned with the flows of social information and are adept at manipulating plant material, no one has ever seen a chimpanzee use plant material as body decoration.

The social use of space

The pervasiveness of social behaviour and thought among modern foragers is also evident from spatial behaviour within their camping sites. When modern foragers make their camp-sites, sit to cook food or repair tools, they are at the same time engaging in complex patterns of social interaction simply by their use of space. They do not place themselves randomly but use the spatial arrangements between themselves and others as a social strategy. This has been demonstrated in considerable detail by Whitelaw's (1991) studies of the spatial layouts of hunter-gatherer camps. To quote him 'spatial organisation is used by different individuals and in different cultures to generate, amplify, facilitate, manipulate and control social interaction and organisation' (1991: 181).

My impression from the literature about chimpanzees is that the spatial behaviour of chimpanzees when engaging in technical tasks is controlled by purely ecological factors, such as the locations of nuts and hammer stones in the Taï forest (Boesch & Boesch 1983, 1984). Similarly, it appears that the spatial positioning of individuals within a group passively reflects pre-existing social relationships, rather than being actively used to manipulate those relationships as among modern humans, although specific studies of the spatial behaviour of chimpanzees and other primates appear to be lacking.

The all-pervasiveness of human social behaviour

In summary, when we look at modern foragers we cannot identify a discrete domain of social behaviour. Interaction with the natural world, the production, the form and the use of tools, the building and placing of hearths and huts are as much social as non-social behaviour. We can perhaps summarize this argument by quoting Ernest Gellner (1988: 45-46): 'The conflation and confusion of functions, aims and criteria, is the normal, original condition of mankind ... it is important to grasp this point fully. A multi functional expression is not one in which a man combines a number of meanings because he is in a hurry and human language has offered him a package deal: on the contrary, the conflated meanings constitute for him, a single and indivisible semantic content'. Gellner was writing about verbally expressed statements; but precisely the same interpretation must hold for the actions and the material culture of hunter-gatherers.

The fact that one action of a human forager may be simultaneously accomplishing ends in multiple domains of activity confounds those who wish to take an ecological approach to hunter-gatherers in which the pigeon-holing of behaviour as either concerned with acquiring food, or making things, or social interaction is desirable so that the costs and benefits of any particular behaviour can be measured. When the consequences of any one activity simultaneously reverberates in multiple domains of behaviour, trying to measure costs, risks and benefits becomes extremely difficult.

While this all pervasiveness of human social behaviour may make life miserable for the behavioural ecologist, it is, however, a godsend to the archaeologist of prehistoric hunter-gatherers. This is because when we try to reconstruct social behaviour in the past we cannot see alliances or kinship groups, let alone deception and social manipulation. All we can see in the archaeological record is the junk that was left behind such as the debris from the animals that were butchered, and the waste from the tools that were made. Yet because human social behaviour is so deeply embedded in, and indeed created by, all activities we can indeed reconstruct some aspects of past social life from such material.

MESOLITHIC FORAGERS IN EUROPE

As the first of two brief case studies which explore this pervasiveness of human social behaviour we can consider the Mesolithic communities of Europe—the prehistoric foragers who lived in the temperate forests of the early post glacial between the end of the Pleistocene, 10,000 years ago and the appearance of Neolithic farming communities (see Mithen 1994a for a review of this period).

Of particular interest during this period is the presence of cemeteries, appearing for the first time in prehistoric Europe. These contain varying number of individuals, many of whom were buried with items such as beads, pendants, stone artefacts and parts of animals (Clark & Neeley 1987). The distribution of grave goods in the cemeteries of southern Scandinavia, Vedbaek (Albrethsen & Petersen 1976) and Skateholm (Larsson 1983), suggests that wealth was acquired during an individual's lifetime, rather than inherited. In contrast at Hoëdic and Téviec in Brittany (Péquart *et al.* 1937; Péquart & Péquart 1954) young children were found with abundant items, suggesting that wealth and status were inherited. At Oleonovstrovski Mogilnik in Karelia, it appears that institutionalized social positions had arisen, as we find the burial of what appears to be a shaman (O'Shea & Zvelebil 1984).

The presence of beads and pendants within these graves illustrates the use of body adornment for social strategies. As these are predominantly made from the teeth of red deer and wild boar we also see the exploitation of the natural world for social ends. This is also evident at Oleonovstrovski Mogilnik where the graves appear to form two clusters, associated with effigies of snakes and elk respectively, suggesting a totemic social structure (O'Shea & Zvelebil 1984). The cognitive fluidity between the social and natural worlds is also evident from the cemetery of Skateholm where dogs were buried with the same type of ritual and grave items as used for people (Larsson 1983, 1990).

While figurative art is not common in the Mesolithic, some of the examples which do exist, such as the faces carved on boulders from Lepenski Vir in the Danube (Srejovic 1972), are anthropomorphic in character. This again reflects the absence of any cognitive barriers between the social and natural worlds.

The use of the natural world for social ends can also be inferred from the hunting practices of Mesolithic groups, especially those in coastal regions where resources were relatively abundant. An analysis of the animal bones from late Mesolithic sites in southern Scandinavia suggests that the hunting of red deer and wild boar was undertaken as much for social prestige as for the supply of food (Mithen 1990). By using a computer simulation of Mesolithic hunting I have argued that the composition of faunal assemblages in terms of the frequencies of different species of terrestrial game imply Mesolithic hunting goals which preferentially killed large animals, notably red deer and wild boar. This hunting pattern is likely to have frequently failed with the Mesolithic hunters returning to their base camps empty handed. This strategy is likely to have been feasible due to the abundance of plant foods and small game, especially fish, caught by untended traps and facilities. Consequently the hunting of large terrestrial game in southern Scandinavia appears to have been principally geared to acquiring prestige and constructing social obligations by providing large carcasses for food sharing. The faunal assemblages from southern Germany, in contrast, indicate that hunting focussed on the small types of game, notably roe deer. These would have been the most reliable to acquire but can be assumed to have carried relatively low social prestige. In this interior region of Europe alternative food resources were less abundant than in coastal regions, requiring the hunting of large game to provide regular supplies of meat. And we have no evidence for the type of social developments that are seen in southern Scandinavia. Mesolithic cemeteries, for instance, are absent from southern Germany.

The hunting of terrestrial game during the Mesolithic was largely undertaken by the use of arrows with tips and barbs made from chipped flint blades, referred to as microliths. While experimental replication and use of such weapons have demonstrated their effectiveness at piercing thick animal hide (e.g. Fris-Hansen 1990), many of these microliths are also likely to have been imbued with social information (Gendel 1984), in a similar manner to the arrow heads of the Kalahari San, as described by Wiessner (1983). Similarly, other types of artefacts display distinct spatial distributions which cannot be explained in purely ecological terms. For instance in Eastern Denmark, flint axes made to different, but functionally equivalent, designs are found in discrete spatial clusters. There are also marked differences in the material culture on either side of the strait separating Eastern Denmark and southern Sweden, even though the environments were broadly comparable (Vang Petersen 1984). These examples seem to reflect the active use of material culture in social strategies, in terms of creating social boundaries between groups. As such, they complement the use of beads and pendants, actively used for the social strategies of individuals within a group.

It is extremely difficult to draw inferences about the social use of space, in terms of the placement of hearths, huts and activities, during the Mesolithic. When a number of dwellings or other features are found on a site it is difficult to determine whether these were precisely contemporary with each other, or indeed to infer the specific form of past constructions. Similarly, while archaeologists have methods to estimate the approximate numbers of people occupying a site (e.g. O'Connell 1987), the methodological tools to infer their social and biological relationships are elusive. This type of data has been critical to Whitelaw's (1991) studies of the social use of space by ethnographically documented foragers. It is nevertheless evident from the archaeological record that it is characteristic of the Mesolithic period that a range of features are found on settlements in spatial relationships which appear similar to those found in the ethnographic record. Sites such as Mount Sandel in Ireland (Woodman 1985) or Vaenget Nord in Denmark (Petersen 1989), provide the remains of dwellings, postholes, pits, hearths and scatters of knapping debris which would appear to reflect a social use of space similar to that documented for modern foragers by Whitelaw, although this cannot be formally demonstrated.

We can see in the Mesolithic period, therefore, the use of nature, material culture and space in the social behaviour of prehistoric foragers. From such data, we can recognize considerable variability in Mesolithic social behaviour across the continent and during the period itself. In southern Scandinavia, for instance, it is likely that during the latter part of the Mesolithic period, the Ertebølle, there were permanently based social groups probably consisting of many hundreds of people who were concerned with marking and defending territories (Price 1985). This is a very different pattern of social behaviour to that commonly associated with foragers: small, egalitarian and highly mobile groups. Such groups may nevertheless have been present in other regions of Europe during the Mesolithic. This returns us to the first difference between the social behaviour of human foragers and non-human primates that I remarked upon above: a veritable gulf in the degree of social complexity and variability.

UPPER PALAEOLITHIC FORAGERS IN EUROPE

For a second brief case study exploring the pervasiveness of human social behaviour we can consider the Upper Palaeolithic communities of Europe, 40–10,000 BP (for a general review see Mellars 1994). This period constituted the final stages of the late Pleistocene including the late glacial maximum at 18,000 BP. Consequently, we are now dealing with prehistoric foragers living in open tundra environments, and a period of climatic deterioration up to 18,000 BP followed by gradual amelioration until the extremely rapid global warming at 10,000 BP. Within these general trends, however, there were many environmental fluctuations, as demonstrated by recent ice cores (e.g., Johnsen *et al.* 1992). There would also have been considerable environmental variation with latitude and the degree of continentality. Whereas the Mesolithic foragers generally stalked individual animals within the thick forests of the postglacial, we are now dealing with foragers who hunted large migratory herds of animals, notably reindeer (Mithen 1990).

The Upper Palaeolithic in Europe is renowned for its art, clustered in south west France and Northern Spain, and predominantly created after 20,000 years ago. The florescence of this art appears to be related to the environmental and economic conditions of the period of the late glacial maximum (Jochim 1983; Mithen 1989, 1991). Several of the images within the art are anthropomorphic, such as the 'sorcerer' from Les Trois Frères, which appears to be a human figure with a bison face and antlers (Bahn & Vertut 1988) and the man/lion figure from Hohlenstein-Stadel, which dates to at least 33,000 years ago (Marshack 1990). Such images indicate a similar cognitive fluidity between the social and natural worlds as we have seen in the Mesolithic and among ethnographically documented foragers. Indeed, although it cannot be demonstrated, it is most likely that the predominance of animal imagery within the art of the Upper Palaeolithic reflects a totemic structure to society.

Material culture appears to have played a major role in social interaction during the Upper Palaeolithic. Gamble (1982, 1991) has argued that items such as Kostienki points and Venus figurines, which show the same basic form across vast expanses of Eurasia and are chronologically restricted to the period of climatic deterioration, were used in the construction of alliance networks. Such networks were critical to the continued occupation of Europe as the glacial maximum approached and constitute a spatial scale of social relations very different to that we see in the Mesolithic. The use of material culture for mediating group interaction is also evident from the discrete distributions of specific motifs in the cave paintings, possibly indicating social territories (Sieveking 1980), and the diversity of imagery on carved bones at sites such as Altamira which appear to have been used for group aggregations (Conkey 1980). The projectile points of the Upper Palaeolithic are widely accepted as having been invested with considerable amounts of social information, in light of the distinct morphologies and their spatial and temporal distributions (Mellars 1994). In some regions, items of material culture are likely to have been exchanged in trade networks, such as amber and fossil sea shells on the Central Russian Plain (Soffer 1985).

Material culture was also used for social interaction in terms of beads, pendants, and bracelets made from ivory, animal teeth and sea shells. Indeed the very start of the Upper Palaeolithic in south west France is marked by a sudden abundance of such items, found in what appear to be domestic and manufacturing contexts (White 1989). Elsewhere such items are found adorning bodies in graves, such as at Dolni Vestonice in Czechoslovakia (Kilma 1988). Perhaps most notable are the remarkably rich graves from Sungir in Russia dating to 28,000 years ago. It is worthwhile here to summarize White's (1993) description of these as they demonstrate the importance of the natural world and material culture for social strategies. There were three particularly rich graves at Sungir. One grave contained an old man (c. 60 years) who was buried with no less than 2936 beads which had once been sewn onto his clothing. He wore painted mammoth ivory bracelets on his arms and a pendant from his neck, painted red with a single black dot. In an adjacent grave there were two adolescents, a boy aged 13 and a girl aged 7–9 years. The body of the boy was covered with 4903 beads. Around his waist there were at least 250 canine teeth of the polar fox which had once formed part of a belt. At his throat there was an ivory pin and under his left side a large ivory sculpture of a mammoth; he also had an ivory disc with a central hole and carved lattice work. On his left side was part of a human femur, which had been polished and packed with red ochre, and on his right side a massive ivory lance, 2.40 m in length. The girl had 5274 beads and fragments. Like the boy, she had a beaded cap, although there were no fox teeth associated with her body. She had small ivory lances at her side and two pierced antler batons. She also had three ivory discs similar to that found with the boy. Both Dennell (1983) and White (1993) have stressed the time investment required to make the beads at Sungir, estimated to have been 2000 hours for the man and 3500 for each of the children.

With regard to making inferences regarding the social use of space, Upper Palaeolithic archaeologists face the same dilemma as those studying the Mesolithic. There are some extremely well preserved settlements, especially from the late glacial period (c. 12–10,000 BP) from which detailed reconstructions of spatial behaviour, such as the seating of flint knappers around hearths and the location of different butchering activities, can be reconstructed (e.g. Pigeot 1990; Enloe et al. 1994; Koetje 1994). But as with the Mesolithic, the specific social and biological relationships between the occupants of these sites are elusive.

Upper Palaeolithic sites on the Central Russian Plain provide the most promising data sets with which to explore the social use of space. Soffer (1985) has argued that the spatial relationships between mammoth bone dwellings and storage pits played an active role in social interaction. At some sites, such as Radomysh'l, there is a single large storage pit surrounded by dwellings. Soffer interprets this as reflecting equal and open access by the inhabitants of the site to stored foodstuffs. At sites such as Dobranichevka, however, approximately equal numbers of pits are found surrounding each dwelling, implying that access was restricted to members of that household. At further sites, such as Mezin and Eliseevichi, storage pits are clustered around one single dwelling, suggesting that a single household controlled access to the stored foodstuffs of the whole group. Moreover, Soffer argues that these three patterns of resource control constitute a chronological sequence reflecting the emergence of a hierarchically structured society during the latter part of the Pleistocene. Whether or not this interpretation is correct, it is readily apparent from the type and distribution of features on Upper Palaeolithic sites that the social use of space was as complex and sophisticated as that seen in the ethnographic record.

In summary, while the specific forms of social behaviour that can be reconstructed for the Upper Palaeolithic in Europe contrast with those for the Mesolithic, both are characterized by the pervasiveness of social behaviour in all domains of activity. In both periods material culture, whether in the form of items for body adornment, hunting weapons, or the positioning of dwellings played an active role in social behaviour. Similarly, while hunting and gathering provided food and raw materials, the acquisition and consumption of these were as much social as non-social activity. Moreover, the anthropomorphic images that we find in both periods suggest the same cognitive fluidity between the social and natural world as we see in the ethnographic record.

SOCIAL BEHAVIOUR DURING THE EARLY PALAEOLITHIC OF EUROPE

I now wish to move to a third case study in which we see a dramatic contrast in the character of the archaeological record and the nature of our inferences about prehistoric social behaviour. This concerns the Early Palaeolithic period in Europe, dating to between the time of first colonization, most probably *c.* 500,000 years ago (Roebroeks & van Kolfschoten 1994), and the start of the Upper Palaeolithic (see Gamble 1994 for a general review). The term 'Early Palaeolithic' combines both the Lower and Middle Palaeolithic, the distinction between which has now become so blurred as to have little utility (Gamble 1986; Stringer & Gamble 1993). My principal concern in this case study will remain with Europe, but practically all of the remarks and interpretations I will be making below are equally applicable to the Old World in general for the period prior to 60–35,000 years ago. This is a period of a global change in the character of the archaeological record, described as the Middle/Upper Palaeolithic transition.

The Middle/Upper Palaeolithic transition

While the transition to the Upper Palaeolithic is well defined in Europe at 40–35,000 years ago, most notably by the appearance of the first art objects, systematic blade technology and bone tools (Mellars 1973, 1989; White 1982) it is rather more fuzzy elsewhere in the world. In the Near East, for instance, there is a clear technological transition at $c.\,40,000$ years ago resulting in the production of blade technologies similar to those found in the Upper Palaeolithic in Europe (Bar-Yosef 1994). But art remains extremely rare until after 20,000 years ago. Similarly in Africa, while the first art objects do indeed date to after 40,000 years, such as the painted slabs from Apollo cave at 27,500 BP, these also remain rare and the technological transitions remain poorly defined (Wadley 1993). In East Asia there are now pieces of art dated to 18,000 BP from China (Bednarik & Yuzhu 1991), although technological changes, if any, are particularly poorly understood (e.g. see Zhonglong 1992). In south east Asia, the colonization of Australasia, most likely prior to 50,000 BP (Roberts et al. 1990, 1994; Allen 1994; Bowdler 1992) is interpreted as an 'Upper Palaeolithic' type behaviour, and it is most likely that art dates to that initial colonization (Bowdler 1992; Davidson & Noble 1992). The stone tools industries of Australia are lacking, however, in any attributes that could be characterised as Upper Palaeolithic in a European sense. In summary, there can be little doubt that there is a global transition in the character of the archaeological record between 60–35,000 years ago which is nevertheless manifest differently in different regions. The most appropriate description for the transition is a cultural mosaic.

In Europe, the transition to the Upper Palaeolithic appears to correlate broadly with the replacement of *H. neanderthalensis* by *H. sapiens sapiens*, although the Chatelperronian industry, which has Upper Palaeolithic attributes, has been claimed to be a product of Neanderthals (Mellars 1989; Harrold 1989; Stringer & Gamble 1993). But this correlation between hominid anatomy and material culture appears to be the exception rather than the rule. In the Near East, for instance, the earliest anatomically modern humans were manufacturing stone industries essentially the same as those produced by Neanderthals (Bar-Yosef 1994), although subtle distinctions in hunting behaviour can be identified (Lieberman & Shea 1994). Similarly in south and north Africa, anatomically modern humans dating to soon after 100,000 years ago appear to remain associated with Middle Palaeolithic technologies until the start of the Upper Palaeolithic (Hublin 1992; Grün & Stringer 1991).

Archaeological evidence for social behaviour in the Early Palaeolithic

The archaeological record for the Early Palaeolithic appears to lack any evidence for complex social behaviour of a type that pervades the archaeology of the Upper Palaeolithic and Mesolithic periods. This is a consistent pattern irrespective of whether it is associated with *H. erectus*, archaic *H. sapiens*, *H. neanderthalensis* or anatomically modern humans.

No objects of art were produced during the Early Palaeolithic. Although a few artefacts are known from this period which have marks of no apparent utilitarian value (Bednarik 1992), these are most likely to be no more than unintended by-products of activities such as cutting grass on a bone support, or products of post-depositional processes, or carnivore gnawing (Davidson 1992; Chase & Dibble 1987, 1992). There is no figurative art, and no regularly repeated images which may have constituted a symbolic code, as found at the start of the Upper Palaeolithic in southwest Europe (Delluc & Delluc 1978). Arguments that the absence of art objects can be explained on taphonomic grounds (e.g. Bednarik 1994) are unconvincing in light of the massive quantities of data from the Early Palaeolithic, and the presence of numerous well preserved sites.

While Neanderthal burials from the Early Palaeolithic are known (such as from Kebara, Bar-Yosef *et al.* 1992), there is no reason to attribute these with social significance. These burials lack grave goods and may simply represent a hygenic means of corpse disposal rather than having the ritual and social significance that is apparent in later periods (Gargett 1989; Gamble 1989). The burials of early anatomically modern humans from the Near East, in the caves of Qafzeh and Skhūl, appear to be similar. These lack any evidence for body adornment. The best candidates for grave goods are the skull and antlers of a large deer buried with a child at Qafzeh (Stringer & Gamble 1993), the significance of which is discussed in Mithen (1996).

With regard to the stone tools of Early Humans, we must first note that these were technically demanding to manufacture. To create items such as the 500,000 year old symmetrical handaxes from Boxgrove (Roberts 1986), or the 60,000 year old levallois points from Kebara (Bar-Yosef *et al.* 1992) technical skills equivalent to those used for the blade technologies of the Upper Palaeolithic were required. This is readily apparent when the manufacture of such tools is replicated (e.g. Pelegrin 1993) or detailed technological studies of lithic assemblages are made (e.g. Bar-Yosef & Meignen 1992). While it is clear that specific forms were being imposed onto some artefacts, such as by producing handaxes in specific shapes and sizes, there is no evidence that these forms carried social and symbolic information in the manner of Upper Palaeolithic and Mesolithic stone tools (Chase 1991; Mithen 1994b; Wynn 1995).

Evidence for complex social behaviour is also lacking from the features and spatial patterns of debris on Early Palaeolithic archaeological sites. As Gamble (1994; Stringer & Gamble 1993) has stressed, the familiar attributes of hunter-gatherer campsites such as hearths, postholes and pits are simply absent from sites of the Early Palaeolithic. This is not simply a factor of preservation, as there are several sites such as Boxgrove (Roberts 1986) and Mastricht-Bélvèdere (Roebroeks 1988) at which large areas of undisturbed knapping and butchery debris survive intact. Yet this debris appears to be distributed in isolated behavioural episodes, lacking the spatial relationships which characterizes modern behaviour at campsites (Farizy 1994). The Early Palaeolithic record lacks, therefore, any evidence for a social use of space as is found in later periods and in the ethnographic record, just as it lacks evidence that material culture and the natural world were used actively in social strategies.

The conventional, indeed practically unanimous, interpretation of this data is that complex social behaviour was indeed absent among these Early Palaeolithic humans. As Mellars (1989: 358) has stated, the most widely held view by archaeologists of social behaviour at this time is that 'local communities ... were generally small ... and largely lacking in any clear social structure or individual social or economic roles'. Similarly, Binford (1989: 33) described Middle Palaeolithic groups as 'uniformly small and with very high mobility whatever the environmental form and dynamic'. Gamble (Stringer & Gamble 1993: 156) characterized the Early Palaeolithic of Europe as a '15 minute culture that lasted in Europe for at least half a million years'.

In effect, the archaeological record implies that social behaviour was not only much less complex than in the Mesolithic or Upper Palaeolithic, but also less complex than that found among chimpanzees, or indeed many other primates, among whom groups are certainly *not* uniformly small, social structure is certainly *not* lacking, mobility *is* responsive to environment and behaviour has a time depth far greater than 15 minutes.

Evidence for social behaviour from the fossil and palaeoenvironmental records

While the archaeological evidence appears to tell us that Early Palaeolithic social behaviour lacked the complexity that is apparent from the archaeological record from the Upper Palaeolithic onwards, this appears to conflict with the evidence from the fossil and palaeoenvironmental records.

With regard to hominid fossils the most significant feature is the large brain size of Neanderthals, and indeed archaic *H. sapiens*. This was not significantly different to the brain size of modern humans (Stringer & Gamble 1993; Aiello & Dunbar 1993). The social implication of this lies in the fact that among primates in general there appears to be a correlation between brain size, as measured by the neocortex ratio, and group size which is a measure of social complexity (Dunbar 1992). Aiello & Dunbar (1993) extrapolated from this correlation to argue that Neanderthals and archaic *H. sapiens* lived in group sizes with a mean of 148. The details of the statistical relationship between brain size and group size may be questioned (Steele 1996), together with the logic of so great an extrapolation (Mithen 1996). Nevertheless, the brain size of these Early Humans implies a degree of social complexity equivalent to that observed among anatomically modern humans such as those of the Upper Palaeolithic or those documented in the ethnographic record. Moreover, the possession of such a large brain has substantial implications for the life history parameters of Early Humans suggesting complex social links between members of the same and different sexes (Foley & Lee 1989).

This is supported by the evident linguistic capacities of Neanderthals and archaic *H. sapiens*, as inferred from their brain size, brain shape and the reconstruction of their vocal tract (Schepartz 1993). The hyoid bone from the Kebara II Neanderthal (Arensburg *et al.* 1989, 1990) and recent reconstructions of the vocal tract (Houghton 1993) indicate that Neanderthals are unlikely to have been constrained to the limited range of vowel sounds as was argued by Lieberman & Crelin (1971). On the contrary, they appear to have had as wide a range of vocalizations as used in modern language. As Aiello & Dunbar (1993; Dunbar 1993) have argued, language is primarily used to communicate social information among modern humans and the need for more efficient exchange of social information is likely to have been the selective pressure for the evolution of the linguistic capacity. Consequently, as with brain size in general, the linguistic capacities of Neanderthals appears to conflict with the inference of a simple and uniform social organization based on small group size as drawn from the archaeological record.

A further source of concern about the inference of a 'simple' social organization for the Neanderthals lies with the diversity of environments which they and their immediate ancestors in Europe inhabited. It is apparent from the marine sediment and ice core records that Europe experienced a continuous sequence of marked environmental changes during the Pleistocene, with no less that eight glacial-interglacial cycles (Shackleton & Opdyke 1973). For the majority of this time, Europe was in neither a full glacial nor interglacial state and Neanderthals inhabited open tundra-like environments (Gamble 1986, fig 3.12). The faunal assemblages from Pleistocene Europe, reviewed by Chase (1986, 1989) and Gamble (1986), indicate that Neanderthals shared these landscapes with a diverse set of carnivores which would have been both competing with Neanderthals for herbivores, and preying on Neanderthals themselves. Moreover, resource distributions in these landscapes are likely to have been extremely patchy and coming in large packages, especially during winter months when early humans may have been dependent upon scavenging carcasses of large herbivores (Gamble 1987). These ecological conditions—high predator risk and patchy food supply—are precisely the conditions which are known to promote large group sizes among primates (Clutton-Brock & Harvey 1977; van Schaik 1983; Dunbar 1988). Moreover, it would contradict much of what we understand about primate social behaviour if Binford was correct and Neanderthals varied neither their social behaviour nor mobility patterns as the Pleistocene environments went through such radical changes as documented in the palaeoenvironmental records.

In effect the evidence from the archaeological record, or at least our interpretation of it, blatantly contradicts our inferences about Neanderthal social behaviour drawn from their fossil remains, the environments they were living in, and what we understand about primate social behaviour in general. We have a palaeoanthropological paradox, not just for Neanderthals but for all types of Early Humans prior to about 50,000 years ago, whether these be archaic *H. sapiens* in Africa or the early anatomically modern humans in the Near East. The archaeological evidence tells us that social behaviour was simple and uniform, the fossil and palaeoenvironmental evidence tell us the opposite, that social behaviour was complex and varied. How can this paradox resolved?

Resolving the paradox: The domain-specific mentality of the Early Human Mind

I think that the answer lies in the nature of the Early Human mind. My contention is that while Neanderthals possessed essentially modern cognitive capacities in the domains of social, technical and natural history intelligence these remained relatively isolated from each other (Mithen 1993; 1994c; 1996). This 'domain-specific mentality' is in marked contrast to the cognitive fluidity characteristic of the modern mind, as I have described above. If we use the terminology of Gardner (1983) we could characterize Neanderthals as having multiple intelligences, which nevertheless lacked the smooth and seamless integration which he argues is characteristic of the modern mind. Indeed Rozin (1976; Rozin & Schull 1988) argued that the evolution of accessibility between cognitive domains/ multiple intelligences is critically important for the evolution of the advanced intelligence of the modern mind. Elsewhere I have argued that this is precisely what occurs at the Middle/Upper Palaeolithic transition (Mithen 1993, 1996).

My notion of a cognitive domain is of a bundle of closely related cognitive processes which might themselves be described as 'Darwinian algorithms' (Cosmides & Tooby 1987) or micro-domains (Karmiloff-Smith 1992). Some of these may be innately hard-wired, such as an intuitive knowledge about certain attributes of physical objects (Spelke *et al.* 1992) or psychology (Whiten 1991). It is clear, however, that the type and character of cognitive domains which arise within a mind are heavily influenced by the context of development (Gardner 1983; Karmiloff-Smith 1992). Consequently those living a hunter-gatherer lifestyle are likely to have a different set of cognitive domains to those of us who live in a western industrial society, although we may share a number of core cognitive building blocks. But the cognitive domains of all behaviourally modern humans appear to have high degrees of accessibility. In contrast, those of Early Humans (i.e. prior to the Middle/Upper Palaeolithic transition) appear to have been relatively isolated, a domain-specific mentality.

Figure 1. Social intelligence as a cognitive domain in the chimpanzee mind (for full discussion of this and other figures see Mithen 1996).

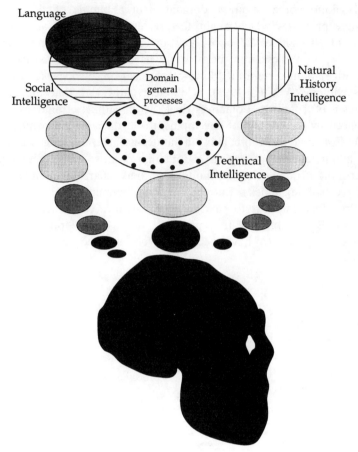

Figure 2. The Early Human Mind c. 100,000 BP.

To support this contention, and to explain how it effects our interpretation of the archaeological record regarding social behaviour, we must momentarily return to the mind of the chimpanzee (Figure 1). As I noted above, an argument can be made that the chimpanzee has a discrete domain of social intelligence which complements a powerful general learning ability. Now this type of mind may well be similar to the mind of the common ancestor of modern humans and apes which lived about six million years ago (Byrne 1995). By the time we reach Neanderthals, say at 100,000 years ago, the archaeological record implies that we have not one but multiple specialized cognitive domains, as illustrated in Figure 2.

In this figure I have proposed a cognitive domain of social intelligence that is more powerful that likely to have existed in the mind of the common ape/human ancestor 6 million years ago. I have appended to this a linguistic capacity. In doing so I have followed Dunbar's (1993) arguments and characterized Early Human language as used predominantly for social purposes, rather than for the wide range of functions for which we use language today.

It is also likely that Early Humans possessed a discrete domain of technical intelligence. Now while domain-general processes such as associative learning appear sufficient to account for the technical abilities of chimpanzees, the complexity of Early Human stone technology implies specialized cognitive processes (Mithen 1996). Generalized learning abilities would have been inadequate to attain the knapping skills required to make artefacts such as handaxes and levallois points. To make these, cognitive process for creating mental templates, planning sequences of knapping actions and mental rotations of artefact form would have been required. I have bundled these together to constitute a domain of technical intelligence.

We must add a further specialized cognitive domain of natural history intelligence to this model of the Early Human mind. Perhaps beginning as early as 1.8 million years ago (Swisher *et al.* 1994) Early Humans moved out of Africa to colonize a vast array of environments demonstrating a capacity to rapidly learn about new types of landscapes and resources. The Neanderthals in Europe, for instance, successfully exploited harsh ice age landscapes without the sophisticated technology used by modern foragers in glaciated landscapes and this suggests that they had a profound understanding of the habits of their game. The archaeological evidence of their subsistence activities, which includes that for big game hunting (Mellars 1989; Chase 1986, 1989), implies abilities to read inanimate secondary cues, such as animal tracks and trails, to build classifications of animals and plants on ecological criteria, and to draw on that understanding for building hypotheses about future resource distribution (Mithen 1996). As with the cognitive processes required to make their stone tools, domain-general learning mechanisms would have been inadequate to such tasks.

Recognising that Early Humans had intellectual abilities in the social, technical and natural history domains of thought little different to those of modern humans is essential for understanding their behaviour and the archaeological record. But of equal importance is to recognize the limited degree of accessibility between these cognitive domains, which constitutes a dramatic contrast to the mind of modern humans. Early Humans appear to have been unable to integrate their thought and knowledge from these multiple cognitive domains.

There was, of course, some degree of cognitive connection; tools were required to exploit the natural world, and hunting is likely to have involved social co-operation. But the archaeological record indicates that the behaviour at the domain-interfaces was markedly less complex and sophisticated than that within the domains themselves. This is most effectively illustrated by Early Human technology. As I have noted above, we cannot doubt that Early Humans had the cognitive skills to make complex artefacts in light of the character of their stone tools; similarly we cannot doubt that they had a profound understanding of the habits of the animals they exploited. Yet we have no evidence that they combined their technical and natural history knowledge to make hunting weapons or traps specialized for specific types of game in specific situations. We find no fine grained correlations between artefact types, environments and faunal assemblages of the kind found in the Upper Palaeolithic (e.g. Clark *et al.* 1986; Peterkin 1993) while Mousterian points, which probably served as spear tips, show a monotonous consistency in form across the Old World (Kuhn 1993). The foragers described in the ethnographic record, and those of the Upper Palaeolithic and Mesolithic, were dependent upon such specialized and dedicated tools, but to make these requires a degree of cognitive fluidity that appears absent from the Early Human mind. And consequently Early Humans were unable to integrate their undoubted technical and natural history intelligences to make specialized hunting weapons. Similarly the absence of beads, pendants and tools carrying social information about ownership or group affiliation can be attributed to an inability to integrate their technical and social intelligences (Mithen 1996).

Perhaps the most compelling piece of evidence for this cognitive constraint is that from the fossil record which indicates that Neanderthals were under severe adaptive stress—90% of Neanderthals were dead by the age of 40 (Trinkaus 1995). In such situations it would seem to have made great ecological sense to have applied their technical skills to making beads and pendants to facilitate social interaction, or to have made specialized and dedicated hunting weapons to have improved foraging efficiency. But they didn't. They appear to have possessed a domain-specific mentality: not for them the confusion and conflation of aims and criteria, but a clear sightedness, and a single mindedness absent from the modern mind.

The presence of this domain-specific mentality appears to resolve the paradox between the degree of social complexity and variability we can infer from the archaeological and the fossil/palaeoenvironmental records. The most accurate picture of Neanderthal social behaviour is likely to come from the second of these. We find no traces of social complexity in the archaeological record not because this was absent, but simply because material culture, the natural world, and space were not actively used in social strategies due to the domain-specific mentality.

THE EVOLUTION OF THE MIND

This domain-specific mentality possessed up to 50,000 years ago by all types of Early Humans, whether they be Neanderthals, archaic *H. sapiens* or anatomically modern humans (but see Mithen 1996: 178–184), appears to be the end point of an ever increasing specialization in cognition that began early in the hominid line (Figure 3). In this figure I have suggested that the long term evolution of the human mind has involved an alternation between a type of mind that can be described as 'specialized' and one that can be described as

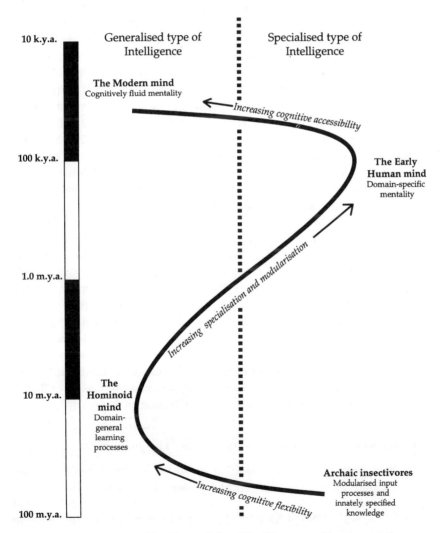

Figure 3. Alternating types of intelligence during 100 million years of human evolution.

'generalized'. If the very earliest primates had a mind composed of specialized hard wired modules for perception and fixed action responses, then it is likely that by 50 million years ago domain-general processes, such as those for associative learning, had evolved providing a new degree of behavioural flexibility. We then begin a period of ever increasing cognitive specialization. As Byrne & Whiten (1992) have argued, soon after 35 million years ago specialized cognitive processes for social interaction had appeared, and then the fossil and archaeological records suggest that specialized cognitive domains for stone technology and natural history appeared about 1.5 million years ago—the timing of the start of the Acheulean and biface manufacture (Asfaw et al. 1992)—and a linguistic capacity perhaps 250,000 years ago (Aiello & Dunbar 1993). In effect an ever increasing specialization of the mind.

This process had ended by 40,000 years ago with a return to a generalized type of intelligence. By this time, the Upper Palaeolithic had begun and the archaeological evidence indicates the cognitive fluidity that is so readily apparent when we consider ethnographically documented hunter-gatherers. I suspect that this transition from a domain-specific to a cognitively fluid mentality was related to the transition of the linguistic capacity from one concerned with social information alone, to one that communicates information about all domains of thought and behaviour, an argument I have expanded upon elsewhere (Mithen 1996). Although we only see this archaeologically at the start of the Upper Palaeolithic, quite when the cognitive architecture for cognitive fluidity arose is unclear. Its universality among all humans today would suggest that the potential for cognitive fluidity was in place prior to the spread of *H. sapiens*, and consequently at least by 100,000 years ago, *if* a replacement scenario for the origins of modern humans is correct. But this leads us into the much debated issue of replacement versus multi-regional evolution which goes beyond the remit of this paper (for these debates see Aiello 1993; Frayer et al. 1993, 1994; Templeton 1993; Stringer & Bräuer 1994). All that is required for my argument is that by 40,000 years ago, all humans possessed a cognitively fluid mentality that can be represented by Figure 4—a mind capable of complex behaviour at the domain interfaces, the mind that has a conflation of aims and criteria into a single and indivisible semantic content.

The mind as software, natural selection as the blind programmer

This long term pattern of cognitive evolution makes sense from our understanding of natural selection and biological evolution. Consider the analogies of the mind as software, a computer program, and natural selection not as a blind watchmaker (Dawkins 1986), but as a blind computer programmer. How could the relatively simple program of the

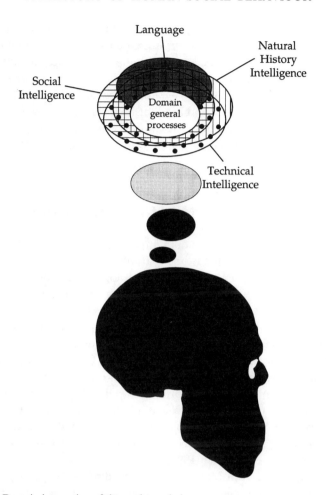

Figure 4. Domain integration of the modern mind.

mind of an early hominoid ancestor be developed to the vastly complex computer program of the modern mind? Note that, as illustrated in Figure 3, both of these have a generalized type of intelligence. Consider how a programmer would do this. Trying to add complexity to all parts of the program at once—which would be represented by moving in a direct vertical line from 'The Hominoid mind' to 'The Modern mind' in Figure 3—would end in a series of untraceable bugs throughout the system. It couldn't be done. A good programmer would follow the curve of Figure 3, i.e. take each routine separately, specialize its function, add the complexity, and test it independently from the other specialized routines. Only finally would these be put back together to make a single complex computer program. This appears to be precisely what natural selection has done when building the modern mind: gradually adding specialized cognitive domains, only to glue these together using language at a very recent date in human evolution. As a consequence it is not surprising that the chimpanzee is often assumed to be so close to humans, such as in the character of its material culture (McGrew 1992) or 'linguistic' capacities (Savage-Rumbaugh & Rumbaugh 1993) because both the chimpanzee and modern humans have a generalized type of intelligence. But these are nevertheless fundamentally different since that of modern humans is based upon an evolved cognitive architecture of specialized cognitive domains.

THE EVOLUTION OF SOCIAL BEHAVIOUR

Attributing Early Humans with a domain-specific mentality is, of course, no more than a hypothesis. Yet it is one that appears to solve the puzzles and paradoxes of the Early Palaeolithic archaeological record, which I have merely touched upon in this paper. It is also a hypothesis which appears compatible with current ideas in evolutionary psychology (e.g. see Barkow *et al.* 1992; Hirschfeld & Gelman 1994). But the domain-specific mentality of Early Humans makes life miserable for archaeologists when trying to reconstruct early prehistoric social behaviour.

With no imposition of social information on tools, no use of body adornment, no use of the natural world for social ends we cannot 'see' social behaviour in the material of the archaeological record before about 50,000 years ago. We can only catch very occasional and blurred glimpses of social variability as this is likely to have been passively reflected in the character of early stone technology (Mithen 1994b).

As a consequence both Mellars (1989) and Binford (1989) appear to be correct: Early Human social groups do indeed look as if their are 'uniformly small'; social structure does indeed appear to be absent. But this is only because the archaeological is interpreted as if Early Humans had cognitively fluid minds. The contradictory inferences about social behaviour that can be drawn from the archaeological and fossil/palaeoenvironmental records suggest that this was not the case, and consequently the nature of Early Human social behaviour remains unclear. All we can be sure of is that Early Human social behaviour must have been something very different from that which we see today among non-human primates, because of the much larger body and brain sizes of Early Humans; it must also have been something very different to that of the hunter-gatherers documented in the ethnographic record, because of the domain-specific mentality of Early Humans. It was a unique type of social behaviour, but one forever lost to us in the darkness of prehistory.

In contrast, our reconstructions of social behaviour during the Upper Palaeolithic and Mesolithic periods are likely to become increasingly refined as further sites are excavated, analyses of existing material undertaken and new methodologies introduced. Similarly, our explanations of the inferred patterns of social behaviour are likely to progress as we understand more fully the relationships between social behaviour and environmental variables, and the specific historical trajectories of social change, and bring theories and models from both behavioural ecology and social anthropology to bear on the data. Bridges between these disciplines can be developed, but the key to this is understanding the major cognitive transition at the start of the Upper Palaeolithic from a domain-specific to a cognitively fluid mentality.

Note. For discussion of some of the issues in this paper I would like to thank Mark Lake, Dick Byrne, Leda Cosmides and Clive Gamble.

REFERENCES

Aiello, L. 1993: The fossil evidence for modern human origins in Africa: A revised view. *American Anthropologist* 95, 73–96.

Aiello, L. & Dunbar, R.I.M. 1993: Neocortex size, group size and the evolution of language. *Current Anthropology* 34, 184–193.

Albrethsen, S.E. & Petersen, E.B. 1976: Excavation of a Mesolithic cemetery at Vedbaek, Denmark. *Acta Archaeologica* 47, 1–28.

Allen, J. 1994: Radiocarbon determinations, luminesence dates and Australian archaeology. *Antiquity* 68, 339–343.

Arensburg, B., Tillier, A.M., Vandermeersch, B., Duday, H. Schepartz, L.A. & Rak, Y. 1989: A Middle Palaeolithic hyoid bone. *Nature* 338, 758–760.

Arensburg, B., Schepartz, L.A., Tillier, A.M., Vandermeersch, B. & Rak, Y. 1990. A reappraisal of the anatomical basis for speech in Middle Palaeolithic hominids. *American Journal of Physical Anthropology* 83, 137–146.

Asfaw, B., Beyene, Y., Suwa, G., Walter, R.C., White, T., Wolde-Gabriel, G. & Yemane, T. 1992: The earliest Acheulean from Konso-Gardula. *Nature* 360, 732–735.

Bahn, P. & Vertut, J. 1988: *Images of the Ice Age*. London: Windward.

Bar-Yosef, O. 1994: The contributions of southwest Asia to the study of the origin of modern humans. In *Origins of Anatomically Modern Humans* (ed. M.H. Nitecki & D.V. Nitecki), pp. 23–66. New York: Plenum Press.

Bar-Yosef, O. & Meignen, L. 1992: Insights into Levantine Middle Palaeolithic cultural variability. In *The Middle Palaeolithic: Adaptation, Behaviour and Variability* (ed. H.L. Dibble & P. Mellars), pp. 163–182. Philadelphia: The University Museum, University of Pennsylvania.

Bar-Yosef, O., Vandermeersch, B., Arensburg, B. *et al.* 1992: The excavations in Kebara Cave, Mt. Carmel. *Current Anthropology* 33, 497–551.

Barkow, J.H., Cosmides, L. & Tooby, J. 1992: *The Adapted Mind: Evolutionary Psychology and the Generation of Culture*. Oxford: Oxford University Press.

Bednarik, R.G. 1992 Palaeoart and archaeological myths. *Cambridge Archaeological Journal* 2, 27–57.

Bednarik, R.G. 1994: A taphonomy of palaeoart. *Antiquity* 68, 68–74.

Bednarik, R.G. & Yuzhu, Y. 1991: Palaeolithic art in China. *Rock Art Research* 8, 119–123.

Binford, L.R. 1989: Isolating the transition to cultural adaptations: an organizational approach. In *The Emergence of Modern Humans: Biocultural Adaptations in the Later Pleistocene* (ed. E. Trinkaus), pp. 18–41, Cambridge: Cambridge University Press.

Birt-David, N. 1990: The 'giving environment': another perspective on the economic system of Gatherer-Hunters. *Current Anthropology* 31, 189–196.

Bleed, P. 1986: The optimal design of hunting weapons. *American Antiquity* 51, 737–747.

Boesch, C. 1993: Aspects of transmission of tool-use in wild chimpanzees. In *Tools, Language and Cognition in Human Evolution* (ed. K.G. Gibson & T. Ingold), pp. 171–183. Cambridge: Cambridge University Press.

Boesch, C. & Boesch, H. 1983: Optimisation of nut-cracking with natural hammers by wild chimpanzees. *Behaviour* 83, 265–286.

Boesch, C. & Boesch, H. 1984: Mental maps in wild chimpanzees: an analysis of hammer transports for nut cracking. *Primates* 25, 160–170.

Boesch, C. & Boesch, H. 1989: Hunting behaviour of wild chimpanzees in the Taï National Park. *American Journal of Physical Anthropology* 78, 547–573.

Boesch, C. & Boesch, H. 1990: Tool use and tool making in wild chimpanzees. *Folia Primatologica* 54, 86–99.

Boesch, C. & Boesch. H. 1993: Diversity of tool use and tool-making in wild chimpanzees. In *The Use of Tools by Human and Non-Human Primates* (ed. A. Berthelet & J. Chavaillon), pp. 158–174. Oxford: Clarendon Press.

Bowdler, S. 1992: Homo sapiens in Southeast Asia and the Antipodes: Archaeological versus biological interpretations. In *The Evolution and Dispersal of Modern Humans in Asia* (ed. T. Akazawa, K. Aoki, & T. Kimura), pp. 559–589. Tokyo: Hokusen-Sha.

Byrne, R.W. 1995. *The Thinking Ape: Evolutionary Origins of Intelligence*. Oxford: Oxford University Press.

Byrne R.W. & Whiten, A. (eds) 1988: *Machiavellian Intelligence: Social Expertise and the Evolution of Intellect in Monkeys, Apes and Humans*. Oxford: Clarendon Press.

Byrne, R.W. & Whiten, A. 1992: Cognitive evolution in primates: evidence from tactical deception. *Man* (N.S.) 27, 609–627.

Chase, P. 1986: *The Hunters of Combe Grenal: Approaches to Middle Palaeolithic Subsistence in Europe*. Oxford: British Archaeological Reports, International Series, S286.

Chase, P. 1989: How different was Middle Palaeolithic subsistence?: A zooarchaeological perspective on the Middle to Upper Palaeolithic transition. In *The Human Revolution* (ed. P. P. Mellars & C. Stringer), pp. 321–337. Edinburgh: Edinburgh University Press.

Chase, P. 1991: Symbols and palaeolithic artefacts: style, standardization and the imposition of arbitrary form. *Journal of Anthropological Archaeology* 10, 193–214.

Chase, P. & Dibble, H. 1987: Middle Palaeolithic symbolism: a review of current evidence and interpretations. *Journal of Anthropological Archaeology* 6, 263–293.

Chase, P. & Dibble, H. 1992: Scientific archaeology and the origins of symbolism: a reply to Bednarik. *Cambridge Archaeological Journal* 2, 43–51.

Cheney, D.L. & Seyfarth, R.S. 1988: Social and non-social knowledge in vervet monkeys. In *Machiavellian Intelligence: Social Expertise and the Evolution of Intellect in Monkeys, Apes and Humans* (ed. R.W. Byrne & A. Whiten), pp. 255–270. Oxford: Clarendon Press.

Cheney, D.L. & Seyfarth, R.S. 1990. *How Monkeys See the World*. Chicago: Chicago University Press.

Churchill, S. 1993: Weapon technology, prey size selection and hunting methods in modern hunter-gatherers: Implications for hunting in the Palaeolithic and Mesolithic. In *Hunting and Animal Exploitation in the Later Palaeolithic and Mesolithic of Eurasia* (ed. G.L. Peterkin, H.M. Bricker & P. Mellars), pp. 11–24. Archaeological Papers of the American Anthropological Association, No. 4.

Clark, G.A. & Neeley, M. 1987: Social differentiation in European Mesolithic burial data. In *Mesolithic Northwest Europe: Recent Trends* (ed. P.A. Rowley-Conwy, M. Zvelebil, H.P. Blankholm), pp. 121–127. Sheffield Department of Archaeology and Prehistory.

Clark, G.A., Young, D., Straus, L.G. & Jewett, R. 1986: Multivariate analysis of La Riera industries and fauna. In *La Riera Cave* (ed. L.G. Straus & G.A. Clark), pp. 325–350. Anthropological Research Papers No. 36. Tempe: Arizona State University.

Clutton-Brock, T.H. & Harvey, P. 1977: Primate ecology and social organisation. *Journal of the Zoological Society of London* 183, 1–39.

Conkey, M. 1980: The identification of prehistoric hunter-gatherer aggregation sites: The case of Altamira. *Current Anthropology* 21, 609–630.

Cosmides, L. & Tooby, J. 1987: From evolution to behaviour: evolutionary psychology as the missing link. In *The Latest on the Best: Essays on Evolution and Optimality* (ed. J. Dupré), pp. 277–306. Cambridge: Cambridge University Press.

Davidson, I. 1992: There's no art—To find the mind's construction—In offence (reply to R. Bednarik). *Cambridge Archaeological Journal* 2, 52–57.

Davidson, I. & Noble, W. 1992: Why the first colonisation of the Australian region is the earliest evidence of modern human behaviour. *Archaeology in Oceania* 27, 113–119.

Dawkins, R. 1986: *The Blind Watchmaker*. Harmondsworth: Penguin.

Delluc, B. & Delluc, G. 1978: Les manifestations graphiques aurignaciennes sur support rocheux des environs des Eyzies (Dordogne). *Gallia Préhistoire* 21, 213–438.

Dennell, R. W. 1983: *European Economic Prehistory*. London: Academic Press.

Dunbar, R.I.M. 1988: *Primate Societies*. London: Chapman & Hall.

Dunbar, R.I.M. 1992: Neocortex size as a constraint on group size in primates. *Journal of Human Evolution* 20, 469–493.

Dunbar, R.I.M. 1993: Coevolution of neocortical size, group size and language in humans. *Behavioural and Brain Sciences* 16, 681–735.

Enloe, J., David, F. & Hare, T.S. 1994: Patterns of faunal processing at section 27 of Pincevent: The use of spatial analysis and ethnoarchaeological data in the interpretation of archaeological site structure. *Journal of Anthropological Archaeology* 13, 105–124.

Farizy, C. 1994: Spatial patterning of Middle Palaeolithic sites. *Journal of Anthropological Archaeology* 13, 153–160.

Foley, R. & Lee, P. 1989: Finite social space, evolutionary pathways, and reconstructing hominid behaviour. *Science* 243, 901–906.

Frayer, D.W., Wolpoff, M.H., Thorne, A.G., Smith, F.H. & Pope, G. 1993: Theories of modern human origins: The paleontological test. *American Anthropologist* 95, 14–50.

Frayer, D.W., Wolpoff, M.H., Thorne, A.G., Smith, F.H. & Pope, G. 1994: Getting it straight. *American Anthropologist* 96, 424–438.

Fris-Hansen, J. 1990: Mesolithic cutting arrows: Functional analysis of arrows used in the hunting of large game. *Antiquity* 64, 494–504.

Gamble, C. 1982: Interaction and alliance in Palaeolithic society. *Man* 17, 92–107.

Gamble, C. 1986: *The Palaeolithic Settlement of Europe*. Cambridge: Cambridge University Press.

Gamble, C. 1987: Man the shoveler: alternative models for Middle Pleistocene colonization and occupation in northern latitudes. In *The Pleistocene Old World* (ed. O. Soffer), pp. 81–98. New York: Plenum Press.

Gamble, C. 1989: Comment on 'Grave shortcomings: the evidence for Neanderthal burial by R. Gargett', *Current Anthropology* 30, 181–82.

Gamble, C. 1991: The social context for European Palaeolithic art. *Proceedings of the Prehistoric Society* 57 (i), 3–15.

Gamble, C. 1994: The peopling of Europe, 700,000–40,000 years before the present. In *The Oxford Illustrated Prehistory of Europe* (ed. B. Cunliffe), pp. 5–41. Oxford: Oxford University Press.

Gardner, H. 1983: *Frames of Mind: The Theory of Multiple Intelligences.* New York: Basic Books.
Gargett, R. 1989: Grave shortcomings: The evidence for Neanderthal burial. *Current Anthropology* 30, 157–190.
Gellner, E. 1988: *Plough, Sword and Book: The Structure of Human History.* London: Collins Harvell.
Gendel, P. 1984: *Mesolithic Social Territories in Northwestern Europe.* Oxford: British Archaeological Reports International Series 218.
Grün, R. & Stringer, C. 1991: Electron spin resonance dating and the evolution of modern humans. *Archaeometry* 33, 153–199.
Harrold, F. 1989: Mousterian, Chatelperronian and early Aurignacian in western Europe: continuity or discontinuity? In *The Human Revolution* (ed. P. Mellars & C. Stringr), pp. 677–713. Edinburgh: Edinburgh University Press.
Hirschfeld, L.A. & Gelman, S.A. (eds.) 1994: *Mapping the Mind: Domain Specificity in Cognition and Culture.* Cambridge: Cambridge University Press.
Hodder, I. 1985: *Symbols in Action.* Cambridge: Cambridge University Press.
Houghton, P. 1993: Neanderthal supralaryngeal vocal tract. *American Journal of Physical Anthropology* 90, 139–146.
Hublin, J.J. 1992: Recent human evolution in northwestern Africa. *Philosophical Transactions of the Royal Society*, Series B, 337, 185–91.
Humphrey, N. 1976: The social function of intellect. In *Gowing Points in Ethology* (ed. P.P.G. Bateson & R.A. Hinde), pp. 303–17. Cambridge: Cambridge University Press.
Ingold, T. 1992: Comment on 'Beyond the original affluent society' by N. Birt-David. *Current Anthropology* 33, 34–47.
Ingold, T. 1993: Tool-use, sociality and intelligence. In *Tools, Language and Cognition in Human Evolution* (ed. K.G. Gibson & T. Ingold), pp. 429–445. Cambridge: Cambridge University Press.
Isaac, G. 1978: The food-sharing behaviour of proto-human hominids. *Scientific American* 238 (April), 90–108.
Johnsen, S.J., Clausen, H.B., Dansgaard, W. *et al.* 1992: Irregular glacial interstadials recorded in a new Greenland ice core. *Nature* 359, 311–313.
Jochim, M. 1983: Palaeolithic cave art in ecological perspective. In *Hunter-Gatherer Economy in Prehistory* (ed. G.N. Bailey), pp. 212–219. Cambridge: Cambridge University Press.
Kaplan, H. & Hill, K. 1985: Hunting ability and reproductive success among male Ache foragers: Preliminary results. *Current Anthropology* 26, 131–133.
Karmiloff-Smith, A. 1992: *Beyond Modularity; A Development Perspective on Cognitive Science.* Cambridge, Mass.: MIT Press.
Kilma, B. 1988: A triple burial from the Upper Palaeolithic of Dolni Vestonice, Czechoslovakia. *Journal of Human Evolution* 16, 831–835.
Koetje, T.A. 1994: Intrasite spatial structure in the European Upper Palaeolithic: Evidence and patterning from the SW of France. *Journal of Anthropological Archaeology* 13, 161–169.
Kuhn, S. 1993: Mousterian technology as adaptive response. In *Hunting and Animal Exploitation in the Later Palaeolithic and Mesolithic of Eurasia* (ed. G.L. Peterkin, H.M. Bricker & P. Mellars), pp. 25–31. Archaeological Papers of the American Anthropological Association, No. 4.
Larsson, L. 1983: The Skateholm Project—A Late Mesolithic Settlement and Cemetery complex at a southern Swedish bay. *Meddelanden fran Lunds Universitets Historiska Museum* 1983–84, 4–38.
Larsson, L. 1990: Dogs in fraction—symbols in action. In *Contributions to the Mesolithic in Europe* (ed. P.M. Vermeersch & P. Van Peer), pp. 153–160. Leuven.
Leacock, E. & Lee, R. 1982: *Politics and History in Band Societies.* Cambridge: Cambridge University Press.

Lee, R.B. 1979: *The !Kung San: Men, Women and Work in a Foraging Society.* Cambridge: Cambridge University Press.
Lieberman, P. & Crelin, E.S. 1971: On the speech of Neanderthal man. *Linguistic Enquiry* 2, 203–222.
Lieberman, D.E. & Shea, J.J. 1994: Behavioural differences between Archaic and Modern Humans in the Levantine Mousterian. *American Anthropologist* 96, 330–332.
McGrew, W.C. 1992: *Chimpanzee Material Culture.* Cambridge: Cambridge University Press.
Marshack, A. 1990: Early hominid symbol and the evolution of human capacity. In *The Emergence of Modern Humans* (ed. P. Mellars), pp. 457–498. Edinburgh: Edinburgh University Press.
Mellars, P. 1973: The character of the Middle-Upper transition in south-west France. In *The Explanation of Culture Change* (ed. C. Renfrew), pp. 255–76. London: Duckworth.
Mellars, P. 1989: Major issues in the emergence of modern humans. *Current Anthropology* 30, 349–385.
Mellars, P. 1994: The Upper Palaeolithic revolution. In *The Oxford Illustrated Prehistory of Europe* (ed. B. Cunliffe), pp. 42–78. Oxford: Oxford University Press.
Mithen, S. 1989: To hunt or to paint? Animals and art in the Upper Palaeolithic. *Man* 23, 671–695.
Mithen, S. 1990: *Thoughtful Foragers: A Study of Prehistoric Decision Making.* Cambridge: Cambridge University Press.
Mithen, S. 1991: Ecological interpretations of Upper Palaeolithic art. *Proceedings of the Prehistoric Society* 57(i), 103–114.
Mithen, S. 1993: From domain-specific to generalised intelligence: A cognitive interpretation of the Middle-Upper Palaeolithic transition. In *The Ancient Mind* (ed. C. Renfrew & E. Zubrow), pp. 29–39, Cambridge: Cambridge University Press.
Mithen, S. 1994a: The Mesolithic Age. In *The Oxford Illustrated Prehistory of Europe* (ed. B. Cunliffe), pp. 79–135. Oxford: Oxford University Press.
Mithen, S. 1994b: Technology and society during the Middle Pleistocene. *Cambridge Archaeological Journal* 4, 3–33.
Mithen, S. 1994c: Domain specific intelligence and the Neanderthal mind. In *The Early Human Mind* (ed. P.Mellars & K.Gibson). Cambridge: MacDonald Institute for Archaeological Research.
Mithen, S. 1996: *The Prehistory of the Mind: A Search for the Origins of Art, Religion and Science.* London: Thames & Hudson.
Mithen, S., Finlayson, B., Finlay, N. & Lake, M. 1992: Excavations at Bolsay Fam, a Mesolithic site on Islay. *Cambridge Archaeological Journal* 2, 242–253.
Morphy, H. (ed) 1989: *Animals into Art.* London: Unwin Hyman.
O'Connell, J. 1987: Alyawara site structure and its archaeological implications. *American Antiquity* 52, 74–108.
O'Shea, J. & Zvelebil, M. 1984: Oleneostrovski Moglinik: reconstructing the social and economic organisation of prehistoric foragers in northern Russia. *Journal of Anthropological Archaeology* 3, 1–40.
Oswalt, W.H. 1976: *An Anthropological Analysis of Food-Getting Technology.* New York: John Wiley.
Pelegrin, J. 1993: A framework for analysing prehistoric stone tool manufacture and a tentative application to some early stone industries. In *The Use of Tools by Human and Non-human Primates* (ed. A. Berthelet & J. Chavaillon), pp. 302–314. Oxford: Clarendon Press.
Péquart, M. & Péquart, St-J. 1954: *Hoëdic: Deuxième Station-Nécropole du Mésolithique Cotier Armoricain.* Anvers: de Sikkel.
Péquart, M., Péquart, St-J., Boule, M. & Vallois, H. 1937: Téviec: Station-Nécropole Mésolithique du Morbihan. *Archives de l'Institut de Paléontologie Humanine*, Mémoire No. 18. Paris: Masson et Cie.

Peterkin, G.L., 1993: Lithic and organic hunting technology in the French Upper Palaeolithic. In *Hunting and Animal Exploitation in the Later Palaeolithic and Mesolithic of Eurasia* (ed. G.L. Peterkin, H.M. Bricker, & P. Mellars), pp. 49–67. Archaeological Papers of the American Anthropological Association, No. 4.

Petersen, E.B. 1989: Vaegnet Nord: Excavation, documentation and interpretation of a Mesolithic site at Vedbaek, Denmark. In *The Mesolithic in Europe* (ed. C. Bonsall), pp. 325–330. Edinburgh: John Donald.

Pigeot, N. 1990: Technical and social actors: Flint knapping specialists and apprentices at Magdalenian Etiolles. *Archaeological Review from Cambridge* 9, 126–141.

Price, T.D. 1985: Affluent foragers of southern Scandinavia. In *Prehistoric Hunter-Gatherers: The Emergence of Cultural Complexity* (ed. T.D. Price & J.A. Brown), pp. 341–63. New York: Academic Press.

Roberts, M.B. 1986: Excavation of the Lower Palaeolithic site at Amey's Eartham Pit, Boxgrove, West Sussex: A preliminary report. *Proceedings of the Prehistoric Society* 52, 215–46.

Roberts, R.G., Jones, R. & Smith, M.A. 1990: Thermoluminesence dating of a 50,000 year old human occupation site in northern Australia. *Nature* 345, 153–6.

Roberts, R.G., Jones, R. & Smith, M.A. 1994: Beyond the radiocarbon barrier in Australian prehistory. *Antiquity* 68, 611–16.

Roebroeks, W. 1988: From flint scatters to early hominid behaviour: a study of Middle Palaeolithic riverside settlements at Mastricht-Bélvèdere. *Analecta Praehistorica Leidensai* 1988.

Roebroeks, W. & van Kolfschoten, T. 1994: The earliest occupation of Europe: a short chronology. *Antiquity* 68, 489–503.

Rozin, P. 1976: The evolution of intelligence and access to the cognitive unconscious. In *Progress in Psychobiology and Physiological Psychology* (ed. J.M. Sprague & A.N. Epstein), pp. 245–77. New York: Academic Press.

Rozin, P. & Schull, J. 1988: The adaptive-evolutionary point of view in experimental psychology. In *Steven's Handbook of Experimental Psychology, Vol 1: Perception and Motivation* (ed. R.C. Atkinson, R.J. Hernstein, G. Lindzey & R.D. Luce), pp. 503–46. New York: John Wiley & Sons.

Saladin D'Anglure, B. 1990: Nanook, super-male; the polar bear in the imaginary space and social time of the Inuit of the Canadian Arctic. In *Signifying Animals: Human Meaning in the Natural World* (ed. R.G. Willis), pp. 173–195. London: Unwin Hyman.

Savage-Rumbaugh, S. & Rumbaugh, D. 1993: The emergence of language. In *Tools, Language and Cognition in Human Evolution* (ed. K. Gibson & T. Ingold), pp. 86–108. Cambridge: Cambridge University Press.

Schepartz, L.A. 1993: Language and modern human origins. *Yearbook of Physical Anthropology* 36, 91–126.

Shackleton, N.J. & Opdyke, N.D. 1973: Oxygen isotope and palaeomagnetic stratigraphy of equatorial Pacific core V28–238. *Quaternary Research* 3, 39–55.

Sieveking, A. 1980: Style and regional grouping in Magdalenian cave art. *Institute of Archaeology Bulletin* 16–17, 95–109.

Silberbauer, G. 1981: *Hunter and Habitat in the Central Kalahari Desert*. Cambridge: Cambridge University Press.

Soffer, O. 1985: *The Upper Palaeolithic of the Central Russian Plain*. New York: Academic Press.

Soffer, O. 1994: Ancestral lifeways in Eurasia—The Middle and Upper Palaeolithic records. In *Origins of Anatomically Modern Humans* (ed. M.H. Nitecki & D.V. Nitecki), pp. 101–120. New York: Plenum Press.

Spelke, E.S, Breinlinger, K., Macomber, J. & Jacobsen, K. 1992: Origins of knowledge. *Psychological Review* 99, 605–632.

Srejovic, D. 1972: *Lepenski Vir.* London: Thames & Hudson.
Steele, J. 1996: On predicting hominid group size. In *The Archaeology of Human Ancestry: Power, Sex and Tradition.* (ed. J. Steele & S. Shennan), pp. 230–252. London: Routledge.
Stringer, C. & Bräuer, G. 1994: Methods, misreading and bias. *American Anthropologist* 96, 416–424.
Stringer, C. & Gamble, C. 1993. *In Search of the Neanderthals.* London: Thames & Hudson.
Swisher, C.C. III., Curtis, G.H., Jacob, T., Getty, A.G., Suprijo, A. & Widiasmoro, 1994. Age of the earliest known hominids in Java, Indonesia. *Science* 263, 1118–1121.
Templeton, A.R. 1993: The 'Eve' hypothesis: A genetic critique and reanalysis. *American Anthropologist* 95, 51–72.
Torrence, R. 1983: Time budgeting and hunter-gatherer technology. In *Hunter-Gatherer Economy in Prehistory* (ed. G.N. Bailey), pp. 11–22, Cambridge: Cambridge University Press.
Trinkaus, E. 1995: Neanderthal mortality patterns. *Journal of Archaeological Science* 22, 121–142.
van Schaik, C.P. 1983: Why are dirunal primates living in large groups? *Behaviour* 87, 120–144.
Vang Petersen, P. 1984: Chronological and regional variation in the late Mesolithic of Eastern Denmark. *Journal of Danish Archaeology* 3, 7–18.
Wadley, L. 1993: The Pleistocene Late Stone Age south of the Limpopo River. *Journal of World Prehistory* 7, 243–296.
White, R. 1982: Rethinking the Middle/Upper Paleolithic Transition. *Current Anthropology* 23, 169–192.
White, R. 1989: Production complexity and standardization in early Aurignacian bead and pendant manufacture: evolutionary implications. In *The Human Revolution* (ed. P. Mellars & C. Stringer), pp. 366–90. Edinburgh: Edinburgh University Press.
White, R. 1992: Beyond art: Toward an understanding of the origins of material representation in Europe. *Annual Review of Anthropology* 21, 537–64.
White, R. 1993: Technological and social dimensions of 'Aurignacian-Age' body ornaments across Europe. In *Before Lascaux: The Complex Record of the Early Upper Palaeolithic* (ed. H. Knecht, A. Pike-Tay & R. White), pp. 2477–299. Boca Raton: CRC Press.
Whiten, A. (ed.) 1991: *Natural Theories of Mind: Evolution, Development and Simulation of Everyday Mindreading.* Oxford: Basil Blackwell.
Whitelaw, T. 1991: Some dimensions of variability in the social organisation of community space among foragers. In *Ethnoarchaeological Approaches to Mobile Campsites* (ed. C. Gamble & W. Boismier), pp. 139–88. Ann Arbor: International Monographs in Prehistory.
Wiessner, P. 1983: Style and social information in Kalahari San projectile points. *American Antiquity* 48, 253–257.
Willis, R.G. (ed.) 1990: *Signifying Animals: Human Meaning in the Natural World.* London: Unwin Hyman.
Wobst, H.M. 1977: Stylistic behaviour and information exchange. In *Papers for the Director: Research Essays in Honour of James B. Griffin* (ed. C.E. Cleland), pp. 317–342. Anthropological papers no. 61, Museum of Anthropology, University of Michigan.
Woodman, P. 1985: *Excavations at Mount Sandel.* Belfast: HMSO.
Wynn, T. 1995: Handaxe enigmas. *World Archaeology* 27, 10–23.
Zhonglong, Q. 1992: The stone industries of H. sapiens from China. In *The Evolution and Dispersal of Modern Humans in Asia* (ed. T. Akazawa, K. Aoki, & T. Kimura), pp. 363–372. Tokyo: Hokusen-Sha.

The Emergence of Biologically Modern Populations in Europe: A Social and Cognitive 'Revolution'?

PAUL MELLARS

Corpus Christi College, Cambridge; Department of Archaeology,
University of Cambridge, Downing Street, Cambridge, CB2 3DZ
Fellow of the British Academy

Keywords: Archaeology; Palaeolithic; human evolution; demography; modern humans; Neanderthal.

Summary. The appearance of anatomically modern populations in Europe around 40–45,000 years ago appears to reflect a major population dispersal, which replaced the preceding Neanderthal populations. Closely associated with this population dispersal there is archaeological evidence for a range of dramatic cultural innovations, including the appearance of more complex forms of stone and bone technology, personal ornaments, larger and more highly structured living sites, and remarkably sophisticated representational art and other forms of visual symbolism. There is also evidence for a major increase in human population densities, marked by an increase in the numbers of occupied sites in many regions. It is argued here that several other social transformations, including the appearance of larger residential group sizes, increased separation and specialization of personal roles within these groups, more sharply bounded territorial and demographic groupings, and more complex forms of descent and kinship structures, may be attributable at least in part to this increase in human population densities. A further critical factor in these social and cultural transformations was almost certainly the appearance of more complex and highly structured language patterns, associated with the dispersal of the anatomically modern populations. While the origins of these changes must be sought outside Europe, it was probably this

© The British Academy 1996.

range of behavioural innovations which allowed the biologically modern populations to compete with, and eventually replace, the pre-existing Neanderthal populations of Europe.

INTRODUCTION

FEW TOPICS HAVE GENERATED MORE DEBATE RECENTLY than the origins of anatomically and biologically 'modern' human populations—that is populations which are anatomically closely similar to ourselves, and which are conventionally assigned to the same sub-species of *Homo sapiens sapiens*. From the spate of research carried out recently there seems to be an increasing consensus that the earliest anatomically and genetically modern populations probably originated in one specific region of the world (most probably Africa) and subsequently dispersed to all other regions. Support for this so-called 'Garden of Eden' or 'Out of Africa' hypothesis has come from extensive studies of both mitochondrial and nuclear DNA patterns in present-day populations, as well as from new discoveries and new dating of a range of human skeletal remains from Africa, Asia and Europe (e.g. Mellars & Stringer 1989; Trinkaus 1989; Aitken *et al.* 1992; Stringer & Gamble 1993; Nitecki & Nitecki 1994). Although still disputed by several workers (e.g. Wolpoff *et al.* 1994; Thorne & Wolpoff 1992), most of the latest research seems to be converging increasingly towards this hypothesis, and away from the alternative scenario of 'multiregional' or 'regional continuity' evolution (Harpending *et al.* 1993; Sherry *et al.* 1994; Cann *et al.* 1994; Stringer & Gamble 1994; Rogers & Jorde 1995).

The implications as far as the more northern latitudes of Asia and Europe are concerned are that anatomically modern populations would seem to have dispersed into the Middle Eastern region by at least 90–100,000 BP (as evidenced by the skeletal remains from Skhul and Qafzeh in Israel) and then, after an interval of perhaps 50,000 years, subsequently dispersed into the much colder and more ecologically demanding environments of eastern, central and western Europe—which at this period were in the grip of an essentially periglacial climate, approximately midway during the last glacial episode. The dispersal of the anatomically modern populations throughout Europe apparently led to the eventual decline and extinction of the preceding Neanderthal populations of the region, which are generally assumed to be the more or less direct descendants of the preceding *Homo erectus* or *Homo Heidelbergensis* populations, which had been present in the continent since at least the earlier stages of the Middle Pleistocene, around 500,000 BP (Stringer & Gamble 1993). According to this scenario, therefore, the anatomically modern populations were replacing populations from whom they had been separated in evolutionary terms over a prolonged period. Whether or not there was any interbreeding between the two populations is currently a matter of lively debate, but the bulk of the available genetic and anatomical evidence would seem to argue against any very significant transfer of Neanderthal genes into subsequent European populations (Stoneking & Cann 1989; Stoneking *et al.* 1992; Stringer & Gamble 1993, 1994).

The critical interest of this episode of population replacement from a cultural or (as I would prefer to say) 'behavioural' point of view is that it allows us to make direct comparisons between the patterns of behaviour of two sharply contrasting biological populations, within precisely the same range of environmental settings. In other words, we can compare the archaeological records of these two successive populations within the different regions of Europe and see how far the replacement of the 'archaic' by the modern populations seems to be reflected in the associated behaviour of the two populations. Exactly how we explain any documented contrasts in behaviour of this kind is of course a separate and far more complicated issue, which may conceivably involve delving not only into the nature of the behavioural patterns themselves, but also into the underlying cognitive and intellectual capacities of the two populations. What I wish to argue in this paper is that many of the documented contrasts between the behaviour of the Neanderthal and modern populations could be related to a number of rather basic and simple changes in the social and demographic organization of the two populations. The separate and much more contentious issue of the deeper cognitive implications of these changes will be touched on more briefly in the final section of the paper.

BEHAVIOURAL CHANGES OVER THE MIDDLE-UPPER PALAEOLITHIC TRANSITION

Establishing exact correlations between the archaeological and anatomical records over the period of the Neanderthal-Modern human transition is still rather difficult, owing to the relatively small proportion of archaeological sites which have produced well preserved skeletal remains over the critical transition period. What can be said with some confidence, however, is that the most striking and dramatic changes in the archaeological records can be shown to coincide fairly closely in a chronological sense with the earliest appearance of typically 'modern' anatomical remains in the different regions of Europe, and that there are strong indications that the *earliest* manifestations of these new behavioural patterns are associated specifically with the dispersal of the anatomically modern populations. Since the evidence for these correlations has been discussed fully elsewhere (Mellars 1992; Kozlowski 1992; Stringer & Gamble 1993, 1994; Howell 1994; Gambier 1993), I will not pursue these arguments in the present context.

The evidence for a major shift in human behavioural patterns at this point in the archaeological succession forms the basis for what archaeologists have traditionally referred to as the 'Middle-to-Upper Palaeolithic transition' or—increasingly over the past few years—the 'Upper Palaeolithic revolution' (Mellars 1989a, 1994). If we focus purely on the most archaeologically visible aspects of this transition, it is possible to document changes in at least seven or eight major behavioural domains, all clearly documented within the archaeological records of Europe within the general time range of $c.$ 35–40,000 BP (Kozlowski 1990; Mellars 1989a, 1989b, 1991, 1996; Stringer & Gamble 1993):

1 A basic shift in the technology of stone-tool production, away from the predominantly 'flake-based' technologies of the Lower and Middle Palaeolithic, to the production of more elongated and technologically efficient 'blade' forms. A possible factor underlying this transition is generally thought to have been the introduction of so-called 'indirect' or 'punch' techniques of blade production.

2 A rapid proliferation in the forms of stone tools—almost certainly reflecting a major increase in the complexity of several other, associated aspects of technology (wood working, skin working, hunting missile technology etc.), and apparently indicating a greatly increased component of visual 'style' and deliberately 'imposed form' in the patterns of tool production (Mellars 1989b, 1991; Chase 1991).

3 An even more striking proliferation of new forms of technologically complex, highly varied and visually standardized forms of bone, antler and ivory tools. This contrasts with the virtual lack of deliberately shaped bone tools in earlier Neanderthal contexts, and again seems to reflect an entirely new emphasis on visual form and standardization in artefact production.

4 A correlated shift in the whole tempo of technological change. While the preceding Middle Palaeolithic/Neanderthal phase was characterized by a remarkable lack of technological innovation (in most spheres) over a time span of around 200,000 years, the succeeding Upper Palaeolithic is marked by a series of rapid and conspicuous changes in both stone and bone tool production occurring at intervals of 3000–5000 years or less (Isaac 1972; Mellars 1989b).

5 The sudden appearance of a wide variety of beads, pendants and other items of 'personal ornament'—ranging from simple perforated animal teeth through to carefully shaped stone and ivory beads, together with a range of decorative sea shells, in many cases apparently transported over distances of 300 km or more (White 1989, 1993; Taborin 1993; Gamble 1986).

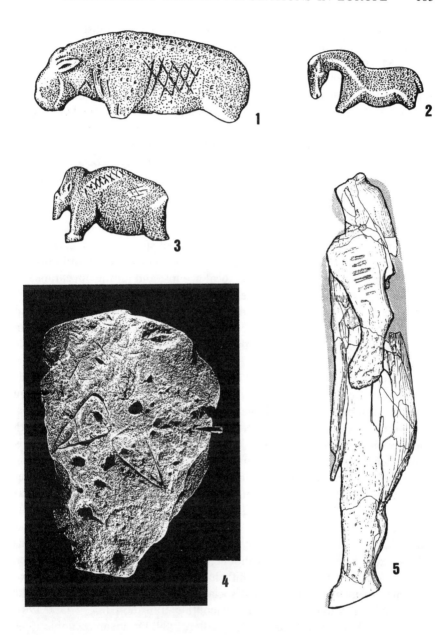

Figure 1. Early Upper Palaeolithic art objects from Aurignacian sites in Europe: 1–3, animal figurines carved from mammoth ivory, Vogelherd, south Germany; 4, female 'vulvar' symbols incised on limestone block, La Ferrassie, southwest France; 5, male human figure with animal's head, of mammoth ivory, Hohlenstein-Stadel, south Germany. Various scales. After Mellars 1989a; Hahn 1977.

6 The appearance of the first incontrovertible musical instruments—in the form of multiple-holed bone flutes (Scothern 1993).

7 Most striking of all, perhaps, the sudden appearance of both extensively incised or 'decorated' bone artefacts and (in certain contexts) remarkably sophisticated and complex forms of representational art. As shown in Figure 1, these range from explicitly vulvar or phallic representations to carefully modelled statuettes of animal, human, or even combined human/animal forms (Hahn 1972, 1993).

8 Lastly, at least strong hints of a number of closely associated changes in both the economic and residential patterns of the human groups. These are inevitably more difficult to document directly from the archaeological records, but nevertheless point strongly to: (a) the appearance of more sharply focused and economically 'specialized' patterns of animal exploitation, targeted on particular, apparently preferred species of game; (b) more efficient and highly organized patterns of procurement and distribution of high quality raw materials between local groups; and (c) the appearance of more highly 'structured' living sites, marked in several cases by deliberately constructed huts or similar structures, suggesting (as in the case of stone and bone tools) a much more explicit component of design and imposed form in the conception and planning of the structures (Mellars 1973, 1989a, 1996; Gamble 1986).

Arguably the most impressive feature of the above list is the wide range of different aspects of behaviour which seem to have been affected—ranging from shifts in basic lithic and bone technology, through subsistence changes, to a veritable explosion of explicitly symbolic expression, reflected in both the obvious fields of art and personal ornamentation, and the new component of increased standardization, 'imposed form' and apparent 'style' reflected in the production of stone and bone artefacts. It is the latter features which have led many authors to talk of a 'symbolic revolution' at this point in the archaeological sequence (Pfeiffer 1982; Chase & Dibble 1987; Mellars 1991; White 1993). To anticipate the discussion in the later part of this paper, it seems almost inconceivable to many workers that this kind of explosion of explicitly symbolic behaviour could have been achieved without at least some associated shifts in the overall complexity, structure or efficiency of language and related social communication patterns over the period of the Middle to Upper Palaeolithic transition (e.g. Clark 1981; Binford 1989; Mellars 1989a, 1991; Whallon 1989; Davidson & Noble 1989, 1993; Bickerton 1990; Lieberman 1991; Donald 1991). Even without this inference, however, it will be clear that we have some justification for speaking of a major 'revolution' in human behavioural patterns at this point in the archaeological sequence, at least as significant in my view as that which heralded the later and more widely publicised 'Neolithic' or 'Urban' revolutions, and apparently reflecting a similar scale of transformations in virtually all of the archaeologically observable aspects of behaviour.

SOCIAL TRANSFORMATIONS

The critical question in the context of the present symposium is what this complex of behavioural changes may tell us about any associated changes in the patterns of social or demographic organization of human communities over the period of the archaic to modern human transition. In other words, does the so-called 'Upper Palaeolithic revolution' also correspond to a social revolution?

The point I wish to argue here is that while there is almost certainly much more to the Upper Palaeolithic revolution than a basic shift in social patterns (see below), it is arguable that at least many of the documented archaeological changes over this transition can be seen most economically as a reflection of some associated social and demographic processes. The main point I will argue is that many of these behavioural changes would fall naturally into place if the process of population dispersal which carried anatomically modern populations across Europe was associated with a major increase in population numbers—and therefore local population densities—in many parts of the continent. Some of the more general components of this model have already been discussed elsewhere, in the context of later developments in social complexity during the course of the Upper Palaeolithic (Mellars 1985). Here I want to expand on these ideas in the particular context of the Neanderthal to modern human transition, and by incorporating some of the recent speculations on the nature of associated cognitive and intellectual changes over this transition.

Population increase

The arguments for a major increase in human population densities associated with the dispersal of modern populations across Europe rest on two kinds of evidence. First, a number of recent studies of the mitochondrial DNA structure of present-day European populations (by the technique known as 'mismatch distributions') seem to point to a sharp increase in population numbers dated approximately to around 40,000 years ago (Harpending *et al.* 1993; Sherry *et al.* 1994). This of course would tally very well with all the current radiocarbon and other dating evidence for the dispersal of anatomically modern populations across Europe, and the associated revolution in behavioural patterns, discussed above. The other line of evidence comes directly from the archaeological records, and lies in the sharp increase in numbers of occupied sites which can be documented in many regions of Europe, coinciding closely with the transition from the Middle to the Upper Palaeolithic. Clear patterns of this kind have been documented for example in the occupation of cave and rock shelter sites in the classic southwest French region (Mellars 1973, 1982), as well as in the adjacent areas of northern Spain and several parts of central and eastern Europe (Straus 1983, 1990: 286; Soffer 1989; Gamble 1986). Exactly what caused or supported this population increase is still a matter of debate, but there can be little doubt that it must have involved some significant increase in the efficiency or productivity of food harvesting strategies. All that needs to be recognized here is that from two quite separate lines of evidence there are now strong indications that human population numbers—and therefore local population densities—did increase sharply in many regions of Europe, at a time corresponding closely with the documented 'revolution' in behavioural patterns reflected in many other aspects of the archaeological evidence.

If population numbers and density did increase sharply with the replacement of Neanderthal by modern populations in Europe, it is reasonable to ask what other shifts in the social structure or organization of local populations we might expect to be associated with this demographic change. It is equally necessary to ask how far such changes can be recognized in the associated archaeological records. The answer, I would suggest, lies in a combination of four major social transformations. To reduce a rather complicated set of arguments to fairly simple terms, the relevant considerations can be summarized fairly briefly as follows:

Size of local residential groupings

The existence of some fairly close relationship between the local density of human populations in particular areas and the sizes of co-residential or co-operating social groups (what in hunter-gatherer societies are often taken to equate broadly with local 'bands') can be argued in at least four different ways:

1 Purely in terms of the basic logistics of economic exploitation and related land-use patterns, it could be argued that any major increase in local population densities would virtually demand the formation of increased co-residential or co-operating social groups, if only to avoid the kinds of recurrent conflicts which would inevitably arise from the uncoordinated activities of a large number of small, independent social units foraging within a relatively small, densely populated area. Arguably, without some degree of co-residence and direct co-operation or communication between group members, this situation would lead to endless confrontations or

conflicts over the exploitation of particular territories or resources, which would seriously undermine the viability of the population as a whole.

2 Second, it could be argued that whatever improvements in the efficiency or productivity of food procurement strategies made possible a significant increase in human population densities over the period of the Middle-Upper Palaeolithic transition would almost inevitably have made possible a corresponding increase in the sizes of local residential groups. The basic reasoning here is that any significant increase in the efficiency, productivity, or (perhaps most significantly) predictability and security of food procurement procedures would automatically increase the amount of food which could be secured on a reliable, day-to-day basis from a particular foraging area, and thereby increase the number of people who could be supported within this area. Exactly what form these improvements in subsistence strategies would have taken is more controversial, but they probably included improved forms of weapon technology, more highly organized and (probably) co-operative hunting methods, better transportation, improved information sharing on the distribution and movements of resources, and perhaps a broadening of the total range of subsistence resources exploited. Improved techniques of food storage could well have been a further critical factor allowing the formation of larger and more stable residential groups in many Upper Palaeolithic contexts (Mellars 1973; Soffer 1985; Peterkin 1993; Testart 1982).

3 Thirdly, it seems likely that at least some of the documented changes in the overall complexity of different kinds of technology and related subsistence strategies which seem to be indicated over the period of the Middle-Upper Palaeolithic transition would have either required, or at least strongly encouraged, the formation of larger groups of co-operating individuals. Obvious examples would be the introduction of more large scale communal hunting strategies; the emergence of wider ranging and more organized systems of procurement and distribution of raw materials; and perhaps the construction of relatively large and complex living structures. Although less easily demonstrable archaeologically, one can probably add the emergence of larger and more complex group ceremonial activities as a further factor favouring at least the temporary formation of relatively large social gatherings in many Upper Palaeolithic contexts.

4 Finally, an increase in the complexity of both social relationships and roles within local groups (as discussed below) and patterns of linguistic or other communication between members of the groups might well have facilitated the integration and co-ordination of larger social groups in the early Upper Palaeolithic. Several studies have emphasized that the capacity of large groups to function effectively—and to persist for long periods without internal conflict—can be critically dependent on the structure of

individual roles, ranking, and authority within the group as a whole (Lee & DeVore 1968; Steward 1972; Johnson 1982; Price & Brown 1985; Cohen 1985; Mellars 1985). In the same context, language could have been an equally critical factor, both in helping to integrate and co-ordinate the activities of large numbers of individuals within the local groups, to formalize rules of social behaviour, and to resolve potential conflicts within the groups.

How far one can support this kind of increase in the size of local residential groups from the archaeological evidence has been debated frequently in the literature. Most authorities seem to agree that local group sizes in most Neanderthal populations were relatively small, and that there is strong evidence for an increase in group sizes in the relatively large dimensions attained by many Upper Palaeolithic settlements (Mellars 1973, 1989a; Binford 1982, 1989; Soffer 1989, 1994; Stringer & Gamble 1993). While the evidence for this is clearest in sites dating from the middle and later stages of the Upper Palaeolithic (as for example in many of the open-air Gravettian sites of central and eastern Europe), there are at least strong indications of these significantly increased site sizes in some of the earliest Upper Palaeolithic settlements—as for example in the Aurignacian levels at Laussel, Abri Caminade and the Abri Pataud in southwest France, and at a number of open-air sites in central and eastern Europe (Mellars 1973; Hahn 1977; Kozlowski 1982). Similarly, there can be little doubt that the internal structure and organization of settlements became more elaborate during the earliest stages of the Upper Palaeolithic (as for example at Arcy-sur-Cure, Cueva Morín, Abri Pataud, Bacho Kiro etc.), which could be seen as a further indication of generally more complex patterns of social organization in early Upper Palaeolithic groups (Freeman & Echegaray 1970; Kozlowski 1982; Mellars 1989a, 1996; Binford 1989; Farizy 1990).

Increased separation and specialization of individual roles within local groups

The emergence of more specialized and sharply defined social and economic roles of individuals within local residential groups could be seen as a largely direct consequence of several of the factors discussed above, for at least three reasons:

1 It is self evident that any increase in the size of local residential groups would automatically create more scope for the increased separation and specialization of individual roles within these groups. It would obviously be difficult, for example, to have much specialization of roles in societies including, say, only 4–5 adults of either sex within a local group; the potential for role-specialization must inevitably be to a large extent contingent on the overall size of the group.

2   By the same token it is clear that any increase in the general complexity of various economic, technological or social activities of the kind which is generally envisaged for the period of the Middle-Upper Palaeolithic transition (for example in the spheres of bone or wood working; the production of artwork or ornaments; more complex hunting strategies, or elaborate ceremonial activities) would automatically create both more *scope* for the work of specialists, and arguably more *need* for individuals to acquire and practise the necessary skills to perform these different roles. As argued further below, similar separation and clear identification of personal roles within local groups may have been necessary to maintain the social integration and cohesion of relatively large numbers of people—at least over extended periods—within these local groups. If, as many of us suspect, there was a significant increase in the relative duration or permanence of occupation in particular settlements over the course of the Middle-Upper Palaeolithic transition, this would have put a further loading on the need for more clearly defined social structures and mechanisms to avoid or resolve conflicts between group members.

3   Finally, one should emphasize the potentially crucial role of advanced, highly structured language in any clear identification and definition of individual social and economic roles within societies. The ability to clearly *categorize* these roles in conceptual and verbal terms, and the ability to *express* potentially complex social relationships between individual roles, could be critical to the emergence of increasingly complex social relationships within human groups (Gellner 1989; Bickerton 1990; Donald 1991; Knight 1991).

Evidence of increased identification and demarcation of personal roles in Upper Palaeolithic societies can be argued from two main aspects of the archaeological data: first, from the dramatic proliferation of various kinds of personal ornaments (perforated animal teeth, stone and ivory beads, transported sea shells etc.) in the earliest stages of the Upper Palaeolithic— which are often seen as a direct reflection of at least the increased potential, if not the increased need, for clearer expression and visual symbolization of the personal roles of individuals within local groups (White 1989, 1993; Soffer 1989; Mellars 1989a, 1991; Wiessner 1983); and second, from the similar emergence of explicitly 'ceremonial' burial practices, which could be taken as a further reflection of the ascription of special roles or status to individuals within the groups (Binford 1968; Harrold 1980). While many of these burials date from the later stages of the Upper Palaeolithic, the discoveries at Sungir in Russia, Dolní Věstonice in Czechoslovakia, Paviland in Wales and (perhaps more tenuously) Cueva Morín in northern Spain, extend back at least to the earlier stages, if not the very beginning, of the Upper Palaeolithic sequence (Gamble 1986; Mellars 1994). Significantly, both well documented personal ornaments and convincing ceremonial burial practices are at present lacking from pre-Upper Palaeolithic Neanderthal contexts in Europe (Chase & Dibble 1987; Mellars 1989a).

More 'bounded' territorial and demographic units

Arguments for the emergence of more sharply prescribed or 'bounded' demographic and territorial units associated with an increase in population densities over the period of the Middle-Upper Palaeolithic transition have been advanced on several occasions (Isaac 1972; Wobst 1974, 1976; Gamble 1986; Mellars 1989a) . The arguments run essentially as follows:

1 Wobst (1974, 1976) and others have argued that the formation of 'closed' demographic networks (i.e. breeding units with distinct boundaries) are only possible where regional population densities are reasonably high. He argues that if groups living in very low population densities were to practise this kind of bounded mating network, it would be necessary for groups occupying the edges of these units to travel over very large distances to maintain contacts with a sufficiently large number of other groups to maintain demographically viable breeding units. Any significant increase in population density would therefore make the formation of more bounded demographic units more viable, if not necessarily more beneficial to the survival prospects of the group.

2 In a related vein, it has often been pointed out that low population densities would largely preclude any clear definition or attempted defence of specific territories among hunter-gatherer groups (e.g. Dyson-Hudson & Smith 1978). As Wobst (1974, 1976) and others have argued, clear definition and defence of social or economic territories is not only generally unnecessary under conditions of low population density, but effectively impossible to operate or 'police' with so few people on the ground.

3 Under these conditions, several factors might well have served to encourage a much stronger separation of demographic units, and sharper definition of territorial boundaries, as a result of increased population densities in the Upper Palaeolithic period. Dyson-Hudson & Smith (1978), for example, argued that in all human societies clear territoriality is most likely to emerge under conditions of direct competition for economic resources, among relatively high density populations. In these situations, the clear definition of territorial boundaries is often beneficial not merely to safeguard essential economic resources for the local groups, but as a legalistic device to avoid recurrent and disruptive conflicts between neighbouring groups for the exploitation of these resources—much in the way that in our own societies 'good fences make good neighbours'.

4 Finally—and perhaps most significantly—one should stress once again the potentially crucial importance of language patterns in the definition and separation of demographic and ethnic units. Almost all of the classic 'tribes' recognized among modern hunter-gatherers (as for example in Australia, the Kalahari or the Arctic) are defined essentially as linguistic units, based on major (and often mutually unintelligible) dialectical differences between adjacent tribes (Lee & DeVore 1968; Damas 1969; Bicchieri 1972; Peterson 1976; Wiessner 1983). If there is any truth in the suspicion that fully developed language emerged only with the appearance of anatomically modern populations in Europe (as discussed above) then at least the *scope* if not the need for the separation and isolation of discrete 'tribal' or 'ethnic' units would presumably have increased commensurately over the period of the Neanderthal to modern human transition.

A direct archaeological reflection of this increased degree of demographic separation over the period of the Neanderthal/Upper Palaeolithic transition has often been seen in the emergence of increasingly localized and sharply defined 'style zones' apparent in many Upper Palaeolithic contexts. Hahn (1972, 1993), for example, has argued for this kind of patterning in the distribution of various artistic and decorative motifs in the early Upper Palaeolithic Aurignacian industries, while similar patterns have been claimed in the distribution of stylistically distinctive technological variants in the later Perigordian and Solutrian periods (David 1973; Smith 1966) and in the distribution of Magdalenian art styles (Jochim 1983; Bosinski 1990 etc.). While there is certainly evidence for some degree of regional patterning in the basic technology of Neanderthal groups (Kozlowski 1992; Mellars 1996), there is general agreement that this kind of patterning is not only very much greater in the Upper than in the Middle Palaeolithic, but almost certainly based on a much more obvious *symbolic* component in effectively all spheres of material culture (Chase & Dibble 1987; Mellars 1991; Knight 1991). The argument is sometimes extended to suggest that much of the impetus for the emergence of a clearly 'stylistic' component in Upper Palaeolithic tool forms may have derived from the need to reflect these increasingly complex demographic and 'ethnic' distinctions in visually symbolic terms (Isaac 1972; Wobst 1977; Close 1978; Gamble 1986; Sackett 1982, 1988).

More complex descent and kinship systems

Finally, and perhaps most importantly, there are strong reasons to suspect that the structure and complexity of social linkages and relationships both within and between individual social groups would have increased sharply over the period of the Middle-Upper Palaeolithic transition. This could be argued from several aspects of the evidence: evidence for the increased size and scale of local residential groups (as discussed above); the apparently clear evidence for the emergence of more wide ranging exchange or 'alliance' networks between widely dispersed communities, in some cases extending over distances of several hundred kilometres (Gamble 1982, 1986); and the virtual inevitability that groups living in high population densities would need to maintain some form of closely structured social relationships between members of the individual local groups, if only to mitigate the potentially disruptive effects of direct competition or conflict for particular economic resources or territories between groups living in close juxtaposition. Once again, the most critical factor in these relationships however is likely to have been the nature and complexity of language patterns. As Donald has recently stressed (1991: 213–5) it is only with the aid of relatively complex linguistic and semantic systems that one can either clearly conceptualize or formally express the kinds of complex social relationships that might be involved in, say, formalized systems of cross-cousin marriage, the formation of male clan groups, or other forms of complex inter-group kinship systems or moieties (see also Gellner 1989). Significantly, it has recently been pointed out that it is these particular between-group patterns of kinship and descent linkages which form the most diagnostic feature of all modern human societies, and which are conspicuously lacking in all of the documented non-human primate groups (Rodseth *et al.* 1991). If language and linguistic complexity *did* change fairly radically over the period of the Middle-Upper Palaeolithic transition, it would be surprising if this were not reflected in the general structure and complexity of social roles and relationships over this period. As I have attempted to argue above, this would be at least consistent with many different dimensions of the archaeological records of the Middle-Upper Palaeolithic transition in Europe.

DISCUSSION

The preceding sections have presented a model of social changes associated with the transition from archaic to modern populations in Europe which reduces in many respects to a question of 'social scale'. The argument, in essence, is that more or less concomitant with the dispersal of anatomically modern populations throughout Europe, there was a major increase in the total numbers of human population, leading to a significant increase in overall population densities in at least many areas of the continent. Partly dependent on this increase in population densities, but also stimulated by other factors such as associated changes in technology, hunting patterns, and probably language and associated symbolic communication, it is suggested that there would have been a corresponding increase in the size of local co-residential and co-operating groups, with its own attendant set of social consequences. Other social adaptations, such as the increased complexity, separation and specialization of roles of individuals within the societies, similar increasing complexity in the structure of local and regional descent and kinship systems and (more hypothetically, but very probably) an increasing trend towards more sharply bounded and territorially defined demographic units (roughly equivalent to the conventional notions of hunter-gatherer 'tribes') can be seen as in many ways directly dependent on these basic changes in the size and scale of social units.

There is no suggestion of course that these changes in the overall size and complexity of social units would necessarily be apparent in *all* the local populations of anatomically modern humans across Europe. Since the most critical factor in these social transformations is assumed to be the *density* of local populations, it is inevitable that these population densities would have varied within fairly wide limits in the different ecological regions of Europe (probably dependent mainly on the character, productivity and long-term dependability of local food resources), which would have led to equally wide variations in the extent of social pressures and constraints arising from these population numbers. Most of the social responses I have described above are therefore likely to be most apparent in the areas with naturally high concentrations of critical food resources (such as concentrations of large herd animals) where the degree of both population crowding and the associated element of *competition* between closely packed human groups for the use of specific territories and economic resources are likely to have been most acute (Mellars 1985; Cohen 1985). As I and several others have pointed out, this is why the most impressive manifestations of social and cultural 'complexity' in the archaeological records of the European Upper Palaeolithic (such as concentrations of cave and portable art, large, rich sites, elaborate dwelling structures, ceremonial burials etc.) seem to be concentrated strongly in certain specific regions, such as the classic Perigord region of southwest France, the adjacent Cantabrian region of northern Spain, and parts of the ecologically productive loessic plains of central and eastern Europe (Jochim 1983; Mellars 1985; Soffer 1985; Straus 1990). In other, ecologically poorer areas, population densities throughout the Upper Palaeolithic sequence may well have remained at relatively low levels, with correspondingly much weaker pressures—and indeed opportunities—for the development of more complex patterns of social organization (Gamble 1982, 1986).

This emphasis on the basic dimensions and scale of social units as a major stimulant of social change is of course by no means new in either the archaeological or ethnographic literature. Much of the discussion of the varying levels of organizational complexity in modern hunting and gathering societies (e.g. Service 1962; Lee & DeVore 1968; Steward 1972; Sahlins 1972; Woodburn 1982; Price & Brown 1985; Cohen 1985; Keeley 1988) has placed a similar emphasis on factors such as the size and scale of local co-residential and co-operating groups, and the extent of competition and interaction maintained between adjacent social groups. Broadly similar processes (though of course on a much larger scale) are generally seen as underlying the major social transformations associated with the so-called 'Neolithic Revolution'—i.e. an increase in local population numbers (ultimately dependent on increased efficiency of food production), leading to larger and more permanent residential groups, which in turn generated further social complexity in both the internal structure and external social relationships of these enlarged social groups (Bar-Yosef 1994). In both cases it is assumed to be shifts in the basic scale of both local population densities and residential group sizes which served as the primary stimulants for further, concomitant patterns of social change.

In many ways the most crucial question in the present context is how far the patterns of social change visualized over the transition from archaic to modern populations were dependent not only on these shifts in the basic scale of social and demographic units, but also on changes in the underlying cognitive structures of the two populations. There is hardly space here to provide a detailed review of the current thorny debates over the nature of cognitive changes over the period of the archaic/modern human transition, but as noted earlier, there is widespread agreement that this involved an effective 'explosion' in most forms of symbolic expression and behaviour, and almost certainly at least some significant changes in the overall structure and complexity of language (Pfeiffer 1982; Gibson 1985; Chase & Dibble 1987; Binford 1989; Mellars 1989a, 1991; Lieberman 1991; Donald 1991; Knight 1991). While it seems highly unlikely that Neanderthals and other archaic populations possessed no forms of language, it has been argued by many authors these are likely to have been far less complex, less structured, and probably less functionally 'efficient' than those which accompanied the spread of biologically and behaviourally modern populations across Europe (e.g. Bickerton 1990; Lieberman 1991).

Exactly how the structure and complexity of language and associated forms of symbolic communication would have impinged on different aspects of social organization can no doubt be argued in several ways. In the preceding sections I have argued that this could have played a crucial role in several kinds of social structures: in the degree to which individual personal roles and identities within societies could be clearly formalized and defined; in the similar conceptualization and formalization of structured descent and kinship relationships—both within and between local groups; in the capacity to integrate increasingly large numbers of individuals into effective interacting and co-operating units; and (above all perhaps) in the potential effects of language and linguistic differentiation on the kinds of social boundaries which would have emerged between neighbouring demographic and territorial groups (Gellner 1989; Whallon 1989; Donald 1991). Whether or not there were any significant contrasts between the innate 'intelligence' of Neanderthal as opposed to anatomically modern populations is of course an entirely separate question, which is notoriously difficult to approach from the standpoint of the archaeological evidence (Gowlett 1984; Gibson 1985; Wynn 1989; Binford 1989; Gibson & Ingold 1993; Mellars & Gibson 1996). Without making any assumptions about changes in intelligence, however, it is clear that any major shift in the character and complexity of language—or indeed other forms of symbolic expression and communication, such as the use of personal ornaments to signify social identity, or the role of 'stylistic' contrasts in tool manufacture as a means of reflecting membership of particular tribal groupings—could have had a potentially profound effect on many different aspects of the social and demographic organization of archaic and early modern human groups.

The final point which must be recognized is that by choosing to focus this study specifically on the evidence from Europe, I have presented what is in effect a 'before and after' scenario for the patterns of social and cognitive change over the period of the archaic-to-modern human transition. As I indicated at the beginning of the paper, all the current evidence points to the conclusion that in Europe the appearance of biologically and behaviourally modern populations was due to a major dispersal of new populations from some region further to the east or south, which eventually replaced the local Neanderthal populations. By comparing the behaviour of these biologically modern populations with that of the preceding Neanderthals, therefore, we are comparing the behaviour of populations who are likely to have been pursuing largely separate lines of both biological and behavioural development over a period of at least 300,000, if not closer to a million years (Stringer & Gamble 1993). The question of exactly how, where and why these new patterns of behaviour and cognition initially evolved is therefore neatly side-stepped by focusing on the evidence from European sites.

As I have discussed at more length elsewhere (Mellars 1989a) the answer to the preceding question almost certainly lies partly in the evidence from western Asia, and partly on the much earlier records from southern Africa (see also Klein 1989, 1994; Deacon 1989; Foley 1989; Clark 1992; Bar-Yosef 1994). At a purely theoretical level it is possible to visualize a wide spectrum of different scenarios whereby complex, multi-dimensional patterns of technological, social and cognitive change could have emerged more or less in parallel with the evolution of biologically modern populations. In theory, changes in many different aspects of behaviour—ranging from technology, through subsistence practices, to demography or even basic cognitive structures such as language—could have served as the initial catalyst for long-term processes of behavioural and social change. One only has to contemplate the wide range of alternative models which has been advanced to account for the so-called 'Neolithic Revolution' (e.g. Cohen 1977; Bender 1978; Gebauer & Price 1992) to appreciate the difficulties of formulating neat, coherent, and archaeologically testable cause-and-effect relationships for these complex, multifactorial processes of cultural change. In archaeological terms the difficulties arise partly from the very patchy and incomplete nature of the available archaeological records in many of the most potentially crucial areas (such as southern Africa, or central Asia) and partly from the difficulties of reconstructing the precise *sequence* in which the different aspects of behavioural change occurred in particular areas. In the present paper I have tried to show how many different dimensions of social and demographic organization appear to be closely interrelated, and to suggest how these might have been related in turn to simultaneous changes in economic, technological and cognitive patterns. But the task of presenting a neat, coherent and easily testable model of exactly how these complex changes originated in the course of the long evolutionary transition from archaic to modern human populations remains, I suspect, a challenge for the next millennium.

REFERENCES

Aitken, M.J., Stringer, C.B. & Mellars, P.A. (eds.) 1992: *The Origin of Modern Humans and the Impact of Chronometric Dating*. London: Royal Society (Philosophical Transactions of the Royal Society, series B, 337, no. 1280).

Bar-Yosef, O. 1994: The contributions of southwest Asia to the study of the origin of modern humans. In *Origins of Anatomically Modern Humans* (ed. M. H. Nitecki & D. V. Nitecki), pp. 24–66. New York: Plenum Press.

Bender, B. 1978: Gatherer-hunter to farmer: a social perspective. *World Archaeology* 10, 204–222.

Bicchieri, M.G. (ed.) 1972: *Hunters and Gatherers Today*. New York: Holt, Rinehart & Winston.

Bickerton, D. 1990: *Language & Species*. Chicago: University of Chicago Press.

Binford, L.R. 1982: Comment on R. White 'Rethinking the Middle/Upper Palaeolithic transition'. *Current Anthropology* 23, 177–181.

Binford, L.R. 1989: Isolating the transition to cultural adaptations: an organizational approach. In *The Emergence of Modern Humans: biocultural adaptations in the later Pleistocene* (ed. E. Trinkaus), pp. 18–41. Cambridge: Cambridge University Press.

Binford, S.R. 1968: A structural comparison of disposal of the dead in the Mousterian and Upper Paleolithic. *Southwestern Journal of Anthropology* 24, 139–154.

Bosinski, G. 1990: *Homo Sapiens*. Paris: Editions Errance.

Cann, R.L., Rickards, O. & Lum, J.K. 1994: Mitochondrial DNA and human evolution: our one lucky mother. In *Origins of Anatomically Modern Humans* (ed. M.H. Nitecki & D.V. Nitecki), pp. 135–148. New York: Plenum Press.

Chase, P.G. 1991: Symbols and Paleolithic artifacts: style, standardization, and the imposition of arbitrary form. *Journal of Anthropological Archeology* 10, 193–214.

Chase, P.G. & Dibble, H.L. 1987: Middle Paleolithic symbolism: a review of current evidence and interpretations. *Journal of Anthropological Archeology* 6, 263–296.

Clark, J.D. 1981: 'New men, strange faces, other minds'. An archaeologist's perspective on recent discoveries relating to the origins and spread of modern man. *Proceedings of the British Academy* 67, 163–192.

Clark, J.D. 1992: African and Asian perspectives on the origins of modern humans. In *The Origin of Modern Humans and the Impact of Chronometric Dating* (ed. M. Aitken, C.B. Stringer & P.A. Mellars), pp. 201–216. London: Royal Society (*Philosophical Transactions of the Royal Society*, series B, 337, no. 1280).

Close, A.E. 1978: The identification of style in lithic artefacts. *World Archaeology* 10, 223–237.

Cohen, M.N. 1977: *The Food Crisis in Prehistory: overpopulation and the origins of agriculture*. New Haven & London: Yale University Press.

Cohen, M.N. 1985: Prehistoric hunter-gatherers: the meaning of social complexity. In *Prehistoric Hunter-Gatherers: the emergence of cultural complexity* (ed. T.D. Price & J.A. Brown), pp. 99–122. Orlando: Academic press.

Damas, D. 1969: *Contributions to Anthropology: ecological essays*. Ottawa: National Museums of Canada, Bulletin 230.

David, N.C. 1973: On Upper Palaeolithic society, ecology and technological change: the Noaillian case. In *The Explanation of Culture Change* (ed. A.C. Renfrew), pp. 277–303. London: Duckworth.

Davidson, I. & Noble, W. 1989: The archaeology of perception: traces of depiction and language. *Current Anthropology* 30, 125–155.

Davidson, I. & Noble, W. 1993: Tools and language in human evolution. In *Tools, Language and Cognition in Human Evolution* (ed. K.R. Gibson & T. Ingold), pp. 363–387. Cambridge: Cambridge University Press.

Deacon, H.J. 1989: Late Pleistocene palaeoecology and archaeology in the southern Cape, South Africa. In *The Human Revolution: behavioural and biological perspectives on the origins of modern humans* (ed. P. Mellars & C. Stringer), pp. 547–564. Princeton: Princeton University Press.

Donald, M. 1991: *Origins of the Modern Mind: three stages in the evolution of culture and cognition*. Cambridge, Massachusetts: Harvard University Press.

Dyson-Hudson, R. & Smith, E.A. 1978: Human territoriality: an ecological reassessment. *American Anthropologist* 80, 21–41.

Farizy, C. 1990: The transition from Middle to Upper Palaeolithic at Arcy-sur-Cure (Yonne, France): technological, economic and social aspects. In *The Emergence of Modern Humans: an archaeological perspective* (ed. P. Mellars), pp. 303–326. Edinburgh: Edinburgh University Press.

Foley, R.A. 1989: The ecological conditions of speciation: a comparative approach to the origins of anatomically modern humans. In *The Human Revolution: behavioural and biological perspectives on the origins of modern humans* (ed. P. Mellars & C. Stringer), pp. 298–320. Princeton: Princeton University Press.

Freeman, L.G. & Echegaray, G. 1970: Aurignacian structural features and burials at Cueva Morín (Santander, Spain). *Nature* 226, 722–726.

Gambier, D. 1993: Les hommes modernes du début du Paléolithique supérieur en France: bilan des données anthropologiques et perspectives. In *El Origen del Hombre Moderno en el Suroeste de Europa* (ed. V. Cabrera Valdès), pp. 409–430. Madrid: Universidad Nacional de Educacion a Distancia.

Gamble, C. 1982: Interaction and alliance in palaeolithic society. *Man* 17, 92–107.
Gamble, C. 1986: *The Palaeolithic Settlement of Europe*. Cambridge: Cambridge University Press.
Gebauer, A.B. & Price, T.D. 1992: *Transitions to Agriculture in Prehistory*. Madison, Wisconsin: Prehistory Press.
Gellner, E. 1989: Culture, constraint and community: semantic and coercive compensations for the genetic under-determination of *Homo sapiens sapiens*. In *The Human Revolution: behavioural and biological perspectives on the origins of modern humans* (ed. P. Mellars & C. Stringer), pp. 514–525. Princeton: Princeton University Press.
Gibson, K.R. 1985: Has the evolution of intelligence stagnated since Neanderthal Man? (Commentary on Parker). In *Evolution and Developmental Psychology* (ed. G. Butterworth, J. Rutkowska & M. Scaife), pp. 102–114. Brighton: Harvester Press.
Gibson, K.R. & Ingold, T. (eds.) 1993: *Tools, Language and Cognition in Human Evolution*. Cambridge: Cambridge University Press.
Gowlett, J.A.J. 1984: Mental abilities of early man: a look at some hard evidence. In *Hominid Evolution and Community Ecology* (ed. R.A. Foley), pp. 167–192. London: Academic Press.
Hahn, J. 1972: Aurignacian signs, pendants, and art objects in Central and Eastern Europe. *World Archaeology* 3, 252–266.
Hahn, J. 1977: *Aurignacien: Das Ältere Jungpaläolithikum in Mittel- und Osteuropa*. Köln: Fundamenta series A9.
Hahn, J. 1993: Aurignacian art in Central Europe. In *Before Lacaux: the complex record of the early Upper Paleolithic* (ed. H. Knecht, A. Pike-Tay & R. White), pp. 229–242. Boca Raton: CRC Press.
Harrold, F. 1980: A comparative analysis of Eurasian Paleolithic burials. *World Archaeology* 12, 195–211.
Harpending, H., Sherry, S., Rogers, A. & Stoneking, M. 1993: The genetic structure of ancient human populations. *Current Anthropology* 34, 483–496.
Howell, F.C. 1994: A chronostratigraphic and taxonomic framework of the origins of modern humans. In *Origins of Anatomically Modern Humans* (ed. M. H. Nitecki & D. V. Nitecki), pp. 253–319 New York: Plenum Press.
Isaac, G.Ll. 1972: Early phases of human behaviour: models in Lower Palaeolithic archaeology. In *Models in Archaeology* (ed. D.L. Clarke), pp. 167–199. London: Methuen.
Jochim, M. 1983: Palaeolithic cave art in ecological perspective. In *Hunter-gatherer Economy in Prehistory: a European Perspective* (ed. G.N. Bailey), pp. 212–219. Cambridge: Cambridge University Press.
Johnson, G. 1982: Organizational structure and scalar stress. In *Theory and Explanation in Archaeology* (ed. C. Renfrew, M. Rowlands & B. Segraves), pp. 389–421. New York: Academic Press.
Keeley, L.G. 1988: Hunter-gatherer economic complexity and 'population pressure': a cross-cultural analysis. *Journal of Anthropological Archaeology* 7, 373–411.
Klein, R.G. 1989: Biological and behavioural perspectives on modern human origins in southern Africa. In *The Human Revolution: behavioural and biological perspectives on the origins of modern humans* (ed. P. Mellars & C. Stringer), pp. 529–546. Princeton: Princeton University Press.
Klein, R.G. 1994: The problem of modern human origins. In *Origins of Anatomically Modern Humans* (ed. M. H. Nitecki & D. V. Nitecki), pp. 3–17. New York: Plenum Press.
Knight, C.D. 1991: *Blood Relations: Menstruation and the Origins of Culture*. New Haven: Yale University Press.
Kozlowski, J.K. 1982: *Excavation in the Bacho Kiro Cave (Bulgaria): Final Report*. Warsaw: Panstwowe Wydawnictwo Naukowe.

Kozlowski, J.K. 1990: A multi-aspectual approach to the origins of the Upper Palaeolithic in Europe. In *The Emergence of Modern Humans: an archaeological perspective* (ed. P. Mellars), pp. 419–437. Edinburgh: Edinburgh University Press.

Kozlowski, J.K. 1992: The Balkans in the Middle and Upper Palaeolithic: the gateway to Europe or a cul de sac? *Proceedings of the Prehistoric Society* 58, 1–20.

Lee, R.B. & DeVore, I. 1968: *Man the Hunter*. Chicago: Aldine.

Lieberman, P. 1991: *Uniquely Human: the evolution of speech, thought, and selfless behavior*. Cambridge, MA: Harvard University Press.

Mellars, P.A. 1973: The character of the Middle-Upper Palaeolithic transition in south-west France. In *The Explanation of Culture Change: Models in Prehistory* (ed. C. Renfrew), pp. 255–276. London: Duckworth.

Mellars, P.A. 1982: On the Middle/Upper Palaeolithic transition: a reply to White. *Current Anthropology* 23, 238–240.

Mellars, P.A. 1985: The ecological basis of social complexity in the Upper Paleolithic of southwestern France. In *Prehistoric Hunter-Gatherers: the emergence of cultural complexity* (ed. T.D. Price & J.A. Brown), pp. 271–297. Orlando: Academic Press.

Mellars, P.A. 1989a: Major issues in the emergence of modern humans. *Current Anthropology* 30, 349–385.

Mellars, P.A. 1989b: Technological changes across the Middle-Upper Paleolithic transition: technological, social, and cognitive perspectives. In *The Human Revolution: behavioural and biological perspectives on the origins of modern humans* (ed. P. Mellars & C. Stringer), pp. 338–365. Princeton: Princeton University Press.

Mellars, P.A. 1991: Cognitive changes and the emergence of modern humans in Europe. *Cambridge Archaeological Journal* 1, 63–76.

Mellars, P.A. 1992: Archaeology and the population-dispersal hypothesis of modern human origins in Europe. In *The Origin of Modern Humans and the Impact of Chronometric Dating* (ed. M. Aitken, C.B. Stringer & P.A. Mellars), pp. 225–234. London: Royal Society (*Philosophical Transactions of the Royal Society*, series B, 337, no. 1280).

Mellars, P.A. 1994: The Upper Palaeolithic Revolution. In *The Oxford Illustrated Prehistory of Europe* (ed. B. Cunliffe), pp. 42–78. Oxford: Oxford University Press.

Mellars, P. 1996: *The Neanderthal Legacy: an Archaeological Perspective from Western Europe*. Princeton: Princeton University Press.

Mellars, P. & Gibson, K. (eds.) 1996: *Modelling the Early Human Mind*. Cambridge: McDonald Institute for Archaeological Research.

Mellars, P.A. & Stringer, C.B. (eds.) 1989: *The Human Revolution: Behavioural and Biological Perspectives on the Origins of Modern Humans*. Princeton: Princeton University Press.

Nitecki, M.H. & Nitecki, D.V. 1994: *Origins of Anatomically Modern Humans*. New York: Plenum Press.

Peterkin, G.L. 1993: Lithic and organic hunting technology in the French Upper Palaeolithic. In *Hunting and Animal Exploitation in the Later Palaeolithic and Mesolithic of Eurasia* (ed. G.L. Peterkin, H.M. Bricker & P. Mellars), pp. 49–68. Archaeological Papers of the American Anthropological Association No. 4.

Peterson, N. 1976: *Tribes and Boundaries in Australia*. Canberra: Australian Institute of Aboriginal Studies.

Pfeiffer, J.E. 1982. *The Creative Explosion: an enquiry into the origins of art and religion*. New York: Harper & Row.

Price, T.D. & Brown, J.A. 1985: Aspects of hunter-gatherer complexity. In *Prehistoric Hunter-Gatherers: the emergence of cultural complexity* (ed.. T.D. Price & J.A. Brown), pp. 3–20. Orlando: Academic Press.

Rodseth, L., Wrangham, R.W., Harrigan, A.M. & Smuts, B.B. 1991: The human community as a primate society. *Current Anthropology* 32, 221–254.

Rogers, A.R. & Jorde, L.B. 1995: Genetic evidence on modern human origins. *Human Biology* 67, 1–36.

Sackett, J.R. 1982: Approaches to style in lithic archaeology. *Journal of Anthropological Archaeology* 1, 59–112.

Sackett, J.R. 1988: The Mousterian and its aftermath: a view from the Upper Paleolithic. In *Upper Pleistocene Prehistory of Western Eurasia* (ed. H. Dibble & A. Montet-White), pp. 413–426. Philadelphia: University of Pennsylvania, University Museum Monographs No. 54.

Sahlins, M.D. 1972: *Stone Age Economics*. New York: Random House.

Scothern, P. 1993: *The Music-archaeology of the Palaeolithic within its Cultural Setting*. PhD Dissertation, University of Cambridge.

Service, E.R. 1962: *Primitive Social Organization*. New York: Random House.

Sherry, S.T., Rogers, A.R., Harpending, H., Soodyall, H., Jenkins, T. & Stoneking, M. 1994: Mismatch distributions of mtDNA reveal recent human population expansions. *Human Biology* 66, 761–775.

Smith, P.E.L. 1966: *Le Solutréen en France*. Bordeaux: Publications de l'Institut de Préhistoire de l'Université de Bordeaux, Mémoire 5.

Soffer, O. 1985: Patterns of intensification as seen from the Upper Paleolithic of the Central Russian Plain. In *Prehistoric Hunter-Gatherers: the emergence of cultural complexity* (ed. T.D. Price & J.A. Brown), pp. 235–270. Orlando: Academic Press.

Soffer, O. 1989: The Middle to Upper Palaeolithic transition on the Russian Plain. In *The Human Revolution: behavioural and biological perspectives on the origins of modern humans* (ed. P. Mellars & C. Stringer), pp. 714–742. Princeton: Princeton University Press.

Soffer, O. 1994: Ancestral lifeways in Eurasia—the Middle and Upper Paleolithic records. In *Origins of Anatomically Modern Humans* (ed. M. H. Nitecki & D. V. Nitecki), pp. 101–109. New York: Plenum Press.

Steward, J.H. 1972: *Theory of Culture Change*. Urbana: University of Illinois Press.

Stoneking, M., & Cann, R.L. 1989: African origin of human mitochondrial DNA. In *The Human Revolution: behavioural and biological perspectives on the origins of modern humans* (ed. P. Mellars & C. Stringer), pp. 17–30. Princeton: Princeton University Press.

Stoneking, M., Sherry, S.T., Redd, A.J. & Vigilant, L. 1992: New approaches to dating suggest a recent age for the human DNA ancestor. In *The Origin of Modern Humans and the Impact of Chronometric Dating* (ed. M. Aitken, C.B. Stringer & P.A. Mellars), pp. 167–176. London: Royal Society (Philosophical Transactions of the Royal Society, series B, 337, no. 1280).

Straus, L.G. 1983: From Mousterian to Magdalenian: cultural evolution viewed from Cantabrian Spain and the Pyrenees. In *The Mousterian Legacy* (ed. E. Trinkaus), pp. 73–111. Oxford: British Archaeological Reports International Series S164.

Straus, L.G. 1990: The early Upper Palaeolithic of southwest Europe: Cro-Magnon adaptations in the Iberian peripheries, 40,000–20,000 BP. In *The Emergence of Modern Humans: an Archaeological Perspective* (ed. P.A. Mellars), pp. 276–302. Edinburgh: Edinburgh Univeristy Press.

Stringer, C. & Gamble, C. 1993: *In Search of the Neanderthals: solving the puzzle of human origins*. London: Thames & Hudson.

Stringer, C. & Gamble, C. 1994: The Neanderthal World: flat earth or new horizon (with accompanying comments). *Cambridge Archaeological Journal* 4, 95–119.

Taborin, Y. 1993: Shells of the French Aurignacian and Perigordian. In *Before Lacaux: the complex record of the early Upper Paleolithic* (ed. H. Knecht, A. Pike-Tay & R. White), pp. 211–229. Boca Raton: CRC Press.

Testart, A. 1982: The significance of food storage among hunter-gatherers: residence patterns, population densities, and social inequalities. *Current Anthropology* 23, 523–537.

Thorne, A.G. & Wolpoff, M.H. 1992: The multiregional evolution of humans. *Scientific American* 266, 28–33.

Trinkaus, E. (ed.) 1989: *The Emergence of Modern Humans: biocultural adaptations in the later Pleistocene*. Cambridge: Cambridge University Press.

Whallon, R. 1989: Elements of cultural change in the later Palaeolithic. In *The Human Revolution: behavioural and biological perspectives on the origins of modern humans* (ed. P. Mellars & C. Stringer), pp. 433–454. Princeton: Princeton University Press.

White, R. 1989: Production complexity and standardization in early Aurignacian bead and pendant manufacture: evolutionary implications. In *The Human Revolution: behavioural and biological perspectives on the origins of modern humans* (ed. P. Mellars & C. Stringer), pp. 366–390. Princeton: Princeton University Press.

White, R. 1993: Technological and social dimensions of "Aurignacian-age" body ornaments across Europe. In *Before Lacaux: the complex record of the early Upper Paleolithic* (ed. H. Knecht, A. Pike-Tay & R. White), pp. 277–300. Boca Raton: CRC Press.

Wiessner, P. 1983: Style and social information in Kalahari San projectile points. *American Antiquity* 48, 253–276.

Wobst, M. 1974: Boundary conditions for Paleolithic social systems: a simulation approach. *American Antiquity* 39, 147–178.

Wobst, M. 1976: Locational relationships in Paleolithic society. *Journal of Human Evolution* 5, 49–58.

Wobst, M. 1977: Stylistic behavior and information exchange. In *Papers for the Director: Research Essays in Honor of James B. Griffin* (ed. C.E. Cleland), pp. 317–342. Ann Arbor: Anthropological Papers, Museum of Anthropology, University of Michigan No. 61.

Wolpoff, M.H., Thorne, A.G., Smith, F.H., Frayer, D.W. & Pope, G.G. 1994: Multiregional evolution: a world-wide source for modern human populations. In *Origins of Anatomically Modern Humans* (ed. M. H. Nitecki & D. V. Nitecki), pp. 175–199. New York: Plenum Press.

Woodburn, J.A. 1982: Egalitarian societies. *Man* 17, 431–451.

Wynn, T. 1989: *The Evolution of Spatial Competence*. Urbana: University of Illinois Press.

Responses to Environmental Novelty: Changes in Men's Marriage Strategies in a Rural Kenyan Community

MONIQUE BORGERHOFF MULDER
Department of Anthropology, University of California, Davis, CA 95616, USA

Keywords: Social change; Kenya; marriage payments; adaptation; behavioural ecology; Kipsigis.

Summary. This chapter examines the use of behavioural ecological models as applied to humans in conditions of environmental novelty. On the assumption that individuals pursue behavioural strategies that maximize their fitness, predictions can be made concerning how social and ecological conditions generate a variety of optimal responses. With environmental novelty, the question arises: For how much of human history (or prehistory) can we assume that the environment remained sufficiently unchanged for appropriate behaviour to be elicited and genuine functional outcomes to be observed? Data from rural Kenya show how Kipsigis men vary their allocations to mating effort, as measured through bridewealth payments, consistent with predictions from an optimality model. The pattern of men paying large bridewealth payments for women of high reproductive value and high labour value disappeared in the 1980s. This shift may reflect the changing reproductive and economic roles of women, contingent on incipient demographic transition in Kenya and an increasing involvement of men in food production and the cash economy. Despite some problems with the interpretation of data such as these, a generally positive appraisal is made of the appropriateness of using behavioural ecological theory in the study of contemporary human populations, both because it provides an empirical measure of the extent to which adaptive responses are

© The British Academy 1996.

still generated, and because it focuses attention on variability. These results challenge the view held by some evolutionary social scientists that there is no a priori reason to suppose that any specific modern cultural or behavioural practice is adaptive. At the same time these findings point to a potentially important area of congruence between behavioural ecology and evolutionary psychology, by highlighting the need to investigate decision-making rules that, on account of their sensitivity to new socioecological conditions, might contibute to the generation of cultural change.

INTRODUCTION

Debates within the evolutionary social sciences

WHILE EVOLUTIONARY APPROACHES find increasing acceptance within the social sciences (e.g., Lieberman 1989), there is still much debate over their appropriate uses and interpretation (Borgerhoff Mulder *et al.*, in press). One hotly contested issue is whether studies of *behavioural variability in contemporary populations* tell us anything about how evolution has shaped human decision making processes and human action. On the one hand, human behavioural ecologists are excited by the prospect of using contemporary ethnographic diversity to identify and test models of how humans strategically respond to varying ecological and social challenges (Smith 1992). On the other hand, evolutionary psychologists offer trenchant criticisms of some of the underlying assumptions of such a research agenda (e.g., Tooby & Cosmides 1990).

At the heart of this critique (examined in more detail in Borgerhoff Mulder *et al.*, in press) lies the issue of novel environments. For how much of human history (or prehistory) can we assume that the environment remained sufficiently unchanged for appropriate behaviour to be elicited and genuine functional outcomes to be observed? Scholars have delineated this period of time in various ways: an unspecified and variable age in the evolution of the mammals and primates, termed the 'environment of evolutionary adaptedness' (Symons 1979; Tooby & Cosmides 1990); the Pleistocene, as bounded by the Neolithic Revolution (Symons 1989); all periods up to the present (Caro & Borgerhoff Mulder 1987) or very recent past (Betzig 1989; Turke 1990). For none of these positions is there a clear rationale for specifying what is, or is not, a 'sufficiently unchanged' environment. Caro & Borgerhoff Mulder (1987), however, argue for the most radical position on methodological considerations: there are no grounds for an *a priori* determination of the environments in which either appropriate responses are unlikely to be elicited or genuine functional outcomes observed *without empirical investigation* (see also Perusse 1992).

This chapter breaks from the somewhat sterile debate that has surrounded the question of environmental novelty, and what it means for evolutionary accounts, by moving towards an empirical test of whether and how humans respond to environmental novelty in adaptive ways; adaptive here is defined as 'likely to be fitness-enhancing in the current environment'. It examines the allocation of male mating effort (in the form of bridewealth payments) in a rapidly changing rural Kenyan community. The full results of this longitudinal study of Kipsigis bridewealth have been published elsewhere (Borgerhoff Mulder 1995), but their significance for theoretical issues within the evolutionary social sciences was not explored in the original (strictly anthropological) treatment. In more general terms, this chapter has two goals. First, it defends a controversial use of evolutionary thinking within the social sciences by demonstrating how behavioural ecology can contribute to an understanding of human behavioural and institutional diversity. Second, it points to a potentially important area of congruence between behavioural ecology and evolutionary psychology, by highlighting the need to investigate decision-making rules that, on account of their sensitivity to features of the social and ecological environment, might shape cultural change (Cosmides & Tooby 1992: 219).

Human behavioural ecology

Human behavioural ecologists explore the function of behaviour in contemporary and past populations by looking at how behaviour is a response to specific ecological and social conditions, insofar as these vary both across societies, within societies, and over time. It has been extensively reviewed in easily available recent sources (Borgerhoff Mulder 1991; Cronk 1991; Smith 1992).

The theoretical underpinnings of behavioural ecology lie in evolutionary ecology, the branch of evolutionary theory that analyzes adaptations in ecological context. Studying behaviour in relation to the ecological and social environment flourished in the 1970s, following on the innovative theoretical work of Hamilton (1964), Williams (1966), Maynard Smith (1974) and Trivers (1971, 1972). These advances opened up new ways of looking at an animal's behaviour, particularly social behaviour. In brief, individuals are viewed as facultative opportunists who assess, either consciously or not, on either the behavioural or the evolutionary time scale, a wide array of environmental conditions (both social and ecological) and determine the optimal fitness-maximizing strategy whereby they can out compete conspecifics in terms of the number of genes transmitted to subsequent generations. As such, behavioural ecology relies on Darwinian theory in a homologous sense, and uses as theoretical tools individual selection, inclusive fitness theory, optimization models and game theory.

Two sets of assumptions are made. First, in order to develop models of optimal behaviour, behavioural ecologists typically ignore the nature of the genetic control of phenotypic design, and adopt a research strategy that has been dubbed 'the phenotypic gambit' (Grafen 1984; Smith 1992). This entails assuming that (a) natural selection can override such conflicting forces as drift, (b) sufficient genetic variation in the past has allowed evolution of the optimal phenotype, and (c) any deviations for simple Mendelian inheritance of phenotypes will not significantly affect the expected evolutionary outcome. Although disregard for the genetic control of adaptations has been criticized (Lewontin 1979), cases where the specifics of inheritance system might make a differences are thought to be rare (Maynard Smith 1982; Grafen 1984). Second, as regards the relationship between genetic differences and phenotypes, behavioural ecologists view behaviour as a flexible phenotype exhibiting reaction norms (Stearns 1989), with behaviour taking different character states in different environments. Hypothetically these different states equilibrate at different local optima, giving rise to the diverse patterns of behaviour that we see both between and within different human populations. Plasticity is a feature of great interest to evolutionists (e.g. Via 1993), and human behaviour can be viewed a prime example of this phenomenon.

The commonest research strategy within behavioural ecology, then, is to rely on the assumption that individuals behave in ways that maximize their fitness, and to develop models of how particular social and ecological configurations impact on an individual's optimal course of action. The goal is not to prove fitness maximization, but rather to use models predicated on this assumption to explain behavioural variability. As in the study of non-humans, predictions for human studies can be imported from general bodies of theory (e.g. Kaplan & Hill 1992), from comparative studies in other species (e.g. Daly & Wilson 1984), or from empirical observations on the determinants of fitness in a given population (e.g. Borgerhoff Mulder 1990).

Studies of humans do, however, raise the problem that most people now live in social and ecological circumstances that are highly altered from their 'traditional' form. While there is much debate over the extent to which any human population ever lived in a state of stability, isolation or equilibrium (Solway & Lee 1990), it is clear that recent social and ecological developments dramatically perturb the vast majority of contemporary human populations. Is it therefore reasonable to expect that men and women do still follow strategies that are fitness maximizing? This is really a double question. Is the environment sufficiently unchanged for appropriate behavioural decisions to be elicited by particular environmental cues? Second, is the environment sufficiently unchanged for expected fitness outcomes to be commonly observed?

As regards the first question, all evolutionary social scientists (by definition) suspect that there is sufficient continuity between ancient and contemporary environments to render legitimate the assumption that humans should still, by and large, reason and behave in ways that are predictable by fitness optimization models. If this were not the case there would be no value in studying contemporary subjects at all. As regards the second question behavioural ecologists and evolutionary psychologists differ. Only the former, and particularly those working in relatively unmodernized populations, are likely to give a positive response. In other words behavioural ecologists are prepared to examine variability from the standpoint of whether or not any given strategy maximizes fitness in *that particular environment* (contemporary or historical), whilst acknowledging that the psychological apparatus whereby adaptive behavioural responses are produced, such as emotions, basic motivations, learning and decision-making abilities, are products of *past* evolutionary pressures (e.g. Irons 1991). The unstated assumption of the behavioural ecologist, then, is that critical environmental features that both affect the expression of the phenotype *and* mediate its fitness consequences have been relatively stable, and are still present (Turke 1990).

One obvious empirical substantiation of the behavioural ecological position lies in investigating how humans respond to rapid conditions of social change. While this topic has for a long time intrigued social (and socio-cultural) anthropologists (e.g. Firth 1959) it has not, at least until recently (e.g. Cronk 1989), been examined by human behavioural ecologists. Voland *et al.* (in manuscript), for example, are trying to determine the lags with which peasant farmers respond to the changing sex-specific opportunities for their sons and daughters. In this study I investigate whether and how rural African men modify their allocations of mating effort in response to changing economic circumstances.

BACKGROUND TO THE KIPSIGIS STUDY

Social organization

Kipsigis are a Kalenjin-speaking Nilotic group living in south western Kenya. Their social organization is very typical of East Africa. Marriages are commonly polygynous, and households consist of related men, typically an elderly man, his wife (wives), his married sons with their wives, and other unmarried dependents. Descent is reckoned exclusively through the paternal line, and all heritable wealth (primarily land and livestock) are passed in equal shares to the sons of each of a man's wives.

Study site

The study area comprises two adjacent clusters of sub-locations within Abosi Location (Kericho District) and Moitanik East Location (Narok District), and is commonly referred to as Abosi, after a hill of that name. It includes land from the former Native Reserve and squatter communities settled during the 1930s. Rainfall is somewhat less than 1000 mm. per annum. Soils are primarily black cotton, of medium and marginal agricultural potential (Kericho District Development Plan 1989–93). In the study site there are three Primary Schools. A government Secondary School and a Catholic Mission Health Centre lie adjacent. Within approximately 45 km of Abosi there are three other secondary schools and two hospitals. The study was initiated in 1982, when 20 months were spent in the field, and then followed up in 1991 for a 3 month period.

Economy

In the pre-colonial period Kipsigis were agro-pastoralists, engaged in the subsistence production of eleusine and some sorghum. Animal husbandry (cattle and small stock) predominated over agriculture both economically and culturally (Peristiany 1939; Manners 1967). With the arrival in 1906/7 of the first European settlers in Kericho District, Kipsigis were relocated into densely-populated 'Native Reserves'. Maize was introduced as a subsistence and cash crop, and became increasingly profitable after World War One on account of the new market among labourers on European estates throughout the Kenya Colony (Manners 1962), and the pattern of individualized land ownership (with title deeds held by men) became legally encoded in the Swynnerton Plan of 1954. After 1960, when the ban against African ownership of European 'grade' cattle was lifted, commercial milk production soared, and was paralleled by a boom in the commercial production of maize (Daniels 1980). Concurrently the District population density escalated from an estimated $58-78/km^2$ to $267/km^2$ for 1993 (for references, see Borgerhoff Mulder 1995).

Women's labour has (until very recently, see below) been critical to production systems such as that of the Kipsigis. Traditionally women were almost exclusively responsible for the production of vegetable food (Peristiany 1939). During the colonial period, as maize became a cash commodity, women continued to maintain responsibility for the labour entailed in its cultivation, even through cash derived from the valuable maize surpluses fell almost exclusively into the hands of men (Sørensen 1990; Davison 1988).

Marriage and marriage payments

Among Kipsigis a single bridewealth payment (from groom's family to bride's family) is made at the time of marriage with no formal expectation of protracted or return payments. The lump sum is constituted of livestock and (since the late 1950s in Abosi) cash. The payments are highly variable, and can be thought of as allocations to mating effort made by men with respect to the kind of wife that they attain. As is common in much of East Africa, first marriages are paid for by the groom's father, and subsequent marriages by the groom himself, although nowadays employed sons contribute some cash to their first marriage payment. The bride's parents are primarily responsible for the negotiation and final acceptance of the bridewealth offer of a potential son-in-law (for details of the negotiation process, see Borgerhoff Mulder 1988).

Bridewealth payments: previous results and new questions

A study of Kipsigis bridewealth payments transferred between 1960 and 1982 showed, amongst other things, that men paid more for women who reached menarche early (Borgerhoff Mulder 1988). Insofar as early maturing Kipsigis women have higher reproductive success than later maturing women, due to their longer reproductive lifespans and higher fertility rates, this result was interpreted as adaptive variation in the exertion of male mating effort, with men paying particularly heavily for women of high reproductive value (Borgerhoff Mulder 1989).

The original study also showed that men paid more heavily for brides who lived outside the small geographically defined local community. In conjunction with the finding that women whose natal homes were distant offered a more reliable source of labour than did women whose natal homes were local (the latter spent disproportionate amounts of time helping out their mother rather than working in their marital home), an adaptive rationale was also placed on this result: high payments for distant brides reflect the tendency for men to make heavy bridewealth payments for women who offered high labour value.

These results, and more precisely their interpretation, lead me to predict that the covariates of bridewealth are likely to vary between different societies, depending on the specifics of the systems of reproduction and production (Figure 1). Specifically, I suggested that 'in societies where it is

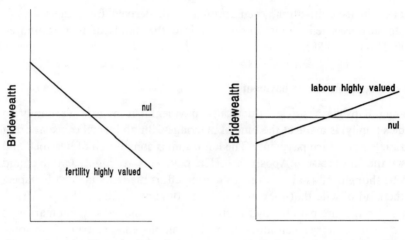

Age at menarche (Reproductive value)   Marital distance (Labour value)

Figure 1. Some hypothetical effects of reproductive value and labour value on bridewealth. (A) Using age of bride at menarche as an indicator of reproductive value the two functions describe how the differential evaluation of reproductive value in different populations might affect negotiated bridewealth outcomes. (B) Using distance between natal and marital home as a measure of the alienation of women's labour services, the two functions describe how differential evaluation of women's labour in different populations might affect negotiated bridewealth; see text for details. (Modified from Borgerhoff Mulder 1988.)

important to produce a large number of children, for example where offspring mortality is high, or where children are not excessively costly to raise, as in labour-intensive systems of production, female reproductive potential will be highly valued' (Borgerhoff Mulder 1988: 78). Alternatively, 'where production of a large family is not feasible, either because children constitute a net cost or because production is limited by capital rather than labour assets, women of high reproductive value may not be more expensive than those of lower reproductive value'.

In similar vein, I predicted that interpopulational differences in the importance of women's economic roles would be associated with the covariates of bridewealth. 'Where women play a critical role in a labour-intensive mode of production, men will pay highly for total control over their wives' labour. Alternatively, where women's labour inputs are less important, differential access to a wife's labour services may not be associated with any variability in bridewealth' (Borgerhoff Mulder 1988: 79).

In the original paper some qualitative supportive evidence was brought from other ethnographic studies to support these propositions. In 1992 I had the opportunity to return to Kenya and test these hypotheses, not across populations as I had originally intended but using longitudinal data on the Kipsigis. Specifically I designed a study by means of which data on bridewealth payments prior to my previous field study could be compared with similar data for the last decade. By way of introduction to these new results, a summary of some of the pertinent economic and social changes in Abosi during the last decade will be outlined.

LONGITUDINAL ANALYSES

Changed historical conditions

Kenya in the 1980s showed signs of reduced economic expansion, with a slowing in the growth rate of real Gross Domestic Product, weakened investment, a hike in inflation rates, and rising costs of agricultural production. Indeed the only boost in the economy was in employment, particularly the informal sector (Economic Survey 1991). The causes of these changes are complex, but are clearly linked to political instability, the retraction of foreign aid programmes to Kenya and the worldwide economic recession. For rural areas such as Abosi, these developments lead to increasing poverty, greater differentiation in wealth holding, and an emergence of a wealthy strata of landholders with sufficient land on which to support one or more families (for details, see Borgerhoff Mulder 1995, Table 1).

Since the early 1980s, for reasons that are unclear, less maize is being sold to the national marketing board, and for a much lower price. Consequently many families are turning to the profitable sale of milk, and a local cooperative has been formed to deliver morning milk to a dairy processing plant 40 km away. Men, retaining their traditional exclusive rights to morning milk, gain directly from the cash thereby generated. In addition, several younger men have begun raising cash through the sale of chickens, eggs and vegetables, commodities traditionally reserved for trade by women; with bicycles, men can take larger quantities to local markets, thereby out competing women (who walk with produce on their backs or in

Table 1. Changes in percentage of income from the economic activities of men and women.[a]

Source of income	1982–83	1991	Sex-specific responsibility
Sale of surplus milk	57	25	Primarily women
Sale of vegetables, etc.	5	12	Men's participation increasing
Sale of surplus milk	21	35	Mainly men
Employment	17	28	Almost exclusively men

[a] Data with interviews with household heads and their wives ($n = 98$, $n = 88$, for 1982–83 and 1991 respectively; 1982 and 1983 scores averaged).

Table 2. Interview responses of women concerning family limitation.

	Date of interview	
	1982–91	1991
Desired family size—modal value		
Women aged:		
20–24 ($n = 18$)	8	7
25–29 ($n = 16$)	8	6
Use of non-traditional birth control methods		
All women		
($n = 114$)	0	2

baskets). Kipsigis of Abosi see this intensification and diversification in cash-producing activities as resulting from increased land shortages. The key social effects of these changes in the Abosi economy over the last decade are that (a) less of the family income depends on women's work in maize production, and (b) men are responsible for a greater proportion of the household's cash income. The data in Table 1 confirm this conclusion: in the 1990s, earnings from maize appear to contribute less income to the household than do earnings from local marketing of other produce and from off-farm employment, both activities pursued almost exclusively by men.

Finally, with respect to demographic patterns, fertility limitation is finally in evidence in rural Kenya. Land already seen as in short supply in the early 1980s is now the focus of intense inter-ethnic conflicts in many parts of Kenya, including Kericho District. Family limitation, not commonly discussed in the early 1980s, is increasingly seen as a necessary response to the daunting challenge of feeding families of seven or more surviving children on farms often averaging little more than 2 or 3 hectares. Though I have no detailed analyses of changes in fertility in the Abosi sample, desired family size has declined (Table 2). More generally, total fertility declined from 8.2 (1973–77) to 6.7 (1984–88), with Kericho District showing a typical 22% decline (Brass & Jolly 1993). In addition it seems as if the rate of decrease accelerated in the late 1980s. The general interpretation of this finding is that Kenyans are producing fewer live births and investing more heavily in the survival and education of those that they do produce, in response to reduced land availability and the increased costs of (and payoffs to) education. Government programmes in health and education, as well as private family planning organizations, have clearly facilitated this transition.

Sample and methods

Data on 248 marriages are divided into four time blocks (1952–61, 1962–71, 1972–81 and 1982–91). Payments for marriages in which the bride was reported pregnant, had already produced a child, or was being married by another woman were excluded ($n = 19$), since these are typically characterized by low bridewealth and constrained negotiation (e.g. Borgerhoff Mulder 1988). From the remaining cases, those for which the full array of independent variables are not available are dropped, leaving a resulting sample of 200 marriage payments, and a reduced sample for those with data on menarcheal age (see below).

Marriages in which the bride was the first, second or subsequent wife are all analyzed together since bridewealth values do not vary according to marital status in either this or the earlier study. Payments for secondary wives are highly variable: some are very expensive on account of the wealth of the suitor; some are very cheap, on account of the higher incidence of pregnant women or unmarried mothers falling into this category (see similarly for the Sebei, Goldschmidt 1974); controlling for these factors shows no overall influence of marital status *per se*.

Statistics were calculated using SPSSpc, with 2-tailed significance values reported. Analyses of variance were conducted, with coefficients of linearity, main effects and interaction effects reported in the figure legends. All main and interaction effects reported are independent of other significant factors affecting bridewealth at the appropriate period in this population; the multivariate analyses showing the independence of these effects are reported elsewhere (Borgerhoff Mulder 1995, Appendix).

Reproductive value

In the original study, menarche was dated to the year preceding clitoridectomy, according to the Kipsigis custom of sending girls for their initiation in the December following first menses. There are increasing difficulties in dating menarche to the year preceding clitoridectomy because of the tendency of young women pursuing education to defer their initiation ceremony until they finish primary school. Thus only girls who failed to proceed beyond Standard 5 were used for this analysis; the typical pattern among these girls was for them to leave school at between 12 and 14 years of age, and then undergo clitoridectomy. In this sub-sample then we can be reasonably certain that age at clitoridectomy provides a good indicator of age at menarche. Furthermore in 33% of this sub-sample I was able to confirm *without exception* that clitoridectomy had been precipitated by first menses (as reported by the subject or her mother) within the preceding 11 months.

The effects of menarcheal age on bridewealth transferred at marriages that occurred between 1952 and 1981 is confirmed in this sub-sample. Prior to 1982 high bridewealth was paid for women who reached menarche at a relatively young age; these women enjoy longer reproductive life spans and

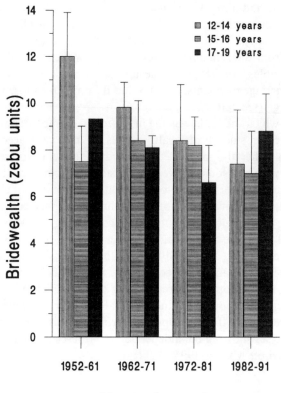

Figure 2. Bridewealth and bride's age at menarche ($n = 101$). Age at menarche classified as 12–14 years, 15–16 years, and 17–19 years. Analysis of variance shows a main effect of age at menarche ($F_{2,95} = 5.83$, $P = 0.004$) and an interaction with time block categorized as 1952–81 and 1982–91 ($F_{2,95} = 4.28$, $P = 0.017$). For explanation for reduced sample size, see text. (From Borgerhoff Mulder 1995.)

higher fertility than do later maturers. The relationship however disappears in the last decade (Figure 2). Grooms are no longer willing to offer large payments for early maturing women.

Labour value

A very similar effect is seen when we consider marital distance. Prior to 1982 higher payments were made for women whose natal homes lay outside of the local community than for women who came from nearby; brides from distant marital homes offer a more reliable labour service to their husband and his kin than do women whose natal homes are close by. In recent times, however, the pattern appears to be changing (Figure 3). Men are no longer

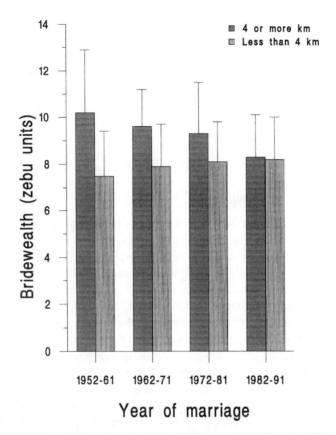

Figure 3. Bridewealth and distance between natal and marital homes ($n = 200$). Marital distance classified as less than 4 km or 4 km or more. Analysis of variance shows a main effect of marital distance ($F_{1,196} = 12.081$, $P < 0.001$) and an interaction with time block categorized as 1952–81 and 1982–91 ($F_{1,196} = 7.75$, $P < 0.006$). (From Borgerhoff Mulder 1995.)

willing to offer large payments for women who are particularly likely to make a reliable labour contribution.

Summary and interpretation

There has been a marked change in how both age at menarche and marital distance are associated with bridewealth. While grooms used to pay highly for brides of who reached menarche early and for brides whose natal home was distant, this pattern disappears in the 1980s. This shift in the covariates of bridewealth may mirror the changing economic and reproductive roles of women. Specifically, the loss of the menarche effect may be associated with a devaluation in the value of women's fertility, indicated by the overall decline in large families at both the national and district level. In similar vein the loss of the marital distance effect may result from a devaluation of women's productive work, reflecting larger structural changes, in particular the drop in the price of maize, the new market orientation in dairying, increased agricultural intensification (particularly by men), and the fact that it is increasingly land and not labour that limits production.

DISCUSSION

Support for the optimality model

The simplest interpretation of these findings is that men modify their mating effort in response to changing circumstances, such as new norms pertaining to fertility and alterations in the division of labour between men and women. These modifications are at least qualitatively consistent with an optimality model. In other words, although such features as birth control technology, a crash in the national maize market, and worldwide economic recession are clearly twentieth century phenomena, the *impact* of these developments on the marriage payments of rural Kenyans are somewhat as one would predict from simple optimality reasoning, predicated on the assumption of fitness maximization. These results therefore challenge the view held by some evolutionary psychologists that there is no a priori reason to suppose that any specific modern cultural or behavioural practice is adaptive (fitness maximizing). From the perspective of behavioural ecology, such a view precludes empirical investigation of interesting, and as yet unanswered questions about the range of human norms of reaction.

Problems of interpretation

Clearly there are many shortcomings to an analysis and interpretation such as this, problems which only far more detailed longitudinal data can remedy. First, the causes of the change in male mating effort allocation are attributed entirely on the basis of correlations between broader socio-economic developments and the changing covariates of bridewealth. Furthermore, there is in effect only a sample size of two periods (1952–1981, and 1982–1991). Conceivably then the decline in bridewealth paid for early maturers may bear no relationship whatsoever with the current demographic transition in Kenya, or with the hypothesized decline in the value of women's fertility. As with all naturalistic studies, we have to rely on inference (and all its inherent shortcomings), rather than experiment. However in the absence of any other obvious explanation for why Kipsigis men should no longer place such high value on early maturing women (see below), consistency with a deductively derived model should lend some credence to the interpretation.

Second, this study does not show whether by modifying their mating effort allocations in this way, Kipsigis men are in fact increasing their fitness (conventionally measured as number of descendants). This raises the much broader question, quite beyond the empirical scope of this paper, of whether or not parents can increase their fitness by decreasing the number of their births. While this is an obvious prediction from Lack's (1968) work on optimal clutch size, and has commonly been adduced as a rationale for demographic transition in human societies (Turke 1989), there is as yet no clear evidence that reductions in the F1 generation serve to enhance overall numbers of descendants, at least as determined from longitudinal studies in western populations (Kaplan *et al.* 1995; no such long term data are available for rural Kenyans, since fertility limitation is only in its very early stages). It is clear, however, that in many rural communities in the developing world, particularly those in which resources are already severely stretched, an offspring's chances in life are negatively affected by large family size (for Kipsigis, unpublished results). Therefore placing value on, for example, a woman's education (see Borgerhoff Mulder 1995) rather than her reproductive value may in the long term prove to be fitness-enhancing; but clearly this is unsubstantiated at present.

Alternative explanations

To what extent can we rule out alternative explanations for the disappearance of the marital distance and age at menarche effects reported here? First, it is possible that the declining wealth of the people of Abosi (Borgerhoff Mulder 1995) leads to a reduction in the *variability* in bridewealth payments, and hence to the likelihood of a correlation being observed. This is not the case: although there has been an overall decline in the size of payments in the last decade (ibid), the variability has remained stable. This therefore seems an inadequate explanation.

Second, it may be that young couples in the 1990s have more autonomy with respect to their marital choices than they did in the past. Under such circumstances, they may have a greater role in the negotiation of their bridewealth transaction, with personal (and perhaps idiosyncratic or unquantifiable) qualities counting for more than, for example, marital distance and age at menarche. While there is some qualitative indication that younger couples do take more initiative in their choice of spouse than formerly, my suspicion is that this freedom is still strongly limited by, in the case of sons, the threat of disinheritance, and in the case of daughters, their almost complete dependence on their mothers and fathers; no unmarried women in the study site had any significant source of independent income. Furthermore, the details of the bridewealth negotiation remain, almost entirely, within the ambit of the senior generation, following many of the same procedures and traditions described over 50 years ago by Peristiany (1939). In addition, the fact that bridewealth payments covary with socioeconomic characteristics of the bride and groom (education, and wealth and education respectively) in ways entirely compatible with the explicit strategies of parents-in-law (Borgerhoff Mulder 1995), indicates that it is still primarily the senior generation that is responsible for negotiated bridewealth outcomes. In short, young couples may have more autonomy these days, but the politics surrounding bridewealth are still primarily the province of parents and elders.

Understanding of mechanism

Even the original finding, that men pay more for women who reach menarche early, raised problems in terms of the *mechanisms* underlying such a decision rule. First, Kipsigis were not aware of the fact that they made higher payments for early maturing girls. Second, even if they had been aware of this tendency, they were rarely cognizant of the exact age of one another; indeed I spent much time ageing women, establishing the dates of their birth and clitoridectomy by means of age ranking, cross-comparisons, and cross-referencing to externally verifiable events. To what mechanisms can the correlation between menarcheal age and bridewealth be attributed, if most people are unaware of exact age?

Detailed analysis of bridewealth payments that were under negotiation while I was in the field (1982–83) showed that other more directly observed phenotypic traits affected the size of the payment. Brides classified as 'skinny' were signficantly cheaper than those classified as 'plump' (Borgerhoff Mulder 1988). Since there is some association, if not direct, between fat deposition and early menarche, I suggested that a woman's shape and body form might act as a cue whereby menarcheal age, and subsequent reproductive success, could be assessed. This could lead to a rule of thumb: pay more for fatter than thinner brides (for detailed support of this argument, see Borgerhoff Mulder 1988).

But what mechanisms might underlie the *changing* covariates in bridewealth payments? Can we really expect that men should be motivated to pay heavily for women classified as fat rather than thin in one decade, and not in the following decade, (and similarly with respect to marital distance)? Even though the empirical findings suggest this is the case (and theory predicts such a response), the mechanistic underpinnings of this behaviour remain totally obscure. This would seem to be a prime example of how studies of function can and should inform studies of mechanism, as recognized by Cosmides & Tooby (1992: 218). The challenge now is to conduct a far more detailed study of mate preferences, or perhaps daughter-in-law preferences, using the methods developed by evolutionary psychologists within the framework of functionally-inspired behavioural ecological anthropology.

The contribution of human behavioural ecology

Evolutionary social sciences consist of three separate though ideally complementary fields: cultural inheritance theory, evolutionary ecology and human behavioural ecology. Each has a distinct and powerful contribution to make towards an understanding of the forces that have shaped human evolution and diversity (Blurton Jones 1990). The particular strength of human behavioural ecology, as argued in this paper, is to explore diversity and change within the paradigm of an optimality approach. On the somewhat onerous assumption that individuals, even in contemporary populations, behave in ways that maximize their fitness, predictions can be made regarding how environment and conspecifics shape different individuals' courses of action in different ways. Under perfect field conditions this methodology allows an empirical determination of the limits of adaptation—how far can the environment be varied before appropriate adaptive responses producing true functions are no longer observed? Under normal (and rather less perfect) fieldwork conditions, such as those reported here, this methodology reveals novel findings that stimulate further questions about fitness and mechanism, questions that probe the validity of the onerous assumption noted above. If this assumption, so widely appropriate for all other life forms on earth, is to be modified, it seems greatly preferable to do so on the basis of *empirical data*, rather than on the basis of some idiosyncratic judgment, as for example denoting the Neolithic revolution as a critical threshold beyond which human behaviour can no longer be explained in terms of its likely fitness consequences. More generally, and with many more studies such as this, we can start to build a more complete picture of how and why humans do, and do not, respond to novel conditions in adapative ways.

A second important contribution from behavioural ecology is that it challenges the somewhat stereotypic characterizations of, for example, mating strategies that have dominated much recent work in evolutionary psychology. Researchers in this area tend to look for the kinds of sex differences that are typical of mammalian patterns of reproduction and parental care (Trivers 1972), and therefore expect men to be particularly concerned with beauty and health, and women with wealth and ambition. In fact there is plenty of evidence that sex specific preferences and strategies are highly variable, both across cultures and over historical periods. The data presented here show subtle variability over only a very short time scale, at least in terms of what men are willing to pay for in women. Exploring this variability remains the principal challenge of human behavioural ecology, and will hopefully become more central to the design of research within evolutionary psychology.

Note. The figures are reproduced from the paper by Borgerhoff Mulder (1955) in *Current Anthropology* published by the University of Chicago: © 1995 by the Wenner-Gren Foundation for Anthropological Research. All rights reserved.

REFERENCES

Betzig L.L. 1989: Rethinking human ethology. *Ethology and Sociobiology* 10, 315–324.

Blurton Jones, N.G. 1990: Three sensible paradigms for research on evolution and human behavior? *Ethology and Sociobiology* 11, 353–359.

Borgerhoff Mulder, M. 1988: Kipsigis bridewealth payments. In *Human Reproductive Behaviour*. (ed. L.L. Betzig, M. Borgerhoff Mulder & P.W. Turke), pp. 65–82. Cambridge: Cambridge University Press.

Borgerhoff Mulder, M. 1989: Early maturing Kipsigis women have higher reproductive success than later maturing women, and cost more to marry. *Behavioral Ecology and Sociobiology* 24, 145–153.

Borgerhoff Mulder, M. 1990: Kipsigis women's preferences for wealthy men: Evidence for female choice in mammals. *Behavioral Ecology and Sociobiology* 27, 255–264.

Borgerhoff Mulder, M. 1991: Human behavioural ecology. In *Behavioral Ecology: An Evolutionary Approach*, 3rd ed. (ed. J.R. Krebs & N.B. Davies), pp. 69–98. London: Blackwell Scientific Publications.

Borgerhoff Mulder, M. 1995: Bridewealth and its correlates: Quantifying changes over time. *Current Anthropology* 36, 573–603.

Borgerhoff Mulder, M., Thornhill, N.W., Voland, E. & Richerson, P.J. (In press). The place of behavioural ecology in the evolutionary social sciences, in *Human by Nature: Between biology and the social sciences* (ed. P.S. Weingart, S.D. Mitchell, P.J. Richerson & S. Maasen). Erlbaum Press.

Brass, W. & Jolly, C.L. (eds.) 1993: *Population Dynamics of Kenya*. Washington: National Academy Press.

Caro, T.M. & Borgerhoff Mulder, M. 1987: The problem of adaptation in the study of human behaviour. *Ethology and Sociobiology* 8, 61–72.

Cosmides, L. & Tooby, J. 1992: Cognitive adaptations for social exchange. In *The Adapated Mind* (ed. J.H. Barkow, L. Cosmides & J. Tooby), pp. 163–228. Oxford: Oxford University Press.

Cronk, L. 1989: From hunters to herders: Subsistence change as a reproductive strategy among the Mukogodo. *Current Anthropology* 30, 224–234.

Cronk, L. 1991: Human behavioral ecology. *Annual Review of Anthropology* 20, 25–53.
Daly, M & Wilson, M. 1984: A sociobiological analysis of human infanticide. In *Infanticide: Comparative and evolutionary perspectives* (ed. G. Hausfater & S.B. Hrdy), pp. 487–502. New York: Aldine de Gruyter.
Daniels, R.E. 1980: Pastoral values among vulnerable peasants. In *Predicting Sociocultural Change* (ed. S. Abbott & J. van Willigen), pp. 57–75. Athens: University of Georgia Press.
Davison, J. 1988: Who own what? Land registration and tensions in gender relations of production in Kenya. In *Agriculture, Women and Land: The African experience* (ed. J. Davison), pp. 157–176. Boulder: Westview Press
Economic Survey 1991: Republic of Kenya. Central Bureau of Statistics/Ministry of Planning and National Development.
Firth, R 1959: *Social Change in Tikopia: Re-study of a Polynesian community after a generation.* London: Allen & Unwin.
Goldschmidt, W. 1974: The economics of bridewealth among the Sebei in East Africa. *Ethnology* 13, 311–33.
Grafen, A. 1984: Natural selection, kin selection and group selection. In *Behavioural Ecology: An evolutionary approach*, 2nd ed. (ed. J.R. Krebs & N.B. Davies), pp. 62–84. Sunderland, MA: Sinauer Associates.
Hamilton, W.D. 1964: The genetical evolution of social behaviour, I and II. *Journal of Theoretical Biology* 7, 1–52.
Irons, W. 1991: Anthropology. In *The Sociobiological Imagination* (ed. M. Maxwell), pp. 79–100. New York: SUNY Press.
Kaplan, H. & Hill, K. 1992: The evolutionary ecology of food acquisition. In *Evolutionary Ecology & Human Behavior* (ed. E.A. Smith & B. Winterhalder), pp.167–201. New York: Aldine de Gruyter.
Kaplan, H.S., Lancaster, J.B., Bock, J.A. & Johnson, S.E. 1995: Fertility and fitness among Albuquerque men: A competitive labour market theory. In *Human Reproductive Decisions: Biological and Social Perspectives* (ed. R.I.M. Dunbar), pp. 96–136. London: Macmillan Press.
Kericho District Development Plan 1989–1993: Ministry of Planning and National Development. Republic of Kenya.
Lack, D. 1968: *Ecological Adaptations for Breeding in Birds.* London: Methuen.
Lewontin, R.C. 1979: Sociobiology as an adaptationist paradigm. *Behavioral Science* 24, 1–10.
Lieberman, L. 1989: A discipline divided: Acceptance of human sociobiological concepts in anthropology. *Current Anthtropology* 30, 676–682.
Manners, R.A. 1962: Land use, labor, and the growth of market economy in Kipsigis country. In *Markets in Africa* (ed. P. Bohannon & G. Dalton), pp. 493–519. Evanston: Northwestern University Press.
Manners, R.A. 1967: The Kipsigis of Kenya: Culture change in a 'model' East African Tribe. In *Contemporary Change in Traditional Societies, Volume I: Introduction and African tribes* (ed. J. Steward), pp. 207–359. Urbana: University of Illinois Press.
Maynard Smith, J. 1974: *Models in Ecology.* Cambridge: Cambridge University Press.
Maynard Smith, J. 1982: *Evolution and the Theory of Games.* Cambridge: Cambridge University Press.
Peristiany, J.G. 1939: *The Social Institutions of the Kipsigis.* London: Routledge & Kegan Paul.
Perusse, D. 1992: Cultural and reproductive success in industrial societies: Testing the relationship at the proximate and ultimate levels. *Behavioral Brain Sciences* 16, 267–322.
Smith, E.A. 1992: Human behavioral ecology. 1 & 2. *Evolutionary Anthropology* 1, 20–25, 50–55.

Solway, J.S & Lee, R.B. 1990: Foragers, genuine or spurious? *Current Anthtropology* 31, 109–146.
Sørensen, A. 1990: The differential effects on women of cash crop production. Centre for Development Research, Copenhagen, Denmark.
Stearns, S.C. 1989: *The Evolution of Life Histories*. Oxford: Oxford University Press.
Symons, D. 1979: *The Evolution of Human Sexuality*. New York: Oxford University Press.
Symons D. 1989: A critique of Darwinian anthropology. *Ethology and Sociobiology* 10, 131–144.
Tooby, J. & Cosmides, L. 1990: The past explains the present: Emotional adaptations and the structure of ancestral environments. *Ethology and Sociobiology* 11, 375–424.
Trivers, R.L. 1971: The evolution of reciprocal altruism. *Quarterly Review of Biology* 46, 35–57.
Trivers, R.L. 1972: Parental investment and sexual selection. In *Sexual Selection and the Descent of Man, 1871–1971* (ed. B. Cambell), pp. 136–179. Chicago: University of Chicago Press.
Turke, P.W. 1989: Evolution and the demand for children. *Population and Development Review* 15, 61–90.
Turke, P.W. 1990: Which humans behave adaptively, and why does it matter? *Ethology and Sociobiology* 11, 305–339.
Via, S. 1993: Adaptive phenotypic plasticity: Target or by-product of selection in a variable environment? *American Naturalist* 42, 352–365.
Voland, E., Dunbar, R.I.M., Engel, C. & Stephan, P. (In manuscript) Population increase and sex-biased parental investment in humans: Evidence from 18th and 19th century Germany. Submitted to *American Naturalist*.
Williams, G.S. 1966: *Adaptation and Natural Selection*. Princeton, NJ: Princeton University Press.

Genetic Language Impairment: Unruly Grammars

M. GOPNIK, J. DALALAKIS, S.E. FUKUDA, S. FUKUDA & E. KEHAYIA

McGill University, Department of Linguistics, 1001 Sherbrooke St. West, Montreal, Quebec, Canada, H3A-1G5

Keywords: specific language impairment; language instinct; compensatory mechanism; automatic procedural rule; cross-linguistic investigation; linguistics.

Summary. Though most children acquire language quickly and easily, some children have great difficulty in acquiring their native language. Over the past several years family studies and epidemiological studies have shown that this disorder aggregates in families. It has also been shown that monozygotic twins are significantly more concordant with respect to this disorder than dizygotic twins. This evidence suggests that at least some cases of this disorder may be heritable.

This paper will report on the linguistic properties of this disorder in English, Japanese and Greek. The cumulative data from several years of testing across several languages show that these subjects do not construct productive rules for such linguistic features as tense or case. They are able to use compensatory mechanisms such as conceptual selection, memorization and explicitly learned rules in place of automatic, procedural rules that normally govern language. These data suggest that this genetic disorder affects normal language learning mechanisms.

© The British Academy 1996.

THE IDEA OF LANGUAGE AS AN INSTINCT

Darwin

CHARLES DARWIN thought that the ability to learn language was a special kind of instinct. '[Language] certainly is not a true instinct, for every language has to be learned. It differs, however, widely from all other arts, for man has an instinctive tendency to speak, as we see in the babble of our young children.' (Darwin, 1874). What we will show here is that this special 'instinct' can be impaired in some individuals. When it is impaired these children have to learn language without the help of their language instinct (Pinker 1994). They can learn by using other cognitive means, but the language that they construct shows characteristic differences from normal language.

Alternative models of language

Darwin's suggestion that language was founded on a biological instinct was lost for a while in a cloud of cultural relativism, which held that all languages (as well as all cultures) were radically different from one another, and behaviourism, which held that scientific explanations must not refer to non-observable entities like minds. The relativist viewpoint led linguists to concentrate on cataloguing the oddities that distinguished one language from another rather than the commonalties which they all shared. In fact Bloomfield specifically said that it would be inappropriate to postulate any universals of language: 'The only useful generalizations are inductive generalizations. Features which we think ought to be universal may be absent from the very next language that becomes accessible' (Bloomfield, 1933: 20). Linguists who adopted the behaviourist stance denied the existence of any such thing as 'mind' and they therefore did not believe in postulating any internalized abstract rules or representations.

Chomsky

Thirty-five years ago Chomsky challenged both relativism and behaviourism as explanations for language and revived the biological story. He said that the ability to acquire language is part of the biological endowment of human beings. He claimed that children come equipped with innate knowledge of the principles that are necessary for constructing human grammars. It is this set of universal principles which provides children with a blueprint for language learning. Children take this general plan and adapt it to fit the particular, specific properties of the language that they hear around them.

Chomsky as a linguist was able to provide empirically testable, specific proposals about the grammatical shape of the language instinct. There are several interrelated consequences to his proposal.

1. All languages built on same principles

If language is a consequence of an innate system then all human languages must be built on the same set of fundamental principles, despite their apparent surface differences. If there were a language that had a structure that violated these universal principles then minds like ours could not have produced it and would not be able to learn it. Linguists have investigated languages all over the world, from remote villages to busy cities, to see if this were true. These empirical studies have led to some revisions in the original proposals for the universal properties of language, but all in all it looks like there is a constrained set of principles that underlies all languages.

2. Modularity

The question is, 'How specific is this set of principles'. Chomskians argue that the principles that govern language design are much more specific than those that are used in general problem solving. They hold that the instinct to learn language is special and modular and is not merely a consequence of general intelligence (Fodor 1983). Others, like some connectionists, have argued that language learning is no different than any other kind of learning. They claim that learning language does not depend on a Universal Grammar, but rather on a Universal Learning Machine that can take anything as input and find the regularities within the system (Rumelhart & McClelland 1986). So far the jury is still out. The connectionists have built computer models that they claim 'learn language' using only general principles of inference. Linguists have shown that these models do not learn in the same way as children do, and besides there are properties of language that cannot be learned by these devices (Pinker & Prince 1988). The connectionists have changed their models to accommodate some of these objections, but some linguists still argue that there are properties of language that even the new connectionist models cannot account for. A more general problem with the connectionist models that have been proposed is whether they really reflect the way in which people think (Holyoak & Thagard 1995). Even if it can be shown that they are not merely interesting computational models, but are in fact accurate models of human cognition it is still an open question as to whether these models can, in principle, account for the complexities of human language. One of the crucial pieces of evidence that will help decide this is the empirical evidence that we can gather from all aspects of language learning, including impairments that interfere with the normal acquisition of language.

3. Acquisition

Studies of the way in which children acquire language has shown that if there is any language around, spoken or signed, children seem to have an instinct that leads them to start building a grammar that follows the principles of language design. They instinctively know what to pay attention to in the language that they hear around them and what to ignore and what kinds of rules are likely to be language rules. For example, Kuhl has shown that when babies are only a few months old they can easily learn to discriminate between a man saying [a] and the same man saying [i] (Marean, et al. 1992). What is surprising is that with no further training the babies can make this discrimination when they hear a woman' s voice or a child's voice saying [a] or [i] even though these same sounds when said by a woman or a child have very different acoustic properties than when they are said by a man. The children appear to distinguish between those acoustic properties of sound that potentially signal linguistic meaning and those that characterize individual variations in voice quality. As children get older they are able to take the limited language that they have observed and infer recursive rules that allow them to produce not just the words and sentences that they have already heard, but a potentially infinite set of words and sentences that they have never heard before. Children do not always get it right the first time. The errors that they make tell us a lot about the rules they are using for grammar building. There are certain kinds of errors that they never make. For example, children learn that you can turn the sentence 'She is playing.' into a question by moving the verb 'is' to the front of the sentence 'Is she playing?'. But children do not try to turn the sentence 'The girl who is happy is playing.' into a question by moving the first 'is' to the front of the sentence. That would give you an ungrammatical sentence 'Is the girl who happy is playing?'. Children all over the world know that rules of language do not operate on the superficial properties of the linear string of words as they actually hear it. The rules that they construct, even the incorrect ones, operate on the hierarchical relations among the elements of the sentence.

Though we all automatically acquire and use these rules, we don't consciously know what they are. If we could bring them to consciousness, describing languages would be an easy task. Native speakers would just have to introspect and then write down the rules of their language. But even after hundreds of linguist-years of work we are still discovering new rules that govern languages.

4. Animal communication

The principles of human language design are radically different from the organizational principles of any animal communication system. Primates or other animals do not naturally develop a human-like language and cannot be explicitly taught the intricacies of such languages as the attempts to teach apes human-like language systems have shown (Terrace et al. 1979). It looks like human language is unrelated to any presently existing animal communication system. Some people have suggested that language is an epiphenomenon of the development of larger, more complex brains (Gould 1981; Lieberman 1992). Others argue that it evolved according to the basic principles of natural selection (Pinker & Bloom 1990). The origins of human language and its evolutionary history are still obscure. It may be the case that 'language' is not a single, unitary object that has a unified evolutionary history. It seems to be more likely that language is a complex system, that has been put together from parts that may have started out from very different beginnings and have followed different evolutionary paths.

5. Neurology

However it arose, it now seems to be the case that there are specified neural structures that subserve the instinct for language. Though the story is very complex, studies of developmental and acquired brain damage have provided evidence for the location and function of some of these structures.

If language is part of the biological endowment of humans that has evolved since the line leading to man split off from those which lead to the other great apes then there must be some genetic properties of humans that build the particular kind of brain circuitry that is specialized for human language. Since the evolution of brain structure is connected with changes in the genes that guide the development of the brain it would not be surprising to find that some change in this genetic endowment can interfere with the way brain circuitry is laid down and thereby impair the ability to acquire or use language in the normal way. There has been some tentative new evidence that appears to indicate that there are anatomical anomalies in the brains of some individuals with language impairments: 'Magnetic resonance imaging scans of specifically language-impaired (SLI) boys were examined ... The distribution of perisylvian asymmetries in SLI subjects was significantly different from the distribution in controls ($P < 0.01$) ... These neuroanatomical findings suggest that a prenatal alteration of brain development underlies specific language impairment' (Plante et al. 1991: 52; Tallal et al. 1991).

Genetic language impairment

This paper will report on a natural experiment that impairs the ability to acquire language and therefore has the potential of giving us some insights into the nature of this genetic endowment. Our team thinks that there is some genetic factor (or factors) that interferes with the establishment of the neurological structures that subserve the acquisition of language. This, in turn, affects the ability to build the kinds of grammars that ordinary children build automatically and unconsciously.

Our work over the last ten years has shown that the language instinct can be impaired in very specific ways. In 1990 when I first reported that an inherited disorder could impair the ability to learn language (Gopnik 1990), the popular press (Associated Press, James Kilpatrick, *et al.*) credited me with having found a gene for grammar. For the record, though it might be nice if it were true, I have not discovered 'a grammar gene' nor do I think that there likely is a grammar gene. Even if there were a single gene that could impair the ability to learn language it would not follow that the good version of that gene would cause good grammar. Complex systems can go dramatically wrong if one small part is defective, as any user of computers knows all too well. Yet no one would say that the one small chip that made the whole system crash accounted for the normal functioning of the computer. Furthermore, if a gene (or genes) interferes with normal neurological development, as evidence seems to suggest, then it is likely that this same disorder may, in some individuals, impair other cognitive functions. It might even turn out that in certain circumstances, this same genetic factor might affect neurological development in areas of the brain that control other cognitive functions, while at the same time sparing the language centres altogether. These are speculative empirical questions, and they can only be resolved when the genes are found and their effects on neurological development are established.

The disorder

It is well established that some children have great difficulty in acquiring their own native language, even though they seem otherwise normal. Specific Language Impairment is defined as a developmental deficit of language in the absence of perceptual, motor, general cognitive, emotional or social problems (Bloom & Lahey 1978; Stark 1980; Wyke 1978; Zangwill 1978). This difficulty can persist into adulthood. Many investigators have reported that these subjects have particular difficulties with inflections such as tense (Miller 1981; Leonard *et al.* 1992; Clahsen 1989, 1992; Rice 1994; Gopnik 1994a; Ullman & Gopnik 1994). This is what we will report on here in some detail: their problems with the linguistic rules that come under the rubric of 'morphology', that is those that mark tense, number, case etc.

Are we claiming that word formation is the only thing that is wrong with their grammar? Not at all. It is just what has been studied cross-linguistically in most detail. It has been documented that they also have difficulties in agreement within sentences, in constructing the rules for the sound system of English, and with more complex syntactic operations like relative clause formation. These other problems have been reported on elsewhere, by our team and others (Miller 1981; Leonard et al. 1992; Piggott & Robb 1994; van der Lely & Harris 1990; Bishop 1992, in press). These other problems appear to be consistent with what has been seen in morphology. They suggest that the impaired subjects build their grammars on different principles than normal children.

EVIDENCE FOR A GENETIC ETIOLOGY

There is converging data from epidemiological studies (Tallal et al. 1989; Tomblin 1989) and family studies (Samples & Lane 1985; Gopnik 1990; Hurst et al. 1990) that show that this disorder aggregates in families. The increased concordance in monozygotic, as compared to dizygotic, twins suggests that this disorder is likely to be heritable (Table 1).

These facts have geneticists all over the world searching for a genetic factor or factors that correlate with this inability to acquire language normally.

Alternative explanations

Several researchers have claimed that this disorder affects the grammar itself (Gopnik 1994a; Clahsen 1989; Rice 1994; van der Lely & Harris 1990) though they differ in the precise details of which particular part of grammar is affected. But not everyone wants to believe that grammar itself can be impaired. Some prefer to think that the problems that these subjects have with language are really caused by problems in the auditory input system (Tallal et al. 1980, 1991; Leonard 1989; Leonard et al. 1992) or in the

Table 1. Per cent of concordance in twin pairs with SLI.

Monozygotic	Dizygotic	Source
80	38	Tomblin & Buckwalter 1994
86	48	Lewis & Thompson 1992
89	48	Bishop et al., in press.

articulatory output system (Fletcher 1990). Others, who doubt that the language faculty is modular, think that the disorder must be caused by some general problem with cognitive functioning (Johnston, in press; Bishop 1992).

These are empirical questions which are resolvable by carefully observing the facts about the natural history of this disorder. It is not enough to show that some of these subjects have articulatory problems, or that some others have auditory processing problems or that some have an assortment of other cognitive problems. The difficulty is that though some of the language impaired subjects have some of these additional problems, others do not have such problems. Furthermore, there are individuals with these other problems who do not have language problems. We know that many disorders can affect more than one system. As we have said above, if this disorder affects neurological development then it would not be surprising if it had broader effects than just language. However, the pattern of double dissociation between the language disorder and these other disorders suggests that in this case co-occurrence is not causation.

If we want to evaluate these alternative explanations then we must look at the specific predictions about what errors language impaired subjects will make that follow from each of these alternatives. Then we have to compare these predictions to the facts of just what these individuals can and cannot learn about language. Only then can we see which explanatory model best accounts for the full range of language facts that are actually observed. (It has been suggested that some non-linguistic factor may affect language development in complex ways that do not result in predictable consequences for language impairment. This, of course, may be the case, but such an explanation would not be testable at present.)

EVIDENCE FOR ABSTRACT GRAMMAR

One of the fundamental assumptions that all linguists make is that the words and sentences that a speaker produces are not simply chosen from a list of appropriate responses but are the product of an internalized system of rules that are capable of producing an infinite number of new sentences, most of which have never been uttered before. Though I have spoken about these issues before I have never used precisely the same set of sentences and you have never heard or read these sentences before. Yet the novelty of the actual utterances for me and for you does not in any way impede communication because we share the system of rules for English that allows us to encode and decode an infinite set of sentences. The same thing is true about words. We can and do make new ones up all the time. I can say:

I was faxing two faxes but one turned out to be unfaxable. Even when I refaxed it they remained faxless. The problems gave me faxophobia so I just e-mailed it.

and though many of the words are new, both for me and for you, you do not have any trouble decoding the message. Therefore the empirically observable data can tell us much more than what sentences or words the speakers actually produce. These data can tell us about the system of rules, the grammar, that produced these responses.

Normal children do not acquire their language skills by mere imitation of what they hear or see around them but by creating a symbolic system of recursive rules that allows them to produce words and sentences that they have never heard before. Children who at 2 and a half say, 'I went to the store,' suddenly at three and a half start saying 'I goed to the store.' None of these children has ever heard the word 'goed' from their parents. So where does it come from? It appears that the child has gone from knowing the meanings of single words e.g. that the word 'went' means 'to move in the past' and that the word 'jumped' means 'to project oneself upward in the past', to knowing that there is a general rule for making past tense forms in English. In the flush of his new discovery the child, like many scientists with a new discovery, applies his new rule everywhere, even to words where it does not apply.

If we want to understand what is going on in either normal or impaired subjects then we must use the observable data as evidence for what is going on in their grammar. This gets complicated because the very concept of observable linguistic data is problematic. Two speakers can produce identical surface forms e.g. 'jumped' and they may be derived by two very different routes from two very different grammars, just as two Turing machines may both produce similar outputs even when they have very different internal rules. The only way you can tell if two forms are the same or different is to see how they fit into all of the other evidence about the system of rules that the subject is using. And that is what we will do here. We will present converging evidence from several different observations that show that language impaired subjects build grammars that are very different in kind than those built by unimpaired speakers and that the same kind of aberrant grammars are built by language impaired subjects in English, in Greek and in Japanese. These three languages have different grammars. By comparing data across these three languages we can determine which problems are the result of the peculiarities of one language and which problems are more general.

These data show that these subjects do not construct unconscious abstract rules for forming words, a skill that four year olds perform automatically. They can learn to compensate for this inability by memorizing complex words like 'walked' in the same way that unimpaired speakers have to memorize an irregular form like 'went'. They can also learn to use conscious explicit rules, which are routinely taught to them in speech therapy, as an imperfect surrogate for the unconscious, implicit rules that are used by unimpaired speakers. Though they have problems understanding the role of grammatical rules for conveying meanings such as 'past', they can rely on the semantic meanings of words and sentences to communicate these concepts effectively (Paradis & Gopnik 1994; Ullman & Gopnik 1994).

We can show that these aberrant grammars cannot be explained in terms of an impairment in auditory perception, articulation or general intelligence. As we have said, some people believe that the ability of children to make these generalizations about language is just a consequence of their general intelligence. Others, following Chomsky, believe that the ability to learn language is not connected to general intelligence, but is the consequence of a specific innate ability that is independent of general intelligence. The evidence from this developmental disorder of language is relevant to this debate. If individuals who have normal intelligence can lose the ability to construct normal grammars then this supports the idea that grammars and general intelligence are independent. On the other hand, if this developmental disorder of language can be shown to be a direct consequence of an impairment in general intelligence then perhaps language is not modular after all.

Evidence

It is widely reported by speech pathologists that this language disorder occurs in children who have very high non-verbal IQ scores, as high as 140. It is also the case, as one would expect if IQ and language impairment were not correlated, that some of the individuals with language impairment have low non-verbal IQ scores. Bishop has reported that monozygotic twins who have similar impairments of language sometimes have quite different non-verbal IQ scores (Bishop, in press a). Our data agree. In a set of monozygotic language impaired twelve year old twins the twin with the more serious language impairment had a higher non-verbal IQ than his brother (PIQ 91 vs. 86, WISC-III). Among the many subjects that we have studied are the members in a large family (Figure 1) in which the pattern of the disorder is indicative of an autosomal dominant gene (Pembrey 1992).

In 1990 Hurst et al. reported that the IQs of the members of this family were within the normal range (Hurst et al. 1990). Pembrey, reporting on the same family said: 'Using the WAIS-R/WISC-R scales, the mean performance IQ of these 13 affected members is 95 (80–112)' (Pembrey 1992: 54).

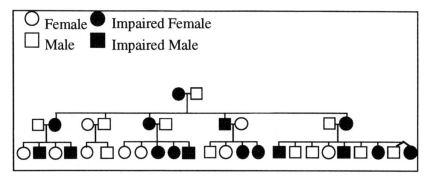

Figure 1. Family tree of affected family

We did not test the IQ scores of these subjects independently, but we cited Hurst and Pembrey's reports in our work and pointed out that in addition to the subjects they reported on, one of the other members of the family in the middle generation was known to have been in a special school for slow learners and as an adult had great difficulty with reading or doing arithmetic which suggested that she had a low performance IQ. Two years later members of the Pembrey research team reported that subjects in this family had '... a mean performance IQ of 86 (range 71–111) (compare the unaffected members' mean score of 104 (range 84–119))' (Vargha-Khadem et al. 1995: 932). In that recent report they do not mention their earlier report. Part of the change is the result of a general revision of the IQ norms that lowered all scores by 4–8 points. The inclusion in this group of the cognitively impaired subject referred to above probably accounts for a portion of the rest of the difference. However, even if the reported means for the affected and unaffected members of the family are different, the IQ scores of the language impaired subjects and their unimpaired relatives overlap substantially. One clearly language impaired subject has a non-verbal IQ of 111, while an unimpaired relative has a non-verbal IQ of 84, almost 30 points lower. Though the IQ scores of the language impaired members of this family are reported to vary widely, they all seem to have similar problems with language, though the severity varies. The impaired Japanese and Greek speakers, who show the same pattern of language deficits, have the same range of performance IQ scores (Japanese, 81–103; Greek, 86–111). So there is no reason to believe that the language disorder in this family is *caused* by a cognitive deficit. On the other hand, if they cannot use their normal instinct for acquiring language and therefore have to use other cognitive skills to learn language then the ones who have better cognitive skills should be better at, for example, learning explicit rules. Therefore while an impairment in intelligence might not *cause* the language disorder it might affect the ability to use various strategies to compensate for the deficit. So you can see that individuals with language impairment run the whole gamut of intelligence, from very bright to not bright at all. And vice-versa. People who have normal language come in all ranges of intelligence. Some people who don't have the cognitive skills to manage simple tasks in the world have fluent, grammatical language. What they say may not be true or make any sense, but their language follows the rules of grammar (Bellugi *et al.* 1991; Smith *et al.* 1994). It looks like general intelligence and the ability to build grammars of a particular kind are independent.

Auditory/articulatory accounts

Could the problems that these people have with language be a result of not being able to process sounds efficiently or not being able to pronounce words accurately? (Tallal *et al.* 1980, 1991; Leonard 1989; Leonard *et al.* 1992; Fletcher 1990.) If that were true then nothing would be wrong with their language ability, only with their input or output systems. At first blush such an explanation might appear to be plausible. Some people with this disorder have difficulty in processing transient auditory signals; other subjects have been shown to have difficulties with sequencing oral-facial movements. So some, though not all, of these language impaired subjects have auditory or articulatory problems. On the language side one of the recurrent problems that is reported is that these subjects have problems with past tense and plural in English. If someone says: 'Last week I jump over the fence,' instead of 'jumped' it could be because they did not process the -ed sound because it went by too quickly and was therefore difficult to process, or because it took too much co-ordination to pronounce.

But when they say 'Two weeks ago I go,' instead of 'went' or 'He did it then he fall,' instead of 'fell', then these anti-grammatical explanations are in trouble. No one could suppose that mistaking 'go' for 'went' or 'fall' for 'fell' can be explained by an auditory or articulatory problem. Furthermore, when these subjects use these same sounds in simple words where they are not being used to signal past, e.g. car/card (similar to mar/marred); ball/bell (similar to fall/fell) they can produce and perceive these sounds quite accurately (Leonard *et al.* 1992; Goad & Gopnik 1994; Gopnik 1994b). There are many similar examples from Japanese and Greek. For example, in Japanese they have difficulties with 'probable-future' marking even though it is encoded by a separate word *deshoo*. The passive is encoded by a two syllable morpheme, *rare*, that occurs in the middle of the word and yet the Japanese subjects omit it and use the past form of eat, *tabeta* (ate) when they should use the past passive *taberareta* (was eaten). In Greek they substitute one complex inflected form for another even though the forms are distinct and they sometimes omit the whole verb. They say:

dhe thelo	*	ki... kimi...	dhe *kimame
NEG want-I	missing compl.	sl... sle... non-continuous verb root only rephrasal (missing affix)	NEG sleep-I continuous root + affix

?*Maria   kato   parea.
(Maria downstairs together.)

when they should say:

Maria   pame   kato   parea.
(Maria   lets-go   downstairs   together.)

They also appear to have trouble with larger syntactic structures like relative clauses which would be hard to account for by any auditory or articulatory deficit.

Table 2. Examples of SLI affixation in Japanese and Greek.

	Japanese	Greek
they say:	tabe-ta (ate)	Maria kato parea (M. downstairs together)
they mean:	tabe-rare-ta (was eaten)	M. pame kato parea (M. lets-go downstairs together)

And they don't have trouble just with producing the correct forms. They have the same problems across the board, in judging sentences to be grammatically correct and in correcting grammatical errors (Table 3). They have the same problems when they hear the sentences or when they read them; in spontaneous speech and in spontaneous letter writing; when they have to respond to a test with a sentence, or with a word, or with a rating number or with 'yes' or 'no'. No matter what the form of the input or output they make the same grammatical mistakes (Matthews 1994). So while a small part of the data might be consistent with an auditory or articulatory problem, there is an overwhelming amount of other data that cannot be explained by either mishearing or misspeaking.

Table 3. Grammaticality judgements of SLI in English, Japanese, and Greek.

	English	Japanese	Greek
Judgement of grammatical errors	57%	43%	34%
Appropriate error corrections	37%	35%	17.2%

Our hypothesis

Evidence for the absence of abstract rules, novels

I made a strong claim before that these subjects cannot construct abstract inflectional rules that allow them to generate, for example, past tense and plural forms. There are two ways that past and plural forms arise. Irregular past forms must be memorized and listed in the lexicon separately from the present form because they are not predictable from the present form. All forms that are not irregular are produced by taking the root form of the word and applying the regular rule (Pinker 1991). These are the kinds of rules that every four year old automatically and unconsciously constructs. Young children know that the regular rule applies equally to existing and novel words. If someone is using a rule to produce the inflected forms then they should be able to apply this rule to novel forms. If, on the other hand, the inflected forms are being stored lexically, that is as separate words in a mental dictionary, then the subjects should have problems with inflecting novel forms. Some investigators (Leonard 1989; Leonard *et al.* 1992; Rice & Oetting 1993; Rice 1994; Bishop, in press b) have argued that the impaired subjects must have these rules because they produce inflected forms like 'walked' in spontaneous speech and in tests. However, even if the subjects produce the right inflected form of an existing word, in spontaneous speech or in test situations, we really cannot tell if they have the rule or if they have simply memorized the word as a whole. The word 'books' could either be generated by rule from the root 'book' and the rule for pluralization or it could be listed in the lexicon as 'books' with the meaning of 'several objects for reading'. But if they know the rule they should be able to apply it to novel words that they have never heard before. The ability to inflect novel forms provides us with an empirical test that can distinguish between a form being retrieved from the lexicon and that same form being generated by a productive rule. We tried several different tests which required the subjects to inflect novel forms in English, Japanese and Greek. The subjects were given novel roots, like 'wug', and then were given a semantic context that required the inflected form of the word. Sometimes this was done by

Figure 2. Sample 'wug' stimuli.

Table 4. Production of inflected novel forms: accuracy of SLI and control populations.

Language	(Morphological operation)	Controls	Impaired
English	Past Tense	95.4%	27.0%
	Plurals	94.9%	9.0%*
Japanese	Past Tense	89.0%	37.0%
	Rendaku†	80.5%	22.1%
Greek	Plurals	78.8%	42.1%
	Compounds	93.6%	12.8%

* After close phonetic analysis.
† *Rendaku*, 'sequential voicing', is a highly productive morphophonological operation in Japanese compound formation. Word initial voiceless obstruents of the second compound member become voiced in the process of compounding.

showing them pictures of imaginary objects (Figure 2). In other cases the context was established within the sentence:

Everyday I crog to John.
Just like everyday, yesterday I _____ to John.

In every one of these tests the language impaired subjects did significantly worse than the controls (Table 4). The data across all three languages shows that the language impaired subjects cannot reliably inflect words that they have never encountered while the controls are able to do so.

The lexical listing hypothesis

These data show that they cannot productively use rules to construct novel inflected words or novel compounds. We still have to account for the fact that they sometimes do produce words that have the same surface form as inflected words. If we want to maintain that they do not have inflectional rules what we need to show is that the words that they produce do not have any internal morphological structure though these same words are represented as a root plus an inflectional ending in the grammars of non-impaired subjects. We have looked at three independent sources of data to test the hypothesis that words that are inflected in the normal grammar are treated by language impaired subjects as unanalysed lexical words: (1) the ability to extract roots from inflected words; (2) on-line processing of inflected and uninflected forms; and (3) frequency effects.

Access to roots

The most direct kind of evidence that can tell us if the impaired subjects know the internal structure of words is their ability to access this internal structure and use these parts in constructing new words. English is not a particularly good language to investigate this hypothesis because words in English can occur without any overt inflections. For example the root underlying 'walks', 'walking' and 'walked' is 'walk'. Unfortunately this root is identical to the first person, singular present form 'walk' which occurs without any overt inflectional marker. Therefore if the subject produces the form 'walk' it is impossible to know if it comes from an analysis of the internal structure of inflected words or is simply the word 'walk'. Greek provides an ideal test case. All nouns in Greek are inflected. For example, the word for 'wolf' has eight different inflected forms.

The root for 'wolf', *lik-*, never occurs by itself in Greek (Table 5). So Greek children never hear the root *lik-* by itself. In order to discover the root of the word they must be able to abstract the root from all of the inflected forms for 'wolf' that they hear. If they can do this then they must be treating these words as made up of a root and an inflection. On the other hand, if they treat each of these forms of 'wolf' as a separate uninflected simple word, then they should not be able to extract the root. One linguistic context which can test for the subjects knowledge of roots is compounding in Greek. Compound words like 'wolfman' are made by concatenating the root for 'wolf', *lik-*, with the root for 'man', *anthrop-*, and then attaching the inflection to the end of the compound stem. This means that the first element of a compound is always a bare root (the second element has the surface form of an inflected word since the inflection for the whole word is attached to the end of the compound). In order to produce a compound correctly the speaker must abstract the bare root form of the first element of the compound from the inflected forms that he has heard. If a speaker can reliably produce the root in a compound then it is reasonable to conclude that the speaker is treating the inflectional forms as if they were made up of a root plus an inflection. On the other hand if they cannot reliably find the roots then it is reasonable to suspect that they do not represent words in Greek as being composed of roots and inflections.

There is one other property of Greek compounding that allows us to investigate the speakers understanding of the internal structure of words. If the second element of a compound noun begins with a consonant then an *o*

Table 5. Full declension of a Greek noun.

	Singular	Plural
Nominative	lik os	lik i
Genitive	lik u	lik on
Accusative	lik o	lik us
Vocative	lik e(!)	lik i

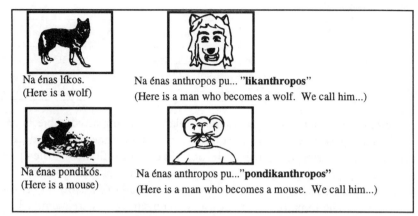

Figure 3. Sample stimuli from Greek compounding task.

is inserted between the two roots. In order to use this rule correctly the speaker must know two different things; first of all the speaker must know where the boundary between the two roots is in order to insert the *o* in the correct location, and secondly must know that the *o* is only inserted before second roots that begin with a consonant.

So compounds in Greek give us two different diagnostic tests of whether language impaired subjects are aware of root boundaries and inflections. To produce a compound they must be able to extract the root for the first element from the inflected words that contain this root. If they choose to insert an *o* in the compound they must insert it at the boundary between the two roots. The knowledge of which compounds require an *o* and which do not, is independent of the speakers knowledge of root boundaries. A speaker could mistakenly believe that a compound requires an *o* when it does not, but still insert the *o* between the two roots, as is the case with younger non-impaired Greek children. In this case the compound is incorrect, but the subject still demonstrates knowledge of the root boundary. The overgeneralization of the *o* in Greek compounds in this case is similar to the overgeneralization of the regular past marker to irregular verbs in English, 'goed' for 'went'.

It should be the case that if the word already exists in the language, like 'wolfman' [Greek: *likanthropos*], the impaired subjects may be able to produce it because they could have it listed as a whole word, though most of these words are infrequent. But if they are asked to produce a novel word like 'mouseman' [Greek: *pondikanthropos*] they should have great difficulty because they will not know how to find the root for 'mouse'. The subjects were visually and aurally presented with the stimuli in Figure 3 and asked to produce novel compounds.

Table 6. Greek error rates.

Error type	FLI subjects	Young controls	Age-matched controls
Sum of Root Errors	83.9%	5.0%	1.25%
Incorrect -o Realization	3.8%	79.0%	12.0%
Total Errors	66.4%	29.3%	1.5%

And that is exactly what happens (Table 6). The impaired subjects are very bad at knowing where the root boundary is. Sometimes they produce a form that is shorter than the real root (like /anthrofàghos/ instead of /anthropofàghos/ for 'maneater') and other times they produce a form that is longer than the real root (/hinothanthropos/ instead of /hinanthropos/ for 'gooseman'). The impaired subjects use *o* significantly less frequently than non-impaired subjects and do not seem to overgeneralize its use. Moreover, when the impaired subjects try to use the *o* that signals the boundary between the two roots they often insert it in positions that are not true root boundaries. In contrast, the younger controls respect root boundaries even though they often overgeneralize the *o* insertion rule and insert an *o* in contexts in which it is not required. For example the young controls produce forms like /pondikoanthropos/ when they should say /pondikanthropos/. They seem to have an *o* insertion rule that marks compounding, but they have not yet learned the constraints on this rule. This tendency of young children to overgeneralize a rule that they have recently learned is a typical property of normal language acquisition. The age-matched controls make virtually no errors at all. They can extract the roots and they know the constraints on the compounding rule.

These data show that the unimpaired children, young and old, know that nouns have a complex internal structure made up of a root and an inflection. The impaired subjects do not appear to recognize that inflected words have this complex internal structure.

Though these native Greek children with language impairment do not appear to acquire the rule for *o* insertion in Greek compounds, you and I, native speakers of English, have unconsciously internalized this Greek rule. Earlier in this paper, in the example about faxes, the word *faxophobia* was used to illustrate the productivity of word formation rules. Native speakers of English judge *faxophobia* to be a natural new coinage and they prefer it to *faxphobia*. What is odd is that the *o* in *faxophobia* follows from the Greek rule for compounding not the English rule. It appears that native speakers of English, on the basis of a few examples from Greek like *claustrophobia* and *acrophobia*, unconsciously construct the Greek *o* insertion rule. This simple, natural ability is what appears to be missing in these children with language impairment.

On-line tests

Another method to see if they process inflected forms differently from uninflected ones is to use on-line lexical decision tasks. These tasks are implicit methods used to tap the underlying mental representations. Kehayia (1994) reports on two different on-line experiments; a simple lexical decision task and a morphological priming task in which the target stimulus was primed by another related or unrelated word. There were five English speaking language impaired subjects and 25 English speaking non-impaired subjects. These tasks measure the amount of time (in milliseconds, ms) it takes for the subject to make that decision and permit us to study the patterns of word recognition and lexical access adopted by the subjects. Finally, these patterns can provide us with valuable insight concerning the organization of underlying representations in the mental lexicon.

In the simple lexical decision task, which we will discuss here briefly, subjects are asked to decide whether a sequence of letters presented to them on a computer screen is a real word of the language. They are told to press a key marked 'yes' if the stimulus is a real word or a key marked 'no' if it is not and to respond as fast as they can, while still being accurate. There were four different kinds of experimental stimuli: uninflected real verbs like 'walk', inflected real verbs, like 'walked', uninflected novel words like 'zash', and inflected novel words like 'zashed'. The stimulus set also included filler and control words to reduce density. The results show that neither the impaired nor control subjects had any difficulty distinguishing between real words and novel words, though both the impaired and the control subjects take longer to process novel words than existing words. However, within these two classes their patterns of word recognition were quite different. Controls, in this experiment and in other similar experiments (e.g.: Kehayia 1993; Kehayia & Jarema 1994) take significantly longer (more than 30 ms, $P < 0.001$) to process the inflected form of the word than to process the uninflected form of the word, i.e. it takes them longer to decide that 'walked' is a word than that 'walk' is a word, or that 'zashed' is not a word than it takes them to decide that 'zash' is not a word (Table 7). These sorts of data have been interpreted to indicate that subjects are sensitive to the presence

Table 7. Lexical decision times (milliseconds) for impaired and controls.

	Impaired	Controls
Non-words (Root)	925 ms	620 ms
Non-words (Inflected)	910 ms	652 ms
Words (Root)	792 ms	450 ms
Words (Inflected)	809 ms	480 ms

of the inflection on either a real or a novel word and that the presence of the inflection requires the speaker to perform some additional processing step. Impaired subjects, on the other hand, were faster, although not significantly, when recognizing inflected novel words, unlike their control counterparts who showed the reverse pattern. With respect to the impaired subjects' recognition of inflected real words even though they appear to take 17 ms longer than when recognizing simple words, this difference is neither significant nor constant across the five subjects. Furthermore results from the priming experiment show a non-differential treatment of inflected and uninflected verbs contrary to results obtained for the control subjects. Findings from this latter experiment also show minimal to non-existent effects of morphological relatedness even in cases of complete root transparency (e.g. wash-washed), unlike findings from the control group that yield strong effects of morphological relatedness (e.g. Stanners *et al.* 1979; Napps 1989; Kehayia 1993; Kehayia & Jarema 1994). Thus, the above results indicate that the language impaired subjects appear to be insensitive to the presence of the inflection when they process words in simple or in primed conditions. Inflected words seem to be recognized as chunks with no processing or decomposition of the inflectional suffix.

Given these results and the possible unavailability of a productive morphological rule for the marking of past tense as has already been suggested, the decomposition of verbs inflected for the past would be impossible. It is thus proposed that the mental lexicon of these subjects contains a full list of all simple and complex lexical items with no internal word structure representation for complex words. Such a proposal would account for the occasional production of past tensed verbs in spontaneous speech or during testing.

The unavailability of rules is further evidenced in the results from four impaired Greek-speaking SLI children tested on real and novel compounds in Greek, using an on-line simple lexical decision task. Their performance was compared to that of 10 normal controls. The overall results show similar recognition patterns for the impaired as those reported in the previous study. The most striking result concerns the unavailability of the compounding rule which led to the swift rejection of novel compounds even though controls appeared to either accept them or reject them with great delays.

Frequency

One way of finding out if the word is generated or is listed in the lexicon is to see if there are any frequency effects. If a word is listed in the lexicon then its frequency should affect whether or not it is retrieved. If, on the other hand,

Table 8. Effect of frequency on morphological production in Japanese.

Task		Frequent	Non-frequent
Rendaku	Impaired	88.8%	33.3%
	Control	94.5% (n.s.)	94.5% ($P = 0.001$)
Past tense	Impaired	73%	68%
(Nas/V)	Control	96% ($P = 0.055$)	97.5% ($P = 0.02$)

it is derived from a rule, then there should be no frequency effect. We know this is true for normals who show frequency effects for irregular verbs, but not for regulars. We looked at the performance of the impaired subjects on existing regular forms. As predicted by the lexical listing hypothesis the likelihood of their producing an existing regular past tense verb in English was dependent on the frequency of the existing past tense form. In Japanese the impaired subjects were much better at voicing existing compounds and inflecting verbs that were judged to be frequent than those which were judged to be infrequent; there was no such effect for the controls (Table 8).

The Greek compounding data, the on-line data and the frequency data all tell the same story; the impaired subjects do not treat inflected words as if they were composed of constituent parts. So it seems reasonable to say that when these subjects produce a word like 'cats', which may look as if is inflected, we cannot assume that this word is composed of a root plus an affix in their mental grammar.

Implicit vs. explicit rules

One of the interesting differences between the impaired and unimpaired subjects is that the unimpaired subjects, even very young ones, acquire the rules of their language unconsciously and with no explicit training. The impaired subjects, even when they have had speech therapy for years that attempts to teach them the rules, never internalize the complex constraints that underlie these rules. They can learn the simple version of the rules, like 'add an -s', but they are oblivious to the wonderful intricacies of language that come for free with the language instinct. The evidence suggests that while the impaired subjects are not able to construct implicit, automatic inflectional rules, some of them are able, on some occasions, to apply explicit rules (Paradis & Gopnik 1994).

There are several empirical differences between the responses from explicit and implicit rules that allow us to distinguish between them. One way is to see if the subject's responses reflect the full complexity of the inflectional rule or if they reflect a much more simplified explicit rule. For

Table 9. Phonological rules for English pluralization.

Singular	Plural	Rule
cat	cats /kæts/	add an /s/ after unvoiced obstruents
dog	dogz /dogz/	add a /z/ after voiced obstruents
bus	buses /mæsiz/	add /Iz/ after sibliants

example, the rule for plurals and pasts in English looks simple. There is a small set of irregular forms that have to be memorized e.g. man/men, go/went. All of the other nouns or verbs in English follow the regular rule. Every English speaker automatically obeys these rules, but very few can describe their workings because they are part of their implicit knowledge. Everybody would say that we add an -s to a noun to make the plural, and that we add an -ed to a verb to make the past, but there is a lot more to it than that (Table 9).

The ending has to agree in voicing with the final sound of the root, *cats* (cats) vs. *dogz* (dogs); *jumpt* (jumped) vs. *jogd* (jogged). If the root ends with a sound that is like the sound of the inflection then you have to insert a vowel, *buses* not *buss*; *loaded* not *loadd*. And there is no stress on the inflectional ending, *buses* not *busES*; *loaded* not *loadED*. This might seem complex for a four year old, but children have help. They do not have to figure out that there is voicing agreement when the root ends in an obstruent like *t* or *g*, or that the inflectional ending is virtually never stressed or that it would be extremely unusual for a language to allow two like sounds together at the end of a word or that the regular rule applies to any word that is not marked as irregular. These are fundamental, widespread properties of language and the child's first guess is that a human grammar should be built like this. Given this advantage it is not hard for children to automatically and implicitly construct the rule for plural with all of its complexities before they are four years old.

The evidence from the performance of these subjects on novel verbs indicates that they do not have the regular rule in English for past or for plural. But they sometimes are able to produce a seemingly inflected form for a novel word and, on occasion, in spontaneous speech they produce overregularizations like *eated*. These data have been used to argue that they have the rule, but just are unable to produce it on all appropriate occasions. A careful phonetic examination of the forms that they produce tells a different story. Some of the impaired adults looked like they were producing

Table 10. A metalinguistic response from a 45-year-old SLI woman.

Stimuli	Produced form	Correct form
vub	/vub-s/	/vub-z/
fen	/fen-s/	/fen-z/
tuss	/tuss-s/	/tuss-iz/
praz	/praz-es/	/præz-iz/

some of the novel plurals and novel pasts correctly. But when you looked closer at the phonetic shape of their attempts it was clear that they were not using the rule used by all four year olds. They were using a much simpler rule that did not have any of the normal constraints.

One 45-year-old who was shown a picture of a crab and given the English word *crab* responded with:

> Crab-S [with an incorrect *s* sound instead of the appropriate *z* sound] you have to put a *s* to it. All the time.

She correctly used the rule that she articulated, but unfortunately it is not the correct rule. She violated the necessary constraints that we discussed above and made precisely those errors that young children do not make. She added an *s* to novel words that ended in a voiced obstruent and produced *vub-s* as the plural of *vub*. She added an *s* directly to words that ended in a sibilant and gave *tusss* as the plural of *tus*. And she put a second stress on the ending and gave *praz-ES* as the plural of *praz* (Table 10). This pattern of providing metalinguistic statements about the rules that they are using and producing errors that show that they are using simple, explicit rules is characteristic of several of the impaired subjects in both the plural and past tense production tasks (Goad & Rebellati 1994; Ullman & Gopnik 1994; Gopnik 1994a). These data indicate that the impaired are not really treating these endings as inflections which are incorporated into the root. The phonetic shape of their productions violates fundamental constraints of English. What they are doing is using an explicit rule to take a word-like element, *s*, which means 'more than one' and simply concatenating it with the root.

SUMMARY

Our observations across three different languages from a wide range of different tests and from naturally occurring language all point in the same direction; that the language impaired subjects do not represent words as

Table 11. Genetic Language Impairment team.

J. Brostoff (R.A.)	E. Kehayia (processing)
J. Dalalakis (Greek)	M. Kessler Robb (phonology)
S. Fukuda (Japanese)	R. Palmour (genetics)
S. E. Fukuda (Japanese)	M. Paradis (neurolinguistics)
H. Goad (plural test)	G. Piggott (phonology)
M. Joanisse (R.A.)	C. Rebellati (plural test)
M. Ullman (past test)	

composed of roots and inflections and they do not construct the same kinds of rules and grammars as normal children do. (There is also evidence, that we do not have time to discuss here, that this cluster of symptoms is neurologically plausible.)

And now we are extending our study to see if we can find out just exactly what the genetic and neurological consequences of this disorder are. We now have a new team that is working on this larger study (Table 11).

It appears that Darwin was right. There is a language instinct. It guides children to pay special attention to the language around them and it tells them how to build a grammar. When it goes wrong they never can just automatically acquire language. They can, with hard work, learn to simulate language behaviour.

REFERENCES

Bellugi, U., Bihrle, A., Jernigan, T., Trauner D. & Doherty, S. 1991: Neuropsychological, neurological, and neuroanatomical profile of Williams syndrome. *American Journal of Medical Genetics Supplement* 6, 115–125.

Bishop, D.V.M. 1992: The Underlying Nature of Specific Language Impairment. *Journal of Child Psychology and Psychiatry* 33, 3–66.

Bishop, D.V.M. (In press, a) Is Specific Language Impairment a valid diagnostic category? Genetic and psycholinguistic evidence. *Philosophical Transactions of the Royal Society, series B.*

Bishop, D.V.M. (In press, b) Grammatical Errors in Specific Language Impairment: Competence or Performance Limitations? *Applied Psycholinguistics.*

Bishop, D.V.M., North, T. & Donlan, C. (In press) Genetic Basis of Specific Language Impairment: Evidence From a Twin Study. *Developmental Medicine & Child Neurology.*

Bloom, L. & Lahey, M. 1978: *Language Development and Language Disorders.* New York: Wiley.

Bloomfield, L. 1933: *Language.* New York: Holt, Reinhart and Winston.

Clahsen, H. 1989: The grammatical characterization of developmental dysphasia. *Linguistics* 27, 897–920.

Clahsen, H. 1992: Linguistic perspectives on Specific Language Impairment. *Theorie des Lexicons Arbeitspapier* 37. Dusseldorf: Heinrich Heine Universitat.

Darwin, C.R. 1874. *The Descent of Man and Selection in Relation to Sex* (2nd edition). New York: Hurst and Company.

Fletcher, P. 1990: Untitled Scientific Correspondence. *Nature* 346.
Fodor, J. 1983: *The Modularity of Mind*. Cambridge, MA: MIT Press.
Goad, H. & Gopnik, M. 1994: Phoneme discrimination in Familial Language Impairment. In *Linguistic Aspects of Familial Language Impairment* (ed. J. Matthews), pp. 10–15. Special Issue of the *McGill Working Papers in Linguistics* 10.
Goad, H. & Rebellati, C. 1994: Pluralization in Specific Language Impairment: Affixation or Compounding? In *Linguistic Aspects of Familial Language Impairment* (ed. J. Matthews), pp. 24–40. Special Issue of the *McGill Working Papers in Linguistics* 10.
Gould, S.J. 1981: *The Mismeasure of Man*. New York: Norton.
Gopnik, M. 1990: Feature-blind grammar and dysphasia. *Nature* 344, 715.
Gopnik, M. 1994a: Impairments of tense in a familial language disorder. *Journal of Neurolinguistics* 8, 109–133.
Gopnik, M. 1994b: The articulatory hypothesis: production final alveolars in monomorphemic words. In *Linguistic Aspects of Familial Language Impairment* (ed. J. Matthews), pp. 129–134. Special Issue of the *McGill Working Papers in Linguistics* 10.
Holyoak, K.J. & Thagard, P. 1995: *Mental Leaps: analogy in creative thought*. Cambridge, MA: MIT Press.
Hurst, J.A., Baraitser, M., Auger, E., Graham, F. & Norell, S. 1990: An extended family with a dominantly inherited speech disorder. *Developmental Medicine and Child Neurology* 32, 352–355.
Johnston, J. (In press) Specific Language Impairment, Cognition, and the Biological Basis of Language. In *The Biological Basis of Language* (ed. M. Gopnik). Oxford University Press.
Kehayia, E. 1993: Morphological priming of inflectionally and derivationally complex words: a neurolinguistic and psycholinguistic study. Presented at the 31st international meeting of the *Academy of Aphasia*, Tucson, Arizona.
Kehayia, E. 1994: Whole-word access or decomposition in Familial Language Impairment: A psycholinguistic study. In *Linguistic Aspects of Familial Language Impairment* (ed. J. Matthews), pp. 10–15, Special Issue of the *McGill Working Papers in Linguistics* 10.
Kehayia, E. & Jarema, G. 1994: Morphological priming (or prim#ing) of inflected verb forms: A comparative study. *Journal of Neurolinguistics* 8(2), 83–94.
Leonard, L.B. 1989: Language learnability and specific language impairment in children. *Applied Psycholinguistics* 10, 179–202.
Leonard, L.B., Bortolini, U., Caselli, M.C., McGregor, K.K. & Sabbadini, L. 1992: Morphological deficits in children with specific language impairment: the status of features in the underlying grammar. *Language Acquisition* 2, 151–179.
Lieberman, P. 1992: Could an autonomous syntax module have evolved? *Brain and Language* 43, 768–774.
Lewis, B.A. & Thompson, L.A. 1992: A study of developmental speech and language disorders in twins. *Journal of Speech and Hearing Research* 35, 1086–1094.
Marean, G.C., Werner, L.A. & Kuhl, P.K. 1992: Vowel catergorization by very young infants. *Developmental Psychology* 28, 396–405.
Matthews, J. (ed.) 1994: *Linguistic Aspects of Familial Language Impairment*. Special Issue of the *McGill Working Papers in Linguistics* 10 (1 & 2).
Miller, J.F. 1981: *Assessing Language Production in Children*. Baltimore, MD: University Park Press.
Napps, S.E. 1989: Morphemic relationships in the lexicon: Are they distinct from semantic and formal relationships? *Memory and Cognition* 17, 729–739.
Paradis, M. & Gopnik, M. 1994: Compensatory strategies in Familial Language Impairment. In *Linguistic Aspects of Familial Language Impairment* (ed. J. Matthews), pp. 142–149. Special Issue of the *McGill Working Papers in Linguistics* 10.

Pembrey, M. 1992: Genetics and language disorder. In *Specific Speech and Language Disorders in Children: Correlates, Characteristics, and Outcomes* (ed. P. Fletcher & D. Hall). San Diego, CA: Singular.

Piggott, G. & Robb, M. 1994: In *Linguistic Aspects of Familial Language Impairment* (ed. J. Matthews), pp. 16–24. Special Issue of the *McGill Working Papers in Linguistics* 10 (1 & 2).

Pinker, S. 1991: Rules of language. *Science* 253, 530–535.

Pinker, S. 1994: *The Language Instinct*. New York: Morrow and Company.

Pinker, S. & Bloom, P. 1990 Natural language and natural selection. *Behavioral and Brain Sciences* 13, 707–784.

Pinker, S. & Prince, A. 1988: On language and connectionism: Analysis of a parallel distributed processing model of language acquisition. *Cognition* 28, 73–193.

Plante, E. 1991: MRI Findings in the Parents and Siblings of Specifically Language-Impaired Boys. *Brain and Language* 41, 67–80.

Plante, E., Swisher, L., Vance, R. & Rapcsak, S. 1991: MRI findings in boys with Specific Language Impairment. *Brain and Language* 41, 52–66.

Rice, M.L. 1994: Grammatical categories of children with Specific Language Impairment. In *Specific Language Impairments in Children* (ed. R. Watkins. & M. Rice), pp. 69–90. Baltimore, MD: Paul H. Brookes.

Rice, M.L. & Oetting, J.B. 1993: Morphological deficits of children with SLI: evaluation of number marking and agreement. *Journal of Speech and Hearing Research* 36, 1249–1257.

Rumelhart, D.E. & McClelland, J.L. 1986: On learning the past tenses of English verbs. In *Parallel Distributed Processing* (ed. J.L. McClelland, D.E. Rumelhart & the PDP Research Group), 2, pp. 216–271. Cambridge, MA: Bradford Books/MIT Press.

Samples, J.M. & Lane, V.W. 1985: Genetic possibilities in six siblings with specific language disorders. *Journal of the American Speech and Hearing Association* 27, 27–31.

Smith, N., Tsimpli, I.-M., & Ouhalla, J. 1994: Learning the impossible: the acquisition of possible and impossible languages by a polyglot savant. *Lingua*.

Stanners, R.F., Neiser, J.J. & Painton, S. 1979: Memory representation of prefixed words. *Journal of Verbal Learning and Verbal Behavior* 18, 733–743.

Stark, J. 1980: Aphasia in children. *Language Development and Aphasia in Children* (ed. R.W. Reiber). New York: Academic Press.

Tallal, P., Ross, R. & Curtiss, S. 1989: Familial aggregation in specific language impairment. *Journal of Speech and Hearing Disorders* 54, 167–173.

Tallal, P., Sainburg, R.L. & Jernigan, T. 1991: The Neuropathology of Developmental Disphasia: Behavioral, Morphological, and Physiological Evidence for a Pervasive Temporal Processing Disorder. *Reading and Writing: An Interdisciplinary Journal* 3, 363–377.

Tallal, P., Stark, R.E., Kallman, C. & Mellits, E.D. 1980: Developmental dysphasia: The relation between acoustic processing deficits and verbal processing. *Neuropsychologia* 18, 273–284.

Terrace, H.S., Pettito, L.A., Sanders, R.J. & Bever, T.G. 1979: Can an ape create a sentence? *Science* 206, 891–902.

Tomblin, J.B. 1989: Familial concentration of developmental language impairment. *Journal of Speech and Hearing Disorders* 54, 287–295.

Tomblin, J.B. & Buckwalter, P.R. 1994: Studies of Genetics of Specific Language Impairment. In *Specific Language Impairments in Children* (ed. R. Watkins & M. Rice), pp. 17–34. Baltimore, MD: Paul H. Brookes.

Ullman, M. & Gopnik, M. 1994: The Production of Inflectional Morphology in Hereditary Specific Language Impairment. In *Linguistic Aspects of Familial Language Impairment* (ed. J. Matthews), pp. 81–118. Special Issue of the *McGill Working Papers in Linguistics* 10, 1 & 2.

van der Lely, H.K.J. & Harris, M. 1990: Comprehension of reversible sentences in Specifically Language Impaired children. *Journal of Speech and Hearing Disorders* 55, 101–117.

Vargha-Khadem, F., Watkins, K., Alcock, K., Fletcher, P. & Passingham, R. 1995: Praxic and cognitive deficits in a large family with a genetically transmitted speech disorder. *Proceedings of the National Academy of Science* 92, 930–933.

Wyke, M.A. 1978: *Developmental Dysphasia*. New York: Academic Press.

Zangwill, O.L. 1978: The concept of developmental dysphasia. In *Developmental Dysphasia* (ed. M.A. Wyke), pp. 1–11. New York: Academic Press.

The Emergence of Cultures among Wild Chimpanzees

CHRISTOPHE BOESCH

Institute of Zoology, University of Basel, Rheinsprung 9, CH-4051 Basel, Switzerland

Keywords: chimpanzees; culture; imitation; social learning; social norms.

Summary. Culture has been granted by primatologists to the chimpanzee, on the base of the many population-specific behaviour patterns they possess. Psychologists tend to disagree arguing that individual learning constrained by ecological factors could produce the same results. After setting up some rigorous criteria to differentiate between these two opposite positions, I show that social canalization, including imitation, is important in explaining the acquisition of nut-cracking behaviour in wild chimpanzees. Then, I argue that a culture requires not only a social learning process to produce a faithful transmission of information, but also a mechanism that guarantees the permanence of the information between transmission events. Leaf-clipping, leaf-grooming, knuckle-knocking and a symbolic drumming communication system are proposed to be examples of behaviour patterns fixed within chimpanzee populations by social norms. The stringent criteria that have to be fulfilled to grant a behaviour cultural properties in an animal species strongly limit the possible candidates. Despite these restrictions, the repertoire of the wild chimpanzee includes many cultural behaviour patterns.

IS CULTURE UNIQUE TO MAN?

THE DEBATE ABOUT WHAT DISTINGUISHES MAN from the other animal species goes on for centuries. And the fact that Darwin proposed in his theory of evolution that human beings had a common ancestor with other primates did not ease the quest for what constitutes our uniqueness. In parallel with the progress of our understanding of the evolutionary processes, paleoanthropologists have discovered an impressive series of early human species showing that hominization was a gradual evolution from very chimp-like ancestors. Therefore, there has been an increasing tendency to search for clear-cut differences between human and other primates in the behavioural domain: tool use and tool making, food sharing and co-operation among others have been proposed to characterize man (Isaac 1978; Johansen & Edey 1981; Leakey 1980; Washburn 1978). But recent observations revealed that wild chimpanzees possess these behaviour patterns too (Boesch & Boesch 1990; Goodall 1986; Nishida 1987), showing that they are not exclusively part of the hominization process (Boesch-Achermann & Boesch 1994). As a consequence, culture and some cognitive abilities underlying it have been proposed to be uniquely human (Galef 1990; Tomasello 1990; Tomasello *et al.* 1993a; Visalberghi & Fragaszy 1990).

Biologists are also interested in culture as it is one of the mechanisms in nature allowing the transmission of information between individuals (Bonner 1980; Maynard-Smith & Szathmary 1995). Contrary to the more common genetical transmission of information, cultural transmission is not genetic. Potentially this non-genetic mechanism could be much quicker than genetic transmission, first because transmission is not dependent upon the reproductive events but could occur at any moment in the lifetime of an individual, and second, because innovation is not dependent upon a rare event such as mutation, and can occur much more often. It is this rapidity that has been proposed to be responsible for the high human social and behavioural diversity (Tomasello *et al.* 1993a; Segall *et al.* 1990).

The chimpanzee challenge

Chimpanzees have been proposed to possess culture (Boesch & Boesch 1990; Goodall 1973, 1986; McGrew 1992; Nishida 1987). The argument of the primatologists is based on the abundant evidence of population-specific behaviour patterns that have been observed. Comparing Figure 1 with Table 1 shows that the distribution of many proposed cultural behaviours in chimpanzees do not follow any obvious ecological differences. For example, the nut-cracking behaviour has been observed in only the western most forest chimpanzee populations in West Africa, whereas those living in forests some 30 km east of the Sassandra river in Côte d'Ivoire or further away do not crack these nuts (Boesch *et al.* 1994). In all the forests where investigations were done for presence or absence of the nut-cracking behaviour, nuts are available as well as the potential hammers and roots to

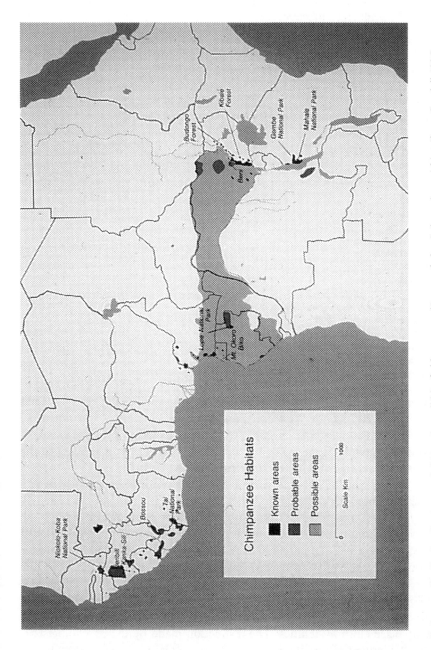

Figure 1. Present distribution (known, probable and possible) of chimpanzees (*Pan troglodytes*) throughout Africa (from Goodall 1986).

Table 1. List of behaviour patterns that have been proposed to be cultural and their distribution within populations of wild chimpanzees. To exclude the most obvious bias (length of the study period), positive and negative results are presented for long-term studies (more than 8 years) in which chimpanzees were directly observed, but only positive observations for shorter studies. [Assirik is located in Senegal (West Africa) and Kibale in Uganda (East Africa).]

Pattern	West Africa		Other sites	East Africa	
	Bossou	Taï		Gombe	Mahale
Ant-dip	+	+	Assirik	+	−
Fly-whisk	+	+		+	−
Leaf-sponge	+	+		+	−
Leaf-clip	+	+		−	+
Nut-crack	+	+		−	−
Play-start	−	+		+	+
Honey-dip	−	+		+	−
Hand-clasp	−	+	Kibale	−	+
Marrow-pick	−	+		−	−
Leaf-groom	−	−	Assirik	+	+
Termite-fish	−	−	Kibale	+	−
Leaf-napkin	−	−		+	−
Self-tickle	−	−		+	−

Bossou: Sugiyama 1981; Sugiyama & Koman 1979
Taï: Boesch & Boesch 1990
Gombe: Goodall 1986; McGrew 1992
Mahale: Nishida 1973; 1987; McGrew 1992
Assirik (Senegal): McGrew et al. 1979
Kibale (Uganda): Wrangham & Isabirye-Busata in McGrew 1992

use as anvils (Boesch et al. 1994). Similarly, the leaf-clipping behaviour has been observed in both West and East African chimpanzees but not in all populations (Boesch 1995; Nishida 1987). It is this seeming independence of cultural behaviour from ecological factors in chimpanzees that have led the primatologists to claim culture in chimpanzees.

Recently, psychologists have challenged this proposition by suggesting a more parsimonious explanation; what we observe in chimpanzees is merely the result of individual learning processes that are constrained by the ecological limitations in which the individuals are learning the task (Galef 1988; Heyes 1993; Tomasello 1990). To take one example, a youngster learning to crack nuts on its own in the Taï forest may end up using the same behaviour as other group members, because the technical and ecological limitation within the forest allows him to solve the problem in only one way. Under such a scheme, we would obviously not grant the chimpanzee cultural abilities. The distinction made by the psychologists is between individual learning and social learning processes and only the second ones could lead to a culture.

ARE CHIMPANZEES CAPABLE OF SOCIAL LEARNING?

What are the social learning processes?

Many different mechanisms have been recognized by different authors (in their review Whiten & Ham 1992 listed 27 of them). It seems relevant to differentiate the part of the task that is copied in the model. At the lowest level, 'local enhancement' is the process in which the naive observer uses an already acquired behaviour in a *new context* used by the model (Thorpe 1956). By 'stimulus enhancement', the observer's attention is drawn to a *stimulus* in the model's performance (Thorpe 1956). For example, if the model is cracking nuts, the observer will use the tool more frequently. In 'emulation', the observer attempts to reproduce or reach the *goal* that the model is pursuing (Tomasello 1990). Finally, in 'imitation', the observer is attempting to copy the *behaviour* of the model and this behaviour was not part of the repertoire of the observer previously (Piaget 1935; Thorpe 1956).

Recently, experiments done with captive primates showed that social transmission of knowledge is less easy for primates than was previously assumed (Galef 1988; Visalberghi & Fragaszy 1990). Even captive chimpanzees were thought to only emulate aspects of the sand-throwing or food-raking behaviour (Tomasello *et al.* 1987). This has strengthened the point of view of some that advocate that imitation is unique to man (Galef 1988; Tomasello 1990). However, other studies working with animals under enriched captive conditions showed that imitation was possible in chimpanzees (Custance & Bard 1994; Tomasello *et al.* 1993b) and in orang-utans (Russon & Galdikas 1995).

How to prove that chimpanzees are capable of social learning?

The bone of contention is to know if what we observe within a given population is the product of individual learning processes constrained by ecological factors or the result of a social learning process. Are termite-fishing, leaf-clipping, ant-dipping or nut-cracking in chimpanzees cultural sets of behaviour as proposed by primatologists working on these populations or are they the product of individual learning as championed by many psychologists?

The key aspect to settle such a point is to show that during the learning process young chimpanzees do not try all the possibilities they have, within the physical and psychological limits of their species, in order to achieve a task, but that they try only a subset of those and that the subset they try is influenced by what they see in social models. Therefore, we need to know *from the chimpanzees themselves* how they would learn a task when not affected by a model and what would be different if they were affected by a model. We, as members of another species with other physical and psychological limits, cannot decide what are these differences. If no difference is expected in the learning of a task, whether it is through individual learning or through social learning, we will not be in a position to differentiate between the two learning processes and we will not be able to test the social learning hypothesis. In other words, we have to use a very limited criterion to test our two alternatives on the learning process and this will, by definition, strongly limit the possible candidates for cultural transmission.

A last criterion has to be fulfilled in the sense that we need this information from a behaviour that has been proposed to be culturally transmitted. Tomasello and his collegues (1987) have shown that captive chimpanzees do not imitate sand-throwing behaviour. Despite the interest of this result, nobody has ever claimed sand-throwing to be a cultural behaviour in chimpanzees. In addition, the absence of imitation in sand-throwing does not say anything about the presence or absence of imitation in the learning processes of ant-dipping, leaf-clipping or nut-cracking behaviour in wild chimpanzees.

To conclude, to test the two alternatives about the learning process of a task, three criteria have to be fulfilled. First, the behaviour under consideration has to be a candidate of a cultural behaviour, which at the present stage mainly means that it is a population-specific behaviour not directly influenced by ecological factors (Boesch, in press). Second, the different ways the behaviour could be acquired or the final form once acquired should differ if learned by individual or social processes. Third, we need to gather the information on the second point from the chimpanzees themselves: What are the individual learning possibilities and what are the socially limited possibilities?

Nut-cracking might be a good candidate as it fulfils the three criteria; it has been proposed to be a cultural behaviour in chimpanzees (Boesch & Boesch 1990; Goodall 1986; McGrew 1992) and an attempt has been made to introduce it to a group of naive chimpanzees in Zürich zoo (Funk 1985), and this allows us to answer the second and third criteria. The chimpanzees in Zürich were offered nuts and hammers, and were observed for two weeks as they manipulated the objects, trying to open the nuts. None of them succeeded, and no reinforcement from other group members operated. If we compare the methods used by the Zürich and the Taï chimpanzees, we see that Taï chimpanzees tried definitively fewer different methods; of the 14 methods used by the Zürich chimpanzees, only 7 were seen in Taï chimpanzees (Table 2). This is intriguing because some of these rarely or unused methods were actually observed in Taï chimpanzees but in other contexts. Stabbing with a stick was observed against a leopard, and rubbing is observed regularly when feeding on other kinds of fruit. Thus, in

Table 2. Social canalization in the learning of nut-cracking behaviour in chimpanzees: list of all methods used to attempt to open nuts by two populations of chimpanzees. The study on the captive chimpanzees of the Zürich zoo was performed by Martina Funk (1985). A + indicates that the method was used in the population, whereas a − indicates that it was never observed.

Method	Zürich chimpanzee	Taï chimpanzee
Hit with a hammer	+	often
Bite the nut	+	often
Pound the nut against hard surface[1]	+	regular
Hit with hand the nut	+	regular
Hit with an object[3]	+	rare
Rub the nut against hard surface[1]	+	rare
Throw the hammer on nut	−	regular
Throw the nut against hard surface[1]	+	−
Hit with other body part[2]	+	−
Shake the nut	+	−
Press nut against teeth[4]	+	−
Sit on nut	+	−
Scratch the nut with fingers	+	−
Press on the nut	+	−
Stab with a stick	+	−

[1] Chimpanzees can rub, pound or throw the nut directly with the hand against the ground, a stone, a tree trunk or a root.
[2] By other body parts is understood the back of the hand or the elbow.
[3] By object, I understand material that could not make a hammer such as a piece of cloth, small twigs or in the Taï forest another nut, a piece of termite mound or a hard-shelled fruit.
[4] Chimpanzees pressed the nut with the hand against the teeth with the mouth kept open.

comparison to the Zürich chimpanzees, a social canalization of the individual learning potentialities is at work in Taï youngsters and this is strongly influenced by the behaviour observed in the model: 5 of the 7 methods used include behavioural movements commonly observed in adults cracking nuts (Boesch, in press). The object used by the models is less of a guidance, since young Taï chimpanzees used as hammers large hard-shelled fruits, pieces of termite mounds and rotten branches. In nut-cracking behaviour, social canalization through imitation is at work and it confines the individual learning possibilities to the different types of objects that could be used to pound the nuts.

WHAT MECHANISM IS NEEDED TO PRODUCE CULTURE?

There are three possible mechanisms:
   1 Imitation, teaching or co-operative learning are the only processes able to produce culture (Galef 1988; Tomasello *et al.* 1993b).

2 All social learning processes (imitation, teaching plus emulation, local enhancement...) produce culture (Whiten & Ham 1992; Russon & Galdikas 1995).

3 Social canalization is the criterion independently of which process produces it (Heyes 1993).

In my opinion, it seems rather arbitrary to single out one process of information transmission as the only one able to produce culture. First, individuals should use all sources of information that could help them to solve a task and information most probably will come from individual experiences and from social partners. For example, in some tasks, some kinds of information cannot be learned through imitation (e.g. non visual properties of tools such as hardness, weight), and without a combination of information from different sources the task will not be acquired. Second, it seems difficult to imagine that an individual imitates a behaviour without being at the same time influenced by the goal to be reached and the object to use, and vice versa. Therefore, I will adhere to the view that social learning processes in general, and not just imitative learning, contribute to cultural transmission. In support of this opinion are the studies showing that in some human populations learning through imitation is rare and sometimes absent (like in the 'Kung bushmen) (Olson & Astington 1993; Rogoff et al. 1993). This shows that in humans culture can develop without much reliance on imitation.

Social transmission of information is necessary to produce culture but is not enough. To use an analogy, in genetical transmission the mechanism itself is not so important and many different kinds are observed, i.e. isogamy or anisogamy, sexual or asexual transmission. However, evolution over generations is only possible because of inheritance that maintains the quality of the genetical information stable over time (heritability of a character has to be high for evolution to take place) (Maynard-Smith 1989). Similarly, culture is only possible if we have a mechanism that guarantees information stability between transmission events, e.g. during the retention period between acquisition and re-transmission. If the information is altered under the influence of individual and ecological factors during this retention period, we will never observe a culture (Heyes 1993). Permanence of the information in cultural transmission requires fidelity during transmission *and* a mechanism to guarantee fidelity between transmissions. In other words, social canalisation should be at work all the time.

Thus, the key to culture is not so much the precise transmission mechanisms, as we saw that many of them could be at work, but a permanence-guaranteeing mechanism. At present, the discussion about culture in animals has been restricted to the transmission mechanisms, and this important aspect of the problem has been forgotten. I shall devote some time to this aspect of culture in chimpanzees.

HOW CAN PERMANENCE OF THE INFORMATION BE GUARANTEED?

Two mechanisms have been proposed (Heyes 1993):
  1 The information could be stored in object or language supports. This could include all extrasomatic artefacts, i.e. technological objects, books, myths and fairy tales about traditions and the past.
  2 Social norms (or social conventions) limit variations in the information, as not all possibles will be socially acceptable.

Logically, we think that these two processes would be likely to operate in conjunction with a symbolic or instructional process of learning, as seen in our own species. However, this might only be an assumption and I think that in the case of the second mechanism, this assumption might be wrong, since this mechanism applies also to animals.

How can we evaluate this point with animals?

Social norms will exist when a strong social canalisation exists in an animal population, that prevents or discourages individuals from modifying the socially acquired information through individual learning and from testing it against all possibilities allowed within a given ecological context. As we can see, such a situation could provide us with a solution to our problem. The lack of testing all possibilities in an ecological context might lead to non-adaptive solutions being retained or to arbitrary solutions being used. I shall review the evidence for the two mechanisms in wild chimpanzees.

Cultural behavioural patterns are maladaptive

The three criteria proposed above still apply, as we need a behaviour that has been proposed to be cultural and we need two different solutions possible from the chimpanzees' point of view. Now, if the solution used by all group members is also the best possible of the two alternatives, we will not be able to differentiate whether what we observed is the result of modifications through individual learning or the result of rigidity resulting from a social norm. This is because we expect individuals to test the possibilities and choose the best ecological solution they find. But if, of the two solutions possible, the group members use the one we know not to be the best one, we could exclude the individual learning alternative. Therefore, for the present test, we need to exclude cases where the animals use the best ecological alternative to a given problem. Please note that such situations might be rare, as we expect in wild populations natural selection to be at work, and there will be a cost related to the selection of non-adaptive behaviours.

The *ant-dipping behaviour* is to my knowledge the best example of such a culturally non-adaptive behaviour of which the maintenance can be explained by different social norms prevailing in different social groups. The ant-dipping behaviour has not been observed in all chimpanzee populations (Table 1), but, more important, the chimpanzees found two different techniques to dip for the ants (Boesch & Boesch 1990; Goodall 1986). Both Gombe and Taï chimpanzees use sticks that they dip into the nest entrance of the driver ants of the species *Dorylus nigricans*, so as to eat them. In Gombe, chimpanzees use one hand to hold the stick among the soldier ants guarding the nest entrance and, once they have swarmed about halfway up the tool, withdraw the stick and sweep it through the closed fingers of the free hand; the mass of insects is then rapidly transferred to the mouth (McGrew 1974). Gombe tools are in average 66 cm long and the dipping is performed 2.6 times per minute. McGrew (1974) estimates that they take 292 ants per dipping movement. In Taï, the chimpanzees hold the stick with one hand among the soldier ants guarding the nest entrance until they have swarmed about 10 cm up the tool. Then, they withdraw it, twist the hand holding it and directly sweep off the ants with the lips. Taï chimpanzees use short sticks of about 24 cm long and perform the dipping movement about 12 times per minute (Boesch & Boesch 1990). We estimate, from our own trials, that they obtain 15 ants per dipping movement. In Gombe, the Taï dipping movement has been observed only sometimes with two individuals, McGregor and Pom (McGrew 1974).

So ants can be dipped by two different techniques, but each of them is seen in only one chimpanzee population. I have tested the two techniques in the two sites and found no ecological factor that would prevent the use of either of them in both sites. The Gombe technique is four times more efficient than the one used in Taï (Gombe, 760 ants/minute; Taï, 180 ants/minute; Boesch & Boesch 1990). Here, Taï chimpanzees restrict themselves to an ecologically sub-optimal solution that must be maintained by a social norm preventing the individuals from testing all possibilities.

Arbitrariness of behaviour is socially dependent

With the same line of argument, the solution selected by group members might have no connection to an ecological solution but be purely socially determined. Such a solution would then be independent of ecological factors and would not present the cost that we expect for non-adaptive behaviours. Therefore, we should expect them to be more frequent.

Leaf-clipping among wild chimpanzees. This behaviour was first described in the Mahale chimpanzees in Tanzania: 'A chimpanzee picks one to five stiff leaves, grasps the petiole between the thumb and the index finger, repeatedly pulls it from side to side while removing the leaf blade with the incisors, and thus bites the leaf to pieces. In removing the leaf blades, a ripping sound is conspicuously and distinctly produced. When only the midrib with tiny pieces of the leaf blade remains, it is dropped and another sequence of ripping a new leaf is often repeated' (Nishida 1987: 466). Note that nothing of the leaves is eaten. This behaviour has also been seen regularly at Bossou (Sugiyama 1981) and Taï (Boesch 1995) but only twice at Gombe (Goodall, personal communication cited in Nishida 1987). The fact that this behaviour is present in three chimpanzee populations but absent in a fourth one could be explained by an ecological difference, although we do not know yet what difference might produce such an irregular distribution of that behaviour.

When present, the function of this behaviour seems arbitrary. In Mahale, the chimpanzees most often use it as a herding or courtship display in sexual contexts (23 of 41 observations; Nishida 1987). Young adult males and adult oestrous females apparently perform it to attract the attention of group members of the other sex (Huffman, personal communication). In Bossou, it occurs mostly in apparent frustration or in play (41 of 44 observations; Sugiyama 1981, personal communication). During the habituation period, individuals surprised in trees would leaf clip while looking at the observer. Once habituation was more advanced, this form of leaf clipping disappeared and is seen now only in youngsters at play. In Taï, leaf-clipping is mainly part of the drumming sequence of the adult males (249 of 319 observations; Boesch 1995) and is seldom seen during a resting period (32 cases) or in frustration situations (34 cases) (Table 3). It seems very difficult to propose ecological reasons to account for the fact that each chimpanzee populations uses leaf clipping in a different context. The arbitrariness in the context of use observed in three chimpanzee populations suggests that leaf clipping is a cultural behaviour whose context of use is locally determined by a social norm fixed among group members.

Conventional behaviour patterns in chimpanzees. Such social norms seem to have influenced other behaviours as well (Table 3). For example, in Mahale young males use leaf clipping to attract the attention of oestrous females (Nishida 1987). This is less conspicuous than the dominant males' way, who routinely shake saplings for the same purpose. Intriguingly, Taï

Table 3. Behavioural variants following population specific norms in wild chimpanzee populations (see text for more explanations).

	Bossou	Gombe	Mahale	Taï
BEHAVIOUR		FUNCTION		
Leaf-clip	Play	–	Courtship	Drumming + Resting
FUNCTION		BEHAVIOUR		
Courtship	–	–	Leaf clip	Knuckle knock
Squash ectoparasite	–	Leaf groom	–	Index hit

low-ranking males also use a less conspicuous way to attract the attention of oestrous females than sapling-shaking behaviour and that is by knocking with their knuckles on the trunks of small saplings (Table 3). Here, we have a class of individuals that have a social problem to solve, 'attract the attention of an oestrus female', and that solve it in two different ways (our criteria 2 and 3). And a different solution is used in each population. Why chimpanzees in Mahale do not knock to attract the attention of females seems to be arbitrarily fixed and determined by a social norm that group members acquire by social learning processes.

Similarly, Gombe chimpanzees have recently started to use leaf grooming in order to squash ectoparasites that they find while grooming somebody else or themselves (Boesch 1995). This has been observed in most members of the group. In Taï, chimpanzees also squash ectoparasites they find during grooming sessions in order to eat them, but they do it in a different way to the Gombe chimpanzees; they place the parasite on one forearm and hit it with the tip of the forefinger until it is squashed, and then eat it. Here, again, we have two different solutions to the same problem, of which only one is used in each population (Table 3). The arbitrariness of the solutions retained as well as of the decision of which one is to be used in one population seems also to indicate that a social norm is at work and that group members acquire it through social learning processes.

Symbolic drumming code in Taï chimpanzees. The last example I want to give of an arbitrary behaviour concerns a case of symbolic communication in forest chimpanzees (Boesch 1991a). Chimpanzees forage typically in ever-fluctuating parties of 7–12 individuals, remaining permanently in auditory contact with the majority (75%) of the community (of 80 chimpanzees), and follow for hours a constant direction even if totally silent. Normally the community splits in at least three major parties that may communicate with one another by vocalizing and drumming. Buttressed trees are abundant in this forest and adult males, after loudly pant-hooting, hit these buttresses powerfully and rapidly with their hands, feet or both. Drumming is a way for males to communicate their position to other group members and it may inform them about the direction in which the drummer progresses, and thus contains information about the group's progression.

However, we suspected that these drummings were more than just an indication of an individual's position, because we tended to lose contact with them just after some drummings were heard. It seemed that the whole chimpanzee community had abruptly and often silently changed direction following these drummings. It took many months to unravel this communicative system. During this time, I learned to differentiate the pant-hootings of the individual adult males. In early 1982, three years after we had initiated the study of the Taï chimpanzees, I began to realize that it was only after Brutus, the alpha male, drummed that the community reacted by abruptly changing the direction of travel. On some occasions, Brutus's drumming sequence appeared to transmit a specific message. There was no audible difference between sequences that did or did not have such a message; rather this message was indicated by the spatial and numerical combination of the sequences. During a 16 month period (January 1983 to May 1984), I studied the information conveyed in Brutus's drummings and was able to identify three messages in Brutus's emissions (Table 4).

1 Change in the travel direction. Brutus, by drumming twice at two different trees, indicated to other community members the direction he was proposing. The direction followed by Brutus when moving between the two drummed trees was used by other group members as indicating the new travel direction he was proposing. In addition, such drummings always occured within a time interval not exceeding two minutes. Individuals that were not part of Brutus's party apparently inferred the direction proposal by mentally visualizing Brutus's displacement between the two trees and then transposed it to their travel direction. Table 4 summarizes the number of occurrences in which I could identify the transfer of information about directions.

2 Indication about resting periods. On other occasions, Brutus seemed to propose a resting period of a specific duration that the community would follow: this was communicated by drumming twice at the same tree within 2 minutes. I was able to identify this message from Brutus in 14 cases (see Table 4) when the community activity stopped for an average of 60 minutes ($N = 12$, range $= 55$ to 65 minutes). Community activity was judged to be resting by the absence of movement and vocalization of parties not observed

Table 4. Symbolic communication in Taï chimpanzees: Brutus's communication system with the frequency of emissions in which communication about travel direction and resting duration could be identified. The number of cases heard correspond to the number of response of the group members in aggreement with my prediction of their response to Brutus's message, except for one case of 1 hour rest, in which Brutus himself canceled his message by drumming farther away 7 minutes later.

Number of drumming	Location of emission	Group response	Number of cases heard
2	same	1 hour rest	8
2	different	Change of direction	8
3	same	–	–
3	different	1 hour rest + change of direction	6
4	same	2 hours rest	1
4	different	–	–

as well as by the behaviour of the party under observation. After this rest, parties sometimes indicated vocally that they had begun to move. A chimpanzees' resting bout in the wild corresponds to an hour and Brutus proposed probably such a resting bout rather than the duration of one hour.

3 *Direction and resting time combined.* By combining both messages, Brutus could propose both a change of direction and an hour's rest; in such a case he would drum once at a tree on the movement axis and then twice at another tree in the direction he was proposing (see Table 4) within a short period of time. Alternatively, Brutus could drum twice on the axis and then once further in the proposed direction. In all cases, the information about time had an immediate effect, whereas that about the direction applied only later. It is worthwhile noting that if Brutus were simply adding information about direction and time, he would have drummed four times (twice for each kind of information). In fact, he really combined them and drummed only three times; thus, one of the drummings contained information on both direction and time.

Brutus stopped using this code rather abruptly, when several of the prime males suddenly disappeared from the community, probably through poaching, and as a consequence the number of travel parties diminished (Boesch 1991a). This mode of symbolic communication has only been observed in the Taï chimpanzees and in this community its use was also limited in time. This is clearly emphasizing the arbitrariness of this form of communication.

In conclusion, the examples given in this section show clearly that social norms exist in wild chimpanzee populations and that they do limit the variation that might be introduced by individual learning. Social norms are thought to bring social advantages that could compensate for the possible costs related to adopting the norm. This is obvious for communicative gestures: if one leaf clips in another context than the one generally used, the risk of being misunderstood exists. Therefore, social norms in the communicative gesture domain are not a surprise. However, this does not seem to apply to domains that represent solutions to ecological problems, like ant dipping or parasite squashing. Why do Taï chimpanzees never dip for ants using the Gombe technique that is so much more efficient? Similarly, why squash parasites only on the forearm, when other methods seem as efficient? It has been argued in humans that one effect of culture is to allow a better identification with a social entity and that part of what we observe merely functions to differentiate individuals from different groups (Segall *et al.* 1990). Would this apply to chimpanzees?

The two mechanisms required to allow a culture have been found in wild chimpanzees: first social learning processes that guarantee the fidelity of the information transfer between individuals, and second, social norms that guarantee the fidelity of the information once it has been acquired by an individual.

IS CULTURE RARE IN ANIMALS?

When studying culture in animals, it is generally required that the behaviour is independent of ecological factors (Bonner 1980; Tomasello 1990). Here, I adopted rigorous criteria to identify the presence of cultural behaviours in chimpanzees. To prove both the existence of social learning processes and of social norms, we need tasks that can be solved in more than one way by the species under consideration, and the individual learning results should differ from the social ones. This excludes many behaviour patterns from the analysis, such as termite fishing or leaf sponging, because we have observed these to be done in only one way by wild chimpanzees. In addition, if individual learning with ecological constraints and social learning give the same result for the same task, we would deny the second explanation on grounds of parsimony. Under such criteria, the appropriate tasks will be difficult to identify and possibly rare. Thus, culture in animals will be rare by definition. If we would apply the same criteria to human cultural behaviours, the list would also be much shorter. Similarly, if we lift the ecological independence criteria for the chimpanzees, culture in this species would then be present in many more aspects of their life (such as hunting or food sharing behaviours).

Nevertheless, we described clear examples of cultures in wild chimpanzees. Does it mean that cultures in chimpanzees and humans are identical and that they could not be used as a criteria to distinguish the two species? I would like first to sound a cautionary note: compared to man, the most studied species on earth, we know impressively little about wild chimpanzees which have been studied for only 30 years in at most 200 different individuals. Any conclusion when comparing the two species has to be tentative. Having said this it seems far-fetched to pretend that human cultures are similar to chimpanzee cultures. But are they qualitatively or only quantitatively different? The comparisons we made showed that the basic mechanisms required to produce culture in humans are present in chimpanzees, whether it be social learning including imitation, teaching and instructional learning (Boesch 1991b), as well as the social norms. Why then is culture not observed in many more aspects of the chimpanzees' behaviour? This brings us to the basic question about the function of culture, which seems to be the possibility for much more rapid adaptation than genetical evolution allows. I would suggest that wild chimpanzees with their suspected limited migratory potentials (Morin et al. 1994) live in stable ecological and social environments and that the need for rapid adaptation is limited. This also seemed to be the case for our early ancestors as is observed in the incredibly stable and rudimentary cultural products for most of the history of *Homo habilis*, *Homo erectus* and for early *Homo sapiens* (Davidson & Noble 1993; Toth & Schick 1993). Only when the number and the products of the cultural behaviours reached a certain threshold did they become part of the environment and require much quicker evolution, thus sparking off the 'human revolution' (Mellars & Stringer 1989).

Note. I thank the 'Ministère de la Recherche Scientifique', the 'Ministère de l'Agriculture et des Ressources Animales' of Côte d'Ivoire, the Tanzania Commission for Science and Technology, the Serengeti Wildlife Research Institute as well as the Tanzania National Parks for permitting this study, and the Swiss National Foundation, the L.S.B. Leakey Foundation and the Jane Goodall Institute for financing it. In Côte d'Ivoire, this project is integrated in the UNESCO project Taï-MAB under the supervision of Dr. Henri Dosso. I am most grateful to A. Aeschlimann, F. Bourlière, J. Goodall, S. Stearns and H. Kummer for their constant encouragement. I thank H. Boesch-Achermann for commenting upon and correcting the manuscript. Figure 1 is reproduced from Goodall (1986) by kind permission of Harvard University Press.

REFERENCES

Boesch, C. 1991a: Symbolic communication in wild chimpanzees? *Human Evolution* 6(1), 81–90.
Boesch, C. 1991b: Teaching in wild chimpanzees. *Animal Behaviour* 41(3), 530–532.
Boesch, C. 1995: Innovation in wild chimpanzees. *International Journal of Primatology* 16(1), 1–16.

Boesch, C. (in press). Three approaches for assessing chimpanzee culture. In *Reaching into Thought* (ed. A. Russon, K. Bard & S. Parker). Cambridge: Cambridge University Press.

Boesch, C. & Boesch, H. 1990: Tool use and tool making in wild chimpanzees. *Folia Primatologica* 54, 86–99.

Boesch, C., Marchesi, P., Marchesi, N., Fruth, B. & Joulian, F. 1994: Is nut cracking in wild chimpanzees a cultural behaviour? *Journal of Human Evolution* 26, 325–338.

Boesch-Achermann, H. & Boesch, C. 1994: Hominization in the rainforest: The chimpanzee's piece of the puzzle. *Evolutionary Anthropology* 3(1), 9–16.

Bonner, J. 1980: *The Evolution of Culture in Animals*. Princeton: Princeton University Press.

Custance, D. & Bard, K. 1994: The comparative and developmental study of self-recognition and imitation: The importance of social factors. In *Self-awareness in Animals and Humans* (ed. S. Parker, R. Mitchell & M. Boccia), pp. 207–226. Cambridge: Cambridge University Press.

Davidson, I. & Noble, W. 1993: Tools and language in human evolution. In *Tools, Language and Intelligence: Evolutionary Implications* (ed. K. Gibson & T. Ingold), pp. 363–388. Cambridge, Cambridge University Press.

Funk, M. 1985: Werkzeuggebrauch beim öffnen von Nüssen: Unterschiedliche Bewältigungen des Problems bei Schimpansen und Orang-Utans. Diplomarbeit: University of Zürich.

Galef, B. 1988: Imitation in animals. In *Social Learning: Psychological and Biological Perspectives* (ed. B. Galef & T. Zentrall), pp. 3–28. Hillsdale: Lawrence Earlbaum.

Galef, B. 1990: Tradition in animals: field observations and laboratory analyses. In *Interpretation and Explanation in the Study of Animal Behavior* (ed. M. Bekoff & I). Jamieson), pp. 74–95. Boulder: Westview Press.

Goodall, J. 1973: Cultural elements in a chimpanzee community. In *Precultural Primate Behaviour* (ed. E.W. Menzel), pp. 195–249. Basel, Karger: Fourth IPC Symposia Proceedings.

Goodall, J. 1986: *The Chimpanzees of Gombe: Patterns of Behavior*. Cambridge: The Belknap Press of Havard University Press

Heyes, C.M. 1993: Imitation, culture and cognition. *Animal Behaviour* 46, 999–1010.

Isaac, G. 1978: The food sharing behavior of protohuman hominids. *Scientific American* 238, 90–108.

Johanson, D. & Edey, M. 1981: *Lucy: The Beginnings of Humankind*. New York: Simon and Schuster.

Leakey, R. 1980: *The Making of Mankind*. London: Book Club Associates.

Maynard-Smith, J. 1989: *Evolutionary Genetics*. Oxford: Oxford University Press.

Maynard-Smith, J. & Szathmary, E. 1995: *The Major Transitions in Evolution*. Oxford: Freeman.

McGrew, W. C. 1974: Tool use by wild chimpanzees in feeding upon driver ants. *Journal of Human Evolution* 3, 501–508.

McGrew, W.C. 1992: *Chimpanzee Material Culture: Implications for Human Evolution*. Cambridge: Cambridge University Press.

McGrew, W.C., Baldwin, P.J. & Tutin, C.E. G. 1979: Chimpanzees, tools and termites: Cross-cultural comparisons of Senegal, Tanzania and Rio Muni. *Man*, 14, 185–214.

Mellars, P. & Stringer, C. 1989: *The Human Revolution*. Oxford, Oxford University Press.

Morin, P.A., Moore, J.J. Chakraborty, R., Jin, L. Goodall, J. & Woodruff, D.S. 1994: Kin selection, social structure, gene flow and the evolution of chimpanzees. *Science*, 265, 1193–1201.

Nishida, T. 1973: The ant-gathering behaviour by use of tools among wild chimpanzee of the Mahale Mountains. *Journal of Human Evolution* 2, 357–370.

Nishida, T. 1987: Local traditions and cultural transmission. In *Primate Societies* (ed. S.S. Smuts, D.L. Cheney, R.M. Seyfarth, R.W. Wrangham & T.T. Strusaker), pp. 462–474. Chicago: University of Chicago Press.

Olson, D. & Astington, J. 1993: Cultural learning and educational process. *Behavioural and Brain Science* 16(3), 531–532.

Piaget, J. 1935: *La Naissance de l'Intelligence chez l'Enfan*t. Neuchâtel: Delachaux et Niestlé.

Rogoff, B., Chavajay, P. & Matusov, E. 1993: Questioning assumptions about culture and individuals. *Behavioural and Brain Science* 16(3), 533–534.

Russon, A.E. & Galdikas, B. 1995: Constraints on great apes' imitation: model and action selectivity in rehabilitant orangutan imitation. *Journal of Comparative Psychology* 109(1), 5–17.

Segall, M., Dasen, P., Berry, J. & Poortinga, Y. 1990: *Human Behavior in Global Perspective: An Introduction to Cross-Cultural Psychology*. New York: Pergamon Press.

Sugiyama, Y. 1981: Observations on the population dynamics and behavior of wild chimpanzees ot Bossou, Guinea, 1979–1980. *Primates* 22, 435–444.

Sugiyama, Y. & Koman, J. 1979: Social structure and dynamics of wild chimpanzees at Bossou, Guinea. *Primates* 20, 323–339.

Thorpe, W. 1956: *Learning and Instinct in Animals*. London: Methuen.

Tomasello, M. 1990: Cultural transmission in tool use and communicatory signalling of chimpanzees? In *Comparative Developmental Psychology of Language and intelligence in Primates* (ed. S. Parker & K. Gibson), pp. 274–311. Cambridge: Cambridge University Press.

Tomasello M., Davis-Dasilva M., Camak L. & Bard K. 1987: Observational learning of tool-use by young chimpanzees. *Journal of Human Evolution* 2, 175–183.

Tomasello, M., Kruger, A. & Ratner, H. 1993a: Cultural learning. *Behavioural and Brain Science* 16(3), 450–488.

Tomasello, M., Savage-Rumbaugh, S., & Kruger, A. 1993b: Imitation of object related actions by chimpanzees and human infant. *Child Development* 64, 1688–1705.

Toth, N. & Schick, K. 1993: Early stone industries and inferences regarding language and cognition. In *Tools, Language and Intelligence: Evolutionary Implications* (ed. K. Gibson & T. Ingold), pp. 346–362. Cambridge, Cambridge University Press.

Visalberghi, E. & Fragaszy, D. 1990: Do monkeys ape? In *Language and intelligence in Monkeys and Apes: Comparative Developmental Perspectives* (ed. S. Parker & K. Gibson), pp. 247–273. Cambridge: Cambridge University Press.

Washburn, S.L. 1978: The evolution of man. *Scientific American* 239, 146–154.

Whiten, A. & Ham, R. 1992: On the nature and evolution of imitation in the animal kingdom: reappraisal of a century of research. In *Advances in the Study of Behavior* (ed. P. Slater, J. Rosenblatt & C. Beer), pp. 239–283. New York: Academic Press.

Terrestriality, Bipedalism and the Origin of Language

LESLIE C. AIELLO

*Department of Anthropology, University College London,
Gower Street, London, WC1E 6BT*

Keywords: evolution of language; cognition; human evolution; evolution of the brain.

Summary. Language is unique to humans, but in the context of the long time span of human evolution it is a fairly recent innovation. All evidence suggests that human brain size and inferred cognitive and linguistic abilities reached their modern norms only within the last quarter of a million years. Foundations for human linguistic and cognitive evolution, however, lie much further back in evolutionary history. Arguments are presented suggesting that these unique human abilities are the legacy of our ancestors' terrestrial and bipedal adaptations. Both terrestriality and bipedalism are directly associated with the unique human propensity for vocal communication, including the range and quality of sound that all humans are capable of producing. Furthermore, terrestriality and bipedalism can also be directly associated with an increase in brain size and cognition. Increases in group size accompanying a committed terrestrial adaptation would have put a premium on social, or Machiavellian intelligence while bipedalism would have been associated with the increased neural circuity involved in enhanced speed and co-ordination of hand and arm movements. The constricted bipedal pelvis would have also necessitated the birth of less mature offspring, exposing them to a rich environment while the brain was still rapidly growing and developing. A larger brain is not without its costs, however. Energetic arguments are also presented which suggest that a large brain can only evolve in concert with a change to a high quality diet, resulting directly in lifestyle changes for our early ancestors.

© The British Academy 1996.

All of these features were in place by the appearance of early *Homo erectus* about 1.8 million years ago and underpinned an apparently stable hominine adaptation that lasted for well over 1.5 million years. Modern human cognitive and linguistic talents are rooted in this earlier *Homo erectus* adaptation and may have begun to develop in response to further need for increased group size. Both the costs and the benefits of this later increase in brain size are considered.

INTRODUCTION

LANGUAGE IS SECOND NATURE TO HUMANS. In everyday life we seldom give serious thought to this phenomenal talent for communication and the fact that normal individuals raised in normal cultural environments acquire language easily, without effort and with little, if any, tuition.

One starting point in the study of the evolution of human language is to understand precisely how it differs from the communication systems of our closest living relatives, the non-human primates. Deacon (1992) has outlined three fundamental areas of difference. The first of these concerns the actual physical production of sound. Only humans have a vocal tract that is shaped in such a way as to permit the production of the range of sounds used in human language. Of particular importance is the fact that humans are able to produce consonants which act as 'stops' in the continuous flow of sound. These consonants are essential to our ability to decode, or make sense of, vocal language. We also have the neurological co-ordination that permits the necessary complex articulatory movements of the mandible, lips and tongue in respect to the teeth, palate and pharynx which allows us to produce not only consonants but also a wide range of vowel sounds. An important point is that these articulatory movements have to be learned, they are not innate.

The second unique aspect of human language is that it involves the use of a finite number of sounds to generate an infinite number of meanings. The order of these sounds, or the syntax, is the source of complex meaning in language. The third unique aspect is the symbolic nature of human language. Combinations of sounds, whether at the level of a word, or a string of words, have complex meanings that are easily recognized by the community of speakers of a particular language but are arbitrary in relation to the object or concept that they represent.

The fact that human language is unique in these three major aspects (the physical production of sound, syntax and symbolic content) does not mean that the non-human primates, and particularly African apes, lack all linguistic ability. Attempts to teach African apes 'non-verbal' language based on either American Sign Language for the Deaf or on computer-based symbol systems have established that they have some of the basic cognitive pre-requisites for language. For example, there seems to be little doubt any longer that African apes, and particularly pygmy chimpanzees, can associate abstract meanings with symbols and use these symbols in novel situations (Savage-Rumbaugh & Lewin 1994; Savage-Rumbaugh et al. 1993). This ability should perhaps not seem too surprising in view of the fact that at least one species of monkey is known to produce calls in the wild that have specific symbolic content (Seyfarth et al. 1980). Claims for syntactic abilities in non-human primates are arguably more controversial, but apes are capable of producing two-symbol combinations that are analogous to the two-word syntax of very young human children. It is also interesting that they appear to be able to understand some of the more complicated syntax of spoken English (Savage-Rumbaugh & Lewin 1994). This is particularly intriguing in view of the fact that, although they are not capable of producing spoken language themselves, they do have both the auditory abilities and the cognitive capacity to interpret at least some of the continuous flow of sound characteristic of human language.

It is clear that the basic symbolic and syntactic talents of living apes exist in these species without the co-existence of symbolically based, vocal (or for that matter, non-vocal), language systems. These abilities must stem form some other aspect of primate life (Povinelli & Preuss, in press). One obvious area for the pre-linguistic use of these talents is social cognition. Work summarized by Tomasello & Call (1994) has shown that there may be little difference between humans and other higher primates in the major aspects of social cognition, and particularly in the basic ability to associate meaning with individuals and to manipulate social situations. The differences that do exit between humans and non-human higher primates more often involve the elaboration and sophistication of basic abilities that exist in other primates than the development of qualitatively new abilities. In particular, Povinelli & Preuss (1995) have recently argued that humans may have specialized in a particular kind of intelligence related to understanding mental states such as desires, intentions, and beliefs.

If the cognitive talents present in living African apes also characterized the last common ancestor of these apes and ourselves, the evolution of human language would have involved a significant elaboration of abilities for sequencing and syntax as well as a significant increase in symbolic capacity. At the same time, it would have involved the development of the vocal basis for language. Any approach to the evolution of human language must take into consideration both of these aspects, the cognitive as well as the vocal. It also must address the questions of why these changes occurred and when in the course of human evolution human language became established. It is the purpose of this contribution to examine these issues.

THE WHY AND THE WHEN OF HUMAN LANGUAGE EVOLUTION

In recent years both the why and the when of language evolution have been under considerable debate. The major issue in relation to the why of language evolution has centred on the question of whether human language appeared as the result of natural selection for linguistic ability (Pinker & Bloom 1990; Pinker 1994). The alternative is that it appeared as a side effect of other evolutionary forces such as the increase in brain size or constraints of brain structure and growth (Piattelli-Palmarini 1989). There are a number of quite compelling theoretical arguments that can be put forward to argue that linguistic ability has specifically been selected for in human evolution (Pinker & Bloom 1990; Pinker 1994). These include the extreme improbability that the complex neurological structures underlying a function as complex as human language could have arisen either entirely by chance or as a by-product of some other unrelated function. The fact that children can acquire language extremely rapidly and with minimal, if any, tuition is strong evidence for the predisposition of the human brain not only for the symbolic requirements of language but also for the sequencing and syntactic structure of all human language.

Perhaps the strongest evidence that language ability was specifically selected for in the course of human evolution is the fact that the prefrontal cortex in humans is the only area of the human brain that is disproportionately large in relation to the brains of other primates (Deacon 1992). The prefrontal cortex is that area of the brain that is specifically responsible for many features of language production and comprehension as well as the unique human ability to reflect on one's own mental states and those of others (Povinelli & Preuss 1995). It is difficult to understand why this particular area of the brain would be the only area that was so disproportionately large if the functions it serves were not interrelated and also the object of continued selection during the course of human evolution. Recent work on the relative sizes of cortical areas in primates support this idea by demonstrating that in animals that are active in daylight, the visual cortex is disproportionately large in relation to its size in nocturnal animals (Barton et al. 1995). There is little reason to have a disproportionately large visual cortex unless the visual acuity conferred by the enlarged cortex gave a specific reproductive advantage to its owner. The same can be said for language, cognition and the prefrontal cortex.

If it is accepted that language did arise by natural selection during human evolution rather than as an accidental by-product of some other process, what were the selective pressures resulting in its appearance? Did both the cognitive and the vocal aspects of human language develop at the same time and for the same reasons during human evolution, or did one significantly precede the other and serve as an exaptation for language? These questions have received surprisingly little attention in the literature on the evolution of language. Rather the main focus has been on when human language first appeared.

It has become popular in recent years to argue that language is a very recent development in human evolution, accompanying the appearance of anatomically modern *Homo sapiens* and/or the Upper Palaeolithic transition (White 1982; Chase & Dibble 1987; Mellars 1991; Noble & Davidson 1991; Davidson & Noble 1993; Milo & Quiatt 1993). These arguments have been made from the points of view of both the ability to produce human vocalizations (e.g. Lieberman *et al.* 1992) and the presence of evidence of the cognitive capacity for symbolization. The vocally based arguments have centred on the conclusion that only anatomically modern humans have a vocal tract capable of producing the full range of human vowel sounds. Over the years this work has been heavily criticized on the basis of the accuracy of the vocal tract reconstructions for the fossil hominines (see particularly Scherpartz 1993; Houghton 1993). Furthermore, the discovery of the first Neanderthal hyoid bone at Kebara, Israel, and the recognition that it is totally modern in form has convinced the great majority of anthropologists that Neanderthals had a vocal apparatus capable of the production of the full range of sounds needed for human language (Arensburg *et al.* 1989, 1990). At least from the physical point of view vocally based language could have characterized pre-modern hominines.

Cognitively based arguments centred on the evidence of symbolism in the fossil record in the form of art or non-utilitarian objects. Noble & Davidson (1991; Davidson & Noble 1993) have been prominent in this debate, arguing that there is no evidence for modern human symbolic, and therefore linguistic, ability until the Upper Palaeolithic transition in Europe (between 30,000 and 40,000 years ago). They do, however, concede that the peopling of Australia at an earlier date of 40,000 years ago (now probably closer to 60,000 years ago, Roberts *et al.* 1990) would mark the most ancient evidence for modern human symbolic ability. The recent discovery of carved harpoons at the Zairian site of Katanda may push the evidence of symbolic behaviour back to about 90,000 years ago (Yellen *et al.* 1995) and the presence of red Ochre and notched bone, engraved ostrich shell and perforated *Conus* shell in southern African sites may push the evidence of symbolic behaviour even further back, to earlier than 100,000 years (Knight *et al.* 1995).

These dates should be taken as the youngest possible occurrence of symbolic language and not the oldest. All that we can be sure of from archaeological evidence is that symbolic behaviour was in place by the time of the particular discovery. It does not tell us how long prior to the date of the discovery that the ability was present. Pinker (1994) makes the very good point that modern human language capability must have been in place by the first appearance of modern humans. This is because we all have the same linguistic abilities and these abilities must have been a feature of the human brain before the spread of modern humans throughout the world. Genetic evidence suggests that the modern human mitrochrondrial ancestor could be no more than about 500,000 years old (Stoneking 1993). The mtDNA ancestor does not necessarily mark the first appearance of modern humans, only the mtDNA characteristic of modern humans. The mtDNA date of 500,000 years together with the archaeological date of about 100,000 years does allow us, however, to bracket the time that modern human symbolic behaviour was most probably evolving.

It is notable, that the period between 500,000 and 100,000 years ago, the second half of the Middle Pleistocene period, corresponds to a period of very rapid brain expansion in our early ancestors (Figure 1). Recent research suggest that this rapid brain expansion may be directly related to the evolution of modern human language (Deacon 1992). Finlay & Darlington (1995) have argued that selection for a particular behavioural ability, such as language, may result in the co-ordinated enlargement of the entire brain (except for the olfactory system which represents a special case). They base this idea on the fact that the primary determinant of the size of a brain structure is the period of time during which the neurons form in early development (the period of neurogenesis). As a result, selection for a particular cognitive function may be easiest achieved by selection for prolonged neurogenesis that would effect the size of the entire neocortex.

THE SIGNIFICANCE OF TERRESTRIALITY TO HUMAN COGNITIVE AND LANGUAGE EVOLUTION

This argument is interesting as an explanation for the explosive increase in the size of the human brain that occurs in the last quarter of a million years or so. It does not tell us, though, why we have an explosive selection for language at this stage and equally importantly why we do not see a similar run-away development of the brain and linguistic capability in other primate species. The conclusion is unavoidable that there is something specific to the course of human evolution that emphasized the development both of verbally based communication and of the phenomenal development of

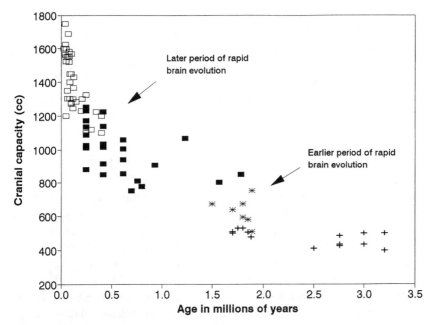

Figure 1. The increase in hominid cranial capacity over the past 3.5 million years (data from Aiello & Dean 1990). Plus = australopithecines, asterisk = early *Homo* (*H. habilis, H. rudolfensis*), filled square = *Homo erectus* (including *Homo ergaster*), open square (archaic and modern *Homo*). Note the two periods of rapid increase in brain size, the first occurring between about 2.0 and 1.5 million years ago and the second after 500,000 years ago.

cognitive capacity that accompanied it. One clue to what this might be can be inferred from the pattern of brain size increase in the hominines over the past 4 million years (Aiello, in press). In addition to the exponential increase in the brain of *Homo* in the second half of the Middle Pleistocene period there is also a marked increase in brain size that accompanies the first appearance of the genus *Homo* at approximately 2 million years ago (Figure 1).

This increase in brain size is correlated with one major feature in hominine adaptation, a transition to a fully terrestrial lifestyle. By the appearance of early *Homo erectus* (*Homo ergaster*) at approximately 1.8 million years ago in East Africa all evidence of any type of committed arboreal adaptation has been lost from the hominine postcranial skeleton. Early *Homo erectus* has modern limb proportions rather than the relatively short legs of the australopithecines and also lacks the specific skeletal morphology of the australopithecines that has been interpreted as indicative of at least a partial arboreal lifestyle (Walker & Leakey 1993). These arboreal adaptations include, among others, a funnel-shaped thorax (Schmid 1983, 1991; Hunt 1994), curved hand and foot phalanges (Stern

& Susman 1983) and features of the upper limb skeleton that are associated with forelimb strength and mobility (Aiello & Dean 1990). *Homo habilis*, an earlier member of the genus *Homo*, may still retain some climbing adaptations in its skeleton (Hartwig-Scherer & Martin 1992), but the palaeoecology of Bed I at Olduvai Gorge where it is found, indicates an adaptation to a mosaic lake-shore environment, which would include open country habitats (Plummer & Bishop 1994).

TERRESTRIALITY, GROUP SIZE AND LANGUAGE

Adaptation to an open-country, terrestrial environment has a number of perhaps unsuspected implications for an increase in brain size in general and for the origin of language in particular. One of the most important of these is that a terrestrial, open-country adaptation is correlated in primates with large group sizes (Foley 1987). This correlation can best be explained by the increase in predator pressure in such open environments and the protection from predators that is gained through larger group numbers (Aiello & Dunbar 1993). Primates that live in large groups also tend to have relatively larger neocortices than those living in smaller groups (Dunbar 1992, 1993). The neocortex is that part of the brain that deals with cognitive (among other) functions. Dunbar explains the relationship between large group size and relatively large neocortex sizes by the increased sophistication in social cognition needed to maintain the increased number and complexity of social relationships in a larger group. Following this line of logic, larger group sizes would not only be associated with larger neocortex sizes but also, and perhaps surprisingly, with a greater reliance on vocally based communication.

The connection between large group sizes and vocally based communication comes in the context of the maintenance of social cohesion. Non-human primates reinforce their social networks through mutual grooming and there is a strong correlation between group size and time spent in such grooming behaviour. The larger the group the more time must be spent in grooming to maintain group cohesion. But if too much time is spent grooming there will be insufficient time for other activities such as feeding, resting or travelling. Dunbar has suggested that the maximum time that can be spent grooming is about 20% of the daily time budget. Any more than this and other activities will suffer and this will ultimately affect individual survival and fitness.

Following this line of reasoning, one can predict when the early hominines might have begun to experience time-budget pressures. Neocortex size, can be predicted from total brain size, group size can be predicted from neocortex size and percentage grooming time can be predicted from group size (Aiello & Dunbar 1993). Although this three-step process of inference compounds error, the results do suggest that early *Homo*, and particular early *Homo erectus*, would have been the first of the hominines to have to find a solution to the problem of maintaining group cohesion. Increased emphasis on vocalization as seen in living Gelada baboons, could serve as a supplement to mutual grooming by reaching more individuals than the specific focus of the grooming activity. At this stage, such vocalizations would not have necessarily had to have any specific symbolic content, but could be seen as something as simple as 'vocal grooming'. Control of the voice in tone and pattern would have been sufficient to spread a feeling of mutual content and well being. At this early stage, such vocalization may have been analogous to the chattering observed today in Gelada baboons, which have some of the largest group sizes of any living primates.

The importance of this to the evolution of human language is that it would have been a means by which conscious and co-ordinated control of vocalization could have become established. And importantly it would have been a way of establishing vocally based communication in our hominine ancestors without presupposing any necessary symbolic or other advanced cognitive capacity.

Is there any specific, empirical evidence that early *Homo* might have had an enhanced reliance on voluntary vocal communication as postulated by this scenario. The answer is yes. Endocranial casts demonstrate that early *Homo* has a reorganization and elaboration of the sulci of the left inferior frontal lobe of the brain in the region of Broca's area (Falk 1983). Broca's area and the adjacent ventral prefrontal cortex are the precise areas of the brain associated with (among other things) voluntary control and co-ordination of the tongue, lips and pharynx in the production of vocalizations (Deacon 1992). Early *Homo* also has brain asymmetries typical of modern humans (Tobias 1987). The most important of these to the current discussion is the relatively low position of the left Sylvian sulcus which separates the parietal from the temporal cortex and seems to suggest an expanded parietal cortex. Work on the neural organization of language functions suggests that the parietal cortex, together with the temporal cortex is involved in the voluntary production of sound as well as in speech comprehension (Deacon 1992).

The important point here is that this type of brain organization is seen for the first time in early *Homo* and can be interpreted in the context of the co-ordination necessary for the voluntary production of complex vocalization which is both a prerequisite for human language and predicted by the group-size model for the origin of language. This reorganization accompanies a modest increase in relative size of the hominine brain, but it is important to note that the size of the early *Homo* brain is still $\frac{1}{2}$ to $\frac{1}{3}$ smaller, both in relative and absolute terms, than the modern human brain (Aiello & Dean 1990). This pattern is precisely what might be expected if the evolution of language proceeded through a phase of voluntary production of relatively elaborate vocalizations, lacking the symbolic, syntactic, or cognitive sophistication of modern human language.

TERRESTRIALITY, BIPEDALISM AND LANGUAGE

The group-size hypothesis for the expansion of the human brain and the evolution of language has many compelling aspects, but it cannot be the whole story. Other terrestrial primates living in large groups have not developed brains the size of modern human brains nor have they developed language. One of the reasons for this may be connected with another unique human adaption, bipedal locomotion. Bipedalism is one of the earliest, if not the earliest, hominine adaptation to appear in the fossil record (White *et al.* 1994; Leakey *et al.* 1995). There is no current consensus over the reasons why bipedalism evolved in the human line (Wheeler 1994; Hunt 1994; Jablonski & Chaplin 1993; also see Rose 1991 for a review), or why the postcranial skeletal and inferred bipedal capabilities were apparently so different in the australopithecines on the one hand and early *Homo erectus* (*Homo ergaster*) on the other. But we can be sure that bipedalism was not an evolutionary option for other primarily terrestrial primates such as baboons or macaques. The reason for this is that these primates are Old World (cercopithecid) moneys that have a limited component of below branch, or suspensory, postures in their repertoires (Rose 1973). As a result, they have skeletons that are very different from those of the apes in features associated with mobility and truncal erectness (Andrews & Aiello 1984). Primates that engage in suspensory postures, and particularly vertical climbers, are very similar to bipedal humans in joint excursion, muscle usage and kinematics (Kimura *et al.* 1979; Prost 1980; Fleagle *et al.* 1981; Ishida *et al.* 1985; Kimura 1985; Okada 1985). This type of positional behaviour is found in chimpanzees and is also inferred for the immediate proto-hominine. It would have made bipedalism a distinct possibility for the early hominines where it was not an option for the Old World monkeys entering the same niche.

Bipedalism may have had at least four direct effects of the evolution of language and cognition, two of these specifically related to the ability to produce human speech sounds and two generally related to the evolution of increased brain size and cognition. In relation to the production of human speech sounds, bipedalism was most probably directly related to the descent of the human larynx (Negus 1929). In humans the larynx lies much lower in the throat than it does in apes and the part of the throat above the larynx, whose anterior border is the rear of the tongue, is necessary for the formation of both human vowels and consonants. In bipedal hominines the spinal cord enters the brain case from below rather than from behind, constricting the space for the larynx between the spinal cord and the mouth. This together with the reduction of the face in early *Homo erectus* (see below) in relation to the australopithecines would be expected to necessitate a lower larynx particularly in these later terrestrially committed bipedal hominines.

Bipedalism may also have been directly related to sound quality (Aiello, in press). Humans have a valvular larynx that allows the airway to be closed off. In speaking this permits pressure to build up below the larynx to be released during phonation. A valvular larynx also may have a locomotor function. It is found in mammals with prehensile forelimbs (Negus 1929). Air pressure below a closed larynx stabilizes the chest to provide a fixed basis for the arm muscles. This is why we tend to hold our breath when exerting ourselves with our arms. Negus (1929) has suggested that one difference between ourselves and climbing mammals is that human vocal folds (vocal cords) are less cartilaginous. The more membraneous human vocal folds allow the production of a less harsh, more melodious sound. This change may have been associated with a relaxed selective pressure on the locomotor function of the valvular larynx in a committed terrestrial bipedal. If this proves to be true it would be possible to speculate that the more arboreally adapted early hominines (including *Australopithecus afarensis*, *Australopithecus africanus* and *Ardipithecus ramidus*), retained the ancestral cartilaginous larynx and a harsher voice than would have characterized later members of the genus *Homo*.

In relation to the general evolution of increased brain size and cognition, committed bipedalism has the obvious effect of freeing the forelimb from locomotor function. Dedicated use of the forelimb for object manipulation would be expected to be associated with enhanced hand-eye co-ordination and an associated increase in neural circuitry. The failure of living apes to produce stone tools as sophisticated as even the earliest Olduwan artifacts (Schick & Toth 1993) may attest to the increased hand-eye co-ordination even in pre-erectus members of the genus *Homo*. Furthermore, Calvin (1983, 1992, 1993) has argued that speed and coordination of arm movement in hammering, throwing and presumedly tool manufacture, would require increased neural capacity and integration. He also argues that the neural adaptations necessary for sequencing which would also be important for accurate throwing could serve as a preadaptation for linguistic sequencing, or syntax (Calvin 1992).

The final way in which bipedalism could be related to general cognitive evolution has to do with the structure of the bipedal pelvis (Wills 1995). In relation to a quadrupedal ape pelvis, a bipedal pelvis (whether australopithecine or *Homo*) structurally has to be much more compact. This results in a considerably more restricted birth canal (Berge *et al.* 1984; Tague & Lovejoy 1986). At the same time the evolution of the large human brain would presuppose a relatively large birth canal through which a large-brained infant could be born. Modern humans bear infants at a relatively premature (or secondary altricial) stage in relation to apes, while the brain is still growing at its rapid fetal rate (Martin 1990). This exposes the rapidly growing brain of the infant to the complex environment outside the womb which would have been a very important selective pressure for brain evolution. There is fossil evidence dating back to approximately 1.6 million years ago that suggests that by this time the human pelvis was so constricted and the brain of the infant potentially so large that it would have had to have been born at a less mature stage than is the case in living apes (Shipman & Walker 1989).

Therefore a terrestrial environment together with bipedalism would have resulted in a larger brain size (larger group size, freeing of the forelimb) and secondarily altricial offspring as well as specific factors which predisposed the early hominines to vocal communication (vocal grooming, membraneous vocal cords and descended larynx). But there is still one other implication of the expanding brain size at this stage of evolution that has direct relevance to the evolution of linguistic ability in our ancestors. This has to do with dietary change.

TERRESTRIALITY, DIET AND LANGUAGE

Brain tissue is energetically among the most expensive tissues in the body, consuming over 22 times more energy than muscle tissue at rest (Aschoff *et al.* 1971). It follows that the larger the relative brain size, the greater the energetic demand would be on the organism and, everything else being equal, the greater difficulty that organism would have in meeting its daily food requirements. In this context, a relatively big brain would potentially be detrimental to an organism because it would significantly increase the total energy budget of that organism. There are a variety of ways in which an organism could compensate for the increased energy demands of an encephalized brain, but the hominines seem to have done this by reducing the size of one of the other energetically expensive tissues in the body, the gastro-intestinal tract. In humans, there are five organs that make up only about 7% of the total body weight but consume over 75% of our basal metabolic rate. These organs are the brain, gastro-intestinal tract, heart, liver and kidneys. Of these the heart, liver and kidneys are tied closely to overall body weight because of their physiological functions. The gastro-intestinal tract is the only 'expensive' organ other than the brain that can vary significantly in size in animals of any given body weight. This is because its overall size, and hence energy requirements, is dependent not only on the size of the organism but also on the digestibility of the food eaten.

Aiello & Wheeler (1995) have demonstrated that humans have hearts, livers and kidneys of a size expected for an average primate of our body weights. They have also demonstrated that human guts have reduced in size by precisely the amount that would balance the energy requirements of our expanded brains. As a result, and in spite of our encephalized brains, the average human basal metabolic rate is at a level that would be expected for an average primate of our body weights. The negative correlation between expanded brains and reduced guts is also apparent in non-human primates. The implication of these results is that no matter what was selecting for a relatively large brain in humans (and in other primates) a high quality, easy to digest diet would be a necessary concomitant of encephalization. For our early ancestors, the most obvious source of such a high quality diet would be animal-based products. This suggests that it is no mere coincidence that early *Homo* is the first of the hominines to be associated with significant amounts of animal bones in the sites in which it is found.

The connection between this necessary change in diet and the origin of language comes through the mechanics of mastication. A diet rich in animal products would not only be easier to digest but also easier to masticate. Early *Homo* is the first of the hominines to show a significant reduction in not only the size of the dentition (McHenry 1988) but also the size of the mandible in relation to body mass. Duchlin (1990) has demonstrated that the geometry of the mandible is crucial to the production of sounds used in human speech. It must be shaped in such a way as to give the muscles that move the tongue proper leverage to position it within the oral cavity in the variety of ways necessary to produce the sounds employed in all human speech. The long and relatively narrow mandible of the chimpanzee precludes such movement of the tongue while the mandible of *Homo erectus* is short enough, deep enough and broad enough to potentially allow the proper muscle leverage. Therefore, the change in diet which accompanies the expansion of the brain in early *Homo* is directly associated with an change in mandibular geometry which facilitates the production of the sounds necessary for human speech. Dietary change can also feed back to brain expansion. High quality diets require increased complexity of foraging behaviour and this would be expected to be another selective pressure for brain enlargement (Aiello & Wheeler 1995; Milton 1995).

THE LINGUISTIC ABILITY OF *HOMO ERECTUS*

By time of the appearance of early *Homo erectus*, the hominine line had experienced a unique set of circumstances that resulted in a series of interconnected adaptations, providing the foundation for both verbally based communication and further cognitive development. But what were these hominines like? How similar or different were they to modern humans in their linguistic and cognitive abilities? The material remains that they left behind suggest they were considerably different from any modern humans. Perhaps most obviously, there is no clear evidence in the fossil record of symbolic behaviour at this stage of human evolution. The only features that have been interpreted in this fashion are the beautifully fashioned bi-facial tools called hand axes that are part of the Acheulian tool tradition that lasted from about 1.4 million (Asfaw *et al.* 1992) until about 150,000 years ago (Gowlett 1992). Schick & Toth (1993) argue that these stereotypic tools may indicate that low levels of symbolic communication or language skills were used to enhance or solidify the ideas that underline their production. However, they also suggest that because of the stylistic uniformity of these tools over almost 1.5 million years *Homo erectus* may have relied primarily on imitation and not verbal instruction. The only thing that is reasonably sure is that there appears to be a marked absence of innovation in material culture throughout the long duration of the Acheulian. Furthermore, although fire was known (Bellomo 1994), there is no evidence of its consistent use and there is only controversial evidence of the construction of shelters. It is perhaps not a coincidence that this period of stasis in material culture correlates with the long period of time where there is little or no significant increase in brain size (Figure 1) (Rightmire 1981; Leigh 1992). It is true that *Homo erectus* was the first hominine to move out of Africa and into Asia and Europe. However other carnivores such as lions, leopards and hyenas also moved out of Africa at about the same time. The geographical expansion may have had more to do with the hunting skills of *Homo erectus* than with any particular increase in intelligence, language or cognitive ability (Schick & Toth 1993).

There are also two features of the anatomy of the most complete *Homo erectus* skeleton that suggest vocally based symbolic communication may not have been developed beyond the most rudimentary stage at this time. This skeleton (KNM-WT 15000) is from the West Turkana region of Kenya and dates to about 1.6 million years ago (Walter & Leakey 1993). Importantly, it lacks the expansion of its neural canal in the mid-thoracic region of its spinal column (MacLarnon 1993). The mid-thoracic expansion is unique to humans and is thought to relate to the local enervation of the thoracic and abdominal muscles which would be associated with fine control of respiration. This is highly important in the context of sustained vocalization associated with human speech. The fact that KNM-WT 15000 lacks this expansion suggests that it had not yet developed the muscular control that would be associated with human language.

KNM-WT 15000 is a juvenile male and its stage and pattern of development also suggest that hominines at this time may not have developed symbolically based verbal communication (Smith 1993). This idea is based on the inference that *Homo erectus* growth and development may have lacked the adolescent growth spurt that is characteristic of modern humans. The inferred dental age at death of KNM-WT 15000 (10–11 years) does not correlate with its inferred age based on epiphyseal closure (13–13.5 years) or stature (15 years). The fact that age inferred from epiphyseal closure (and stature) is in advance of dental age suggests that the growth of the skeleton is in advance of dental development, a pattern found in living chimpanzees which lack the adolescent growth spurt. Bogin (1988, 1990) has suggested that the function of the adolescent growth spurt is primarily to reduce the rate of growth of children to keep them in a greater state of dependency for a longer period of time to facilitate the transfer of symbolically-based cultural knowledge. The inferred absence of the growth spurt in *Homo erectus* may therefore also suggest the absence of symbolically based learning in this species (Smith 1993).

These speculations suggest that although better pre-adapted to symbolically based verbal communication than any other primate, *Homo erectus* was yet to develop true human language. Its terrestrial and bipedal heritage may have preadapted it to increased vocalization, but evidence of syntactic and symbolic skills, with all of their cultural manifestations, is lacking. It is a fallacy to view *Homo erectus* simply as a transition to ourselves. This stage of human evolution represents a long standing, and highly successful hominine adaptation. Perhaps the most significant unanswered question in human evolution is why *Homo erectus* ultimately gave way to larger brained and cognitive more advanced hominines.

COGNITIVE AND LINGUISTIC EVOLUTION AFTER *HOMO ERECTUS*

Increased cognitive ability accompanied by symbolic and syntactic verbal language has generally been viewed as being such an advantage to human adaptation that the evolution of these features need little explanation. But the considerable costs of larger brain sizes and verbally based symbolic communication need to be weighed against the benefits of information transfer and increased social cohesion. When this is done the increase in brain size (and by inference cognitive capacity and linguistic ability) that gets underway midway through the Middle Pleistocene period seems all the more remarkable.

There is one major cost of increased brain size and another of symbolically based verbal communication that undoubtedly not only radically changed the social organization of the hominines but also set up a feed-back loop which continued to select for increased cognitive ability throughout this time period. Increased brain size would place an increasing metabolic cost particularly on the females (Foley & Lee 1991; Leonard & Robertson 1992). Power & Aiello (in press) have argued that the Middle Pleistocene increase in hominine brain size would have increased this stress to the point that it would have required considerable paternal investment in the offspring to insure survival. This necessity for long term paternal investment could have marked the beginning of modern human family social organization. It also could have been one of the important factors underlying the origin of human ritual symbolism (Knight *et al.* 1995; Power & Aiello, in press).

Whether symbolically based verbal language evolved in response to the postulated changes in hominine inter-gender social organization (Power & Aiello, in press) to other factors involving general social cohesion and information transfer (Aiello & Dunbar 1993) or to a combination of these, language also had a major cost. It escalated the possibility of cheating which potentially could have significantly lowered individual reproductive fitness. As a result, it would have increased the selection pressure on the development of cognitive ability and particularly on the development of one's ability to reflect on one's own mental states and those of others. From this perspective it may be no coincidence that the human prefrontal cortex is responsible not only for this ability which is considered by some researchers to be unique to humans (Povinelli & Preuss 1995) but also for many features of language production and comprehension.

We have very few clues as to why brain size began to rapidly increase in the middle part of the Middle Pleistocene and why, by inference, the linguistic and cognitive abilities of our ancestors also began to change. The only thing that is apparent from the fossil record is that the long stasis in the Acheulian culture also begins to break down about 250,000 years ago when improvements in manufacturing techniques and specialization of flake tools that foreshadowed the later Mousterian and Middle Stone Age traditions began to appear (Gowlett 1992). This is undoubtedly a consequence of increased brain size (and inferred intelligence) rather than a cause. One of the best explanations at the present time for the increase in brain size is that it was in response to the necessity for increased group size, the same explanation put forward for the earlier Lower Pleistocene increase in brain size (Aiello & Dunbar 1993). Over the long period of *Homo erectus* existence a virtually imperceptible annual increase in population numbers could produce a seeming population explosion (Foley, in press). As a result, living group size may have been forced to increase in response to population numbers in an 'evolutionary arms race' to provide protection against other human populations (Alexander 1989; Aiello & Dunbar 1993). A related possibility might have to do with the dispersed nature of human populations and the advantages of a nomadic or migrating lifestyle. This may have been increasingly important as the hominines adapted to more severe habitats in the later Middle and early Upper Pleistocene.

CONCLUSIONS

In the context of the long time span of human evolution, the last quarter of a million years in which human brain size, cognitive and linguistic abilities reached their modern norms is a relatively very short time. The arguments present here suggest that although the foundations for human language and cognition extend back over the past 4 to 5 million years of human evolutionary history, during the great majority of this period hominines did not possess either the cognitive or the linguistic talents that we would recognize as human. A terrestrial lifestyle and bipedal locomotion provided our evolutionary ancestors with unique preadaptions to human cognitive and linguistic ability, but it was not until the middle part of the Middle Pleistocene period that the increase in absolute brain size and escalating change in material culture suggest that human abilities may have at last begun to appear.

In trying to understand the reasons behind human cognitive and linguistic evolution we also have to recognize that anatomically modern humans were not the only large brained hominine in existence. Up until as recently as 29,000 years ago Neanderthals were still present in Spain at the site of Zafarraya. Neanderthals were therefore co-existing with anatomically modern *Homo sapiens* that first appeared in Europe about 40,000 years ago and in Africa and the Near East about 100,000 years ago (Aiello 1993). The inferred behavioural differences between Neanderthals and modern humans attest to the fact that selection for increased brain size alone did not necessarily presuppose the development of human cognitive and symbolic abilities. Although there are no physical reasons why Neanderthals could not have produced human language, the difference in material culture between Neanderthals and Upper Palaeolithic modern humans suggests that there were marked differences in their cognitive abilities. The conclusion that can be drawn from this is that, building on the exaptations of our *Homo erectus* ancestors, human cognitive and linguistic abilities most probably had as much to do with the particular social environment in which brain expansion was taking place as with the brain expansion itself.

REFERENCES

Aiello, L.C. 1993: The fossil evidence for modern human origins in Africa: a revised view. *American Anthropologist* 95, 73–96.

Aiello, L.C. (In press) Hominine preadaptations for language and cognition. In *Modelling the Early Human Mind* (ed. P. Mellars & K. Gibson). Cambridge: McDonald Institute Monograph Series.

Aiello, L.C. & Dean, M.C. 1990: *An Introduction to Human Evolutionary Anatomy*. London: Academic Press.

Aiello, L.C. & Dunbar, R.I.M. 1993: Neocortex size, group size, and the evolution of language. *Current Anthropology* 34, 184–193.

Aiello, L.C. & Wheeler, P. 1995: The expensive-tissue hypothesis: the brain and the digestive system in human and primate evolution. *Current Anthropology* 36, 199–221.

Alexander, R.D. 1989: The evolution of the human psyche. In *The Human Revolution* (ed. P. Mellars & C.B. Stringer), pp. 455–513. Edinburgh: Edinburgh University Press.

Andrews, P. & Aiello, L.C. 1984: An evolutionary model for feeding and positional behaviour. In *Food Acquisition and Processing in Primates* (ed. D.J. Chivers, B.A. Wood & A. Bilsborough), pp. 429–466. New York: Plenum Press.

Arensburg, B., Schepartz, L.A., Tillier, A.M., Vandermeersch, B. & Rak, Y. 1990: A reappraisal of the anatomical basis for speech in Middle Paleolithic hominids. *American Journal of Physical Anthropology* 83, 137–146.

Arensburg, B., Tillier, A.M., Vandermeersch, B., Duday, H., Schepartz, L.A. & Rak, Y. 1989: A Middle Paleolithic human hyoid bone. *Nature* 338, 758–760.

Aschoff, J., Gunther, B. & Kramer, K. 1971. *Energiehaushalt und Termperaturregulation*. Munich: Urban and Schwarzenberg.

Asfaw, B., Beyene, Y., Suwa, G., Walter, R.C., White, T.D., WoldeGabriel, G. & Yemane, T. 1992: The earliest Acheulean from Konso-Gardula. *Nature* 360, 732–735.

Barton, R., Purvis. A., & Harvey, P. 1995: *Philosophical Transactions of the Royal Society of London, B* 348, 381

Bellomo, R.V. 1994: Methods of determining early hominid behavioral activities associated with the controlled use of fire at FxJj 20 Main, Koobi Fora, Kenya. *Journal of Human Evolution* 27, 173–196.

Berge, C., Orban-Segebarth, R. & Schmid, P. 1984: Obstetrical interpretation of the australopithecine pelvic cavity. *Journal of Human Evolution* 13, 573–587.

Bogin, B. 1988: *Patterns of Human Growth*. Cambridge: Cambridge University Press.

Bogin, B. 1990: The evolution of human childhood. *BioScience* 40, 16–25.

Calvin, W.H. 1983: A stone's throw and its launch window: timing precision and its implications for language and hominid brains. *Journal of Theoretical Biology* 104, 121–135.

Calvin, W.H. 1992: Evolving mixed-media messages and grammatical language: secondary uses of the neural sequencing machinery needed for ballistic movements. In *Language Origin: A Multidisciplinary Approach* (ed. J. Wind, B. Chiarelli, B. Bichakjian & A. Nocentini). Dordrecht: Kluwer Academic Publishers (NATO ASI Series–Series D: Behavioural and Social Sciences, vol. 61, pp. 163–179).

Calvin, W.H. 1993: The unitary hypothesis; a common neural circuitry for novel manipulations, language, plan-ahead, and throwing? In *Tools, Language and Cognition in Human Evolution* (ed. K.R. Gibson & T. Ingold), pp. 230–250. Cambridge: Cambridge University Press.

Chase, P.G. & Dibble, H.L. 1987: Middle Paleolithic symbolism: a review of current evidence and interpretations. *Journal of Anthropological Archaeology* 6, 263–296.
Davidson, I. & Noble,W. 1993: Tools and language in human evolution. In *Tools, Language and Cognition in Human Evolution* (ed. K.R. Gibson & T. Ingold), pp. 363–3B8. Cambridge: Cambridge University Press.
Deacon, T. 1992: The neural circuitry underlying primate calls and human language. In *Language Origin: A Multidisciplinary Approach* (ed. J. Wind, B. Chiarelli, B. Bichakjian & A. Nocentini) Dordrecht: Kluwer Academic Publishers (NATO ASI Serie—Series D: Behavioural and Social Sciences, vol. 61, pp. 121–162).
Duchlin, L.E. 1990: The evolution of articulate speech: comparative anatomy of the oral cavity in *Pan* and *Homo*. *Journal of Human Evolution* 19, 687–698.
Dunbar, R.I.M. 1992: Neocortex size as a constraint on group size in primates. *Journal of Human Evolution* 22, 469–493.
Dunbar, R.I.M. 1993: Co-evolution of neocortex size, group size, and language in humans. *Behavioral and Brain Sciences* 16, 681–735.
Falk, D. 1983: Cerebral cortices of east African early hominids. *Science* 221, 1072–1074.
Finlay, B.L. & Darlington, R. B. 1995: Linked regularities in the development and evolution of mammalian brains. *Science* 268, 1578–1584.
Fleagle, J.G., Stern, J.T., Jungers, W.L., Susman, R.L., Vangor A.K. & Wells, J. 1981: Climbing, a biomechanical link with brachiation and with bipedalism. In *Vertebrate Locomotion* (ed. M.H. Day), pp. 359–375. *Symposium of the Zoological Society of London* 48. London: Zoological Society.
Foley, R. 1987: *Another Unique Species*. London: Longman Scientific & Technical.
Foley, R. (In press) The evolutionary ecology of hominid dispersals. In *The Human Tide* (ed. O. Soffer).
Foley, R. & Lee, P. 1991: Ecology and energetics of encephalization in hominid evolution. *Philosophical Transactions of the Royal Society of London, B* 334, 223–232.
Gowlett, J.A.J. 1992: Tools—the Palaeolithic record. In *Cambridge Encyclopedia of Human Evolution* (ed. S. Jones, R. Martin & D. Pilbeam), pp. 350–364. Cambridge: Cambridge University Press.
Hartwig-Scherer, S. & Martin, R.D. 1992: Allometry and prediction in hominids: a solution to the problem of intervening variables. *American Journal of Physical Anthropology* 88, 37–57.
Houghton, P. 1993: Neandertal supralaryngeal vocal tract. *American Journal of Physical Anthropology* 90, 139–146.
Hunt, K. 1994: The evolution of human bipedality: ecology and functional morphology. *Journal of Human Evolution* 26, 183–202.
Ishida, H., Kumakura, H. & Kondo, S. 1985: Kinesiological aspects of bipedal walking in gibbons. In *Primate Morphophysiology: Locomotor Analysis and Human Bipedalism* (ed. S. Kondo), pp. 59–80. Tokyo: University of Tokyo Press.
Jablonski, N.G. & Chaplin, G. 1993: Origin of habitual bipedalism in the ancestor of the Hominidae. *Journal of Human Evolution* 24, 259–280.
Kimura, T. 1985: Bipedal and quadrupedal walking of primates: comparative dynamics. In *Primate Morphophysiology: Locomotor Analysis and Human Bipedalism* (ed. S. Kondo), pp. 81–104. Tokyo: University of Tokyo Press.
Kimura, T., Okada, M. & Ishida, H. 1979: Kinesiological characteristics of primate walking: its significance in human walking. In *Environment, Behavior and Morphology: Dynamic Interactions in Primates* (ed. M.E. Morbeck, H. Preuschoft & N. Gomberg), pp. 297–311. New York: Gustav Fischer.
Knight, C., Power C. & Watts, I. 1995: The human symbolic revolution: a Darwinian account. *Cambridge Archaeological Journal* 5, 75–114.
Leakey, M.G., Feibel, C.S., McDougall I. & Walker, A. 1995: New four-million-year-old hominid species from Kanapoi and Allia Bay, Kenya. *Nature* 376, 565–571.

Leigh, S.R. 1992: Cranial capacity evolution in Homo erectus and early *Homo sapiens*. *American Journal of Physical Anthropology* 87, 1–14.

Leonard, R.W. & Robertson, M.L. 1992: Nutritional requirements and human evolution: a bioenergetics model. *American Journal of Human Biology* 4, 179–195.

Lieberman, P., Laitman, J.T., Reidenberg, J.S. & Gannon, P.J. 1992: The anatomy, physiology, acoustics and perception of speech: essential elements in analysis of the evolution of human speech. *Journal of Human Evolution* 23, 447–468.

McHenry, H.M. 1988: New estimates of body weight in early hominids and their significance to encephalization and megadontia in 'robust' australopithecines. In *Evolutionary History of the 'Robust' Australopithecines* (ed. F.E. Grine), pp. 133–148. New York: Aldine de Gruyter.

MacLarnon, A. 1993: The vertebral canal. In *The Nariokotome* Homo erectus *Skeleton* (ed. A. Walker & R.E. Leakey), pp. 359–390. Cambridge: Harvard University Press.

Martin, R.D. 1990: *Primate Origins and Evolution*. London: Chapman and Hall.

Mellars, P. 1991: Cognitive changes in the emergence of modern humans. *Cambridge Archaeological Journal* 1, 63–76.

Milo, R.G. & Quiatt, D. 1993: Glottogenesis and anatomically modern *Homo sapiens*: the evidence for and implications of a late origin of vocal language. *Current Anthropology* 34, 569–598.

Milton, K. 1995: Reply to Aiello & Wheeler: The expensive tissue hypothesis: the brain and the digestive system in primate and human evolution. *Current Anthropology* 36, 214–216.

Negus, V.E. 1929: *The Mechanism of the Larynx*. London: Wm. Heinemann Medical Books Ltd.

Noble, W. & Davidson, I. 1991: The evolutionary emergence of modern human behaviour: language and its archaeology. *Man* 26, 222–253.

Okada, M. 1985: Primate bipedal walking: comparative kinematics. In *Primate Morphophysiology: Locomotor Analysis and Human Bipedalism* (ed. S. Kondo), pp. 47–58. Tokyo: University of Tokyo Press.

Piattelli-Palmarini, M. 1989: Evolution, selection, and cognition: from 'learning' to parameter setting in biology and the study of language. *Cognition* 31, 1–44.

Pinker, S. 1994: *The Language Instinct*. William Morrow & Company, Inc.

Pinker, S. & Bloom, P. 1990: Natural language and natural selection. *Behavioral and Brain Sciences* 13, 707–784.

Plummer, T.W. & Bishop, L.C. 1994: Hominid paleoecology at Olduvai Gorge, Tanzania as indicated by antelope remains. *Journal of Human Evolution* 27, 47–75.

Power, C. & Aiello, L.C. (In press) Female proto-symbolic strategies. In*Women in Human Origins* (ed. L. Hager). London: Routledge.

Povinelli, D.J. & Preuss, T.M. 1995: Theory of mind: Evolutionary history of a cognitive specialization. *Trends in Neuroscience* 18, 418–424.

Prost, J.H. 1980: Origin of bipedalism. *American Journal of Physical Anthropology* 52, 175–189.

Rightmire, G.P. 1981: Patterns in the evolution of *Homo erectus*. *Paleobiology* 7, 241–246.

Roberts, R.G., Jones, R. & Smith, M.A. 1990: Thermoluminescence dating of a 50,000 year old human occupation site in northern Australia. *Nature* 345, 153–156.

Rose, M.D. 1973: Quadrupedalism in primates. *Primates* 14, 337–358.

Rose, M.D. 1991: The process of bipedalization in hominids. In *Origines de la Bipédie chez les Hominidés* (ed. Y. Coppens & B. Senut), pp. 37–48. Paris: Éditions du centre national de la recherche scientifique.

Savage-Rumbaugh, S. & Lewin, R. 1994: *Kanzi: The Ape at the Brink of the Human Mind*. London: Doubleday.

Savage-Rumbaugh, E.S., Murphy, J., Sevoik, R.A., Brakke, K.E., Williams, S.L. & Rumbaugh, D.M. 1993: Language comprehension in ape and child. *Monographs on Social Research into Child Development* 58(3–4), 1–221.

Schepartz, L.A. 1993: Language and modern human origins. *Yearbook of Physical Anthropology* 36, 91–126.

Schick, K.D. & Toth, N. 1993: *Making Silent Stones Speak*. London: Orion Books Ltd.
Schmid, P. 1983: Eine Rekonstruktion des Skelettes von A.L. 288-1 Hadar und deren Konsequenzen. *Folia primatologica* 40, 283-306.
Schmid, P. 1991: The trunk of the australopithecines. In *Origines de la Bipédie chez les Hominidés* (ed. Y. Coppens & B. Senut), pp. 225-234. Paris: Éditions du centre national de la recherche scinetifique.
Seyfarth, R.M., Cheney, D. & Marler, P. 1980: Monkey responses to three different alarm calls: evidence for predator classification and semantic communication. *Science* 210, 801-803.
Shipman, P. & Walker, A. 1989: The costs of being a predator. *Journal of Human Evolution* 18, 373-392.
Smith, B.H. 1993: The physiological age of KNM-WT 15000. In *The Nariokotome* Homo erectus *Skeleton* (ed. A. Walker & R.E. Leakey), pp. 196-220. Cambridge: Harvard University Press.
Stern, J.T. & Susman, R.L. 1983: The locomotor anatomy of *Australopithecus afarensis*. *American Journal of Physical Anthropology* 60, 279-317.
Stoneking, M. 1993: DNA and recent human evolution. *Evolutionary Anthropology* 2, 60-73.
Tague, R.G. & Lovejoy, C.O. 1986: The obstetric pelvis of A.L. 288-1 Lucy. *Journal of Human Evolution* 15, 237-255.
Tobias, P.V. 1987. The brain of *Homo habilis*: a new level of organization in cerebral evolution. *Journal of Human Evolution* 16, 741-761.
Tomasello, M. & Call, J. 1994: Social Cognition of Monkeys and Apes. *Yearbook of Physical Anthropology* 37, 373-305.
Walker, A. & Leakey, R.E. 1993: *The Nariokotome* Homo erectus *Skeleton*. Cambridge: Harvard University Press.
Wheeler, P. 1994: The thermoregulatory advantages of heat storage and shade-seeking behaviour to hominids foraging in equatorial savannah environments. *Journal of Human Evolution* 26, 339-350.
White, R. 1982: Rethinking the Middle/Upper Paleolithic transition. *Current Anthropology* 23, 169-192.
White, T.D., Suwa, G. & Asfaw, B. 1994: *Australopithecus ramidus*, a new species of early hominid from Aramis, Ethiopia. *Nature* 371, 280-281.
Wills, C. 1995: *The Runaway Brain*. London: Flamingo.
Yellen, J.E., Brooks, A.S., Cornelissen, E., Mehlman, M.J. & Steward, K. 1995: Middle Stone Age worked bone industry from Katanda, Upper Semliki Valley, Zaire. *Science* 268, 553-6.

Conclusions

JOHN MAYNARD SMITH
School of Biological Sciences, University of Sussex, Lewes Road, Brighton, BN1 9QG
Fellow of the Royal Society

THIS WAS AN UNUSUAL MEETING between biologists and human scientists. In these concluding remarks, I will concentrate on two topics which were looked at from both sides of the divide. First, rather briefly, the use of optimization theory to analyse behaviour, and second, at greater length, the nature of mind, and in particular the difference between human and primate minds.

Optimization theory, and game theory, which is simply the extension of optimization to cases in which different participants have conflicting interests, has been widely used both in biology and economics, but rather little used in the rest of the human sciences. The use of optimization is easier to justify in biology than in economics, because natural selection provides a dynamics which will, subject to constraints, cause a population to evolve towards an optimum, and specifies the quantity—Darwinian fitness, or, crudely, expected number of offspring—that will be optimized. In economics, the dynamics of natural selection is replaced by an assumption of rationality, which often does not hold, and the quantity maximized—'utility'—is hard to define or measure. For these reasons, optimization and game theory have, I think, proved to have greater explanatory power in biology.

One feature of optimization models in biology that can give rise to misunderstandings is that, often, little is said about the perceptual and cognitive processes determining the behaviour of individual animals. Indeed, optimization models are often a guide to understanding cognitive processes, rather than based on a prior knowledge of them. The point is important in understanding the use of optimization in several of the papers in the symposium, so I will say a few words about it. One of the earliest applications of game theory to animals (Brockmann & Dawkins 1979)

© The British Academy 1996.

analysed the behaviour of digger wasps. The first model formulated by the authors predicted behaviour rather badly. It turned out that this was because they had assumed, reasonably enough, that a wasp knows whether another wasp is using the same burrow. Further observations showed that wasps do not know: when this was allowed for, the model gave good predictions. This story illustrates two points: optimal behaviour does indeed depend on the perceptual and cognitive abilities of the participants, and optimization models can be a good starting point for a study of behavioural mechanisms.

These points are likely to be relevant to the papers by Dunbar, Borgerhoff Mulder, and van Schaik at the symposium. Dunbar uses an optimization approach to predict group size in several genera of primates as a function of three variables—climate, predation pressure and cognitive ability. Von Schaik discusses the importance of infanticide by males in determining group size and structure. Neither author has attempted to explain how the behaviour of individual animals generates the optimal group size. But this is not intended as a criticism: optimization models are a stimulus to mechanistic studies of behaviour, not an alternative.

> The absence of behavioural mechanisms may seem even odder when the object of study is the human animal. Borgerhoff Mulder describes changes in marriage strategies in changing circumstances in rural Kenya. She does so by asking whether male behaviour is optimal in the new circumstances, although, as she makes clear, she is not able to show that it is Darwinian fitness that is being optimized, but only traits that may well be correlated with fitness. It is not the case that men are consciously trying to maximize their fitness, or even that they are conscious of the criteria that they are using in choosing marriage partners. Yet her approach is very much in the spirit of behavioural ecology as applied to other animals: first find out what is being maximized, and only then ask how it is being done. I must confess that, when I first met this approach to human behaviour, I was unsympathetic. Surely we have reasons for what we do. But I have been impressed by at least some of the earlier work along these lines, notably by Irons (1979) and Dickemann (1979). The great virtue of the approach is that it makes very specific predictions, and so is testable. Borgerhoff Mulder's data on marriage strategies are promising: if optimization methods are inappropriate, the data surely ought to show it.

There is, however, a possible criticism of Borgerhoff Mulder, other than the simple objection that is strange to imagine that people are maximizing their fitness without being aware of it. This is that the adaptation she observes is to a social environment that has existed for a time that is far too short to permit the evolution of a genetic adaptation to that environment. Maximization of fitness, therefore, requires that humans possess a behavioural repertoire that will maximize fitness even in a new and unfamiliar environment.

Cosmides & Tooby take a different view of the relation between evolution by natural selection and human nature. The theory of evolution, they argue, predicts that universal features of human nature should be such as to maximize fitness in the 'environment of evolutionary adaptedness': that is, the environment in which most of our behavioural evolution occurred. There is, therefore, no reason to expect our behaviour to be adaptive in the modern world. Since evolutionary change is slow, the argument is persuasive. But I also see some force in Borgerhoff Mulder's argument that the question is an empirical one, best investigated by applying straightforward optimization methods to existing populations in existing environments. If, as her own work on marriage strategies and earlier on food-gathering suggests, human behaviour can be explained as optimal in the current environment, there is no need to invoke adaptation to an 'environment of evolutionary adaptation'. In contrast, Tooby & Cosmides prefer to rely on the theoretical prediction that genetic adaptation is possible only to an environment that lasts for many generations. Fortunately, perhaps, it is not my job to decide between these alternative research strategies.

Returning to primate societies, there is now a lot that can be said about the kinds of social systems found, and the reasons for them. Dunbar suggests that it should be possible to reconstruct the past history of human social systems, given knowledge of past climates and past dietary habits. Foley draws on our knowledge of present primate societies to make an ambitious reconstruction of the past 50 million years of human evolution. I was particularly attracted by his use of modern phylogenetic methods to deduce ancestral states. His paper is clear, and I see little point in trying to summarize it. But it is interesting that he identifies, not one crucial event in our ancestry, but a succession of such events.

I turn now to the nature of the human mind, and to the differences between the minds of men and apes. It is conventional, and probably correct, in the human sciences to assume that differences between cultures, in time or space, are not caused by genetic differences between peoples. I take it, however, that we agree that the differences between human and ape societies do depend, in large part, on genetic differences between us. But what is the nature of these differences, and when did they occur? A number of views, not necessarily mutually exclusive, were expressed on the nature of the differences.

I will start with the largely theoretical paper by Boyd. Although theoretical, his paper is based on a rather surprising observation. By and large, animals do not learn by copying one another. There are exceptions:

for example, birds learn the details of their songs by listening to adults, so that local song dialects arise. But in general, if two populations of a species behave differently, and if the difference can be shown to be culturally, not genetically, inherited, it is not because young individuals copy their elders, but because their elders create an environment in which individual learning is easy. If this is so, Boyd argues, continuous cultural change will not occur. But why is copying so rare in animals? Boyd suggests that it is because copying only pays an individual if it is already common in the population: in a population of non-copiers, copying is a bad strategy. Thus a genetic tendency to copy one's elders is difficult to establish, but, once present, likely to be maintained.

How far is it true that copying is rare in primates? Boesch argues that at least some of the behavioural differences between groups of chimpanzees in different places depend on imitation. The methods of using a stick to dip for ants are different in Taï and Gombe. Since the Gombe method is four times as efficient, it is hard to see how the Taï method could be explained solely by individual learning. A second example is the use of the same signal, leaf-clipping, in different contexts in different places. Although it is not strictly a matter of imitation, I cannot resist mentioning what was for me the most remarkable fact reported at the meeting. An individual chimpanzee, Brutus, can, by drumming on trees, send the message 'time for a rest', which is perhaps not too surprising, and 'time for a rest, and then move off in a specified new direction', which surely is. Lovers of social insects might claim that the latter is no more than a honeybee can do, but there is a difference: it is unlikely that Brutus was genetically programmed to signal a direction. It seems, then, that chimps do sometimes imitate. The same may be true of other primates. Discussing alarm calls, Cheney & Seyfarth remark, rather in passing, that 'infant vervets seem to learn appropriate usage simply by observing adults'. Thus vervet monkeys may inherit a tendency to respond to flying objects with a particular alarm call, but must learn by copying adults to make the call to eagles but not to pigeons. True, adult vervets do not correct mistaken signals by their offspring, but we rarely correct speech errors in our children either.

If it is strange that imitation should be so rare in animals, even stranger is the apparent absence of intentionality. Cheney & Seyfarth argue that, although both baboons and vervets make sounds that carry useful information, it would be a mistake to think that the signaller makes the call with the intention of altering the behaviour of the hearer. Indeed, there is no reason to think that the signaller conceives of the existence of a hearer. Thus vervets continue to give alarm calls after all other monkeys have taken cover, and baboons are more likely to answer their own barks than the barks of others. The point is important. In humans, co-operation depends in part on 'social contracts': it is hard to form a contract with someone unless you can conceive of them as being an individual like yourself.

The ability to conceive of others as having a mind like one's own has sometimes been referred to as a 'theory of mind'. Tooby & Cosmides discuss the evolution of human altruism, co-operation and friendship in a way that implies such a theory of mind. They start by making the reasonable point that we need to regard altruistic behaviour as an adaptation only if, like the structure of the eye, it is too complex to have arisen as the unselected by-product of something else. Human altruism, they suggest, does possess the degree of complexity, and adaptedness for conferring benefit, that requires a selective explanation (the same is certainly true of altruistic behaviour in social insects, although the evolutionary mechanism is different). A simplified version of their argument goes as follows. Suppose that individual X repeatedly helps Y. This may be because Y has 'trained' X to help, by rewarding him when he did so. Such training would work only if X can classify his actions into those that help Y, and those that do not. If the helping actions performed are various, and sometimes novel, this requires that X has a theory of mind.

One difference between humans and other primates that arose frequently during the symposium is the competence for language. Chomsky (1957, and subsequently) has persuaded most of us that linguistic competence is peculiar, both in the sense of not being merely an aspect of general intelligence, and of being confined to humans. I accept this, but am less willing to go along with Chomsky's insistence that it is fruitless to speculate about the evolutionary origin of language. Perhaps our best hope of making progress on this topic is through the studies described by Gopnik on 'specific language impairment'. The existence of inherited defects in linguistic competence, not necessarily associated with any general cognitive disability, is a confirmation of Chomsky's claim that language is peculiar. But it also holds out hope that genetic analysis will help us to understand the nature and evolution of the 'language organ', just as it is now helping us to understand the nature and evolution of animal and plant development. Gopnik insists that she has not discovered 'the gene for grammar', but it seems that she has discovered at least one gene which, if mutated, causes a specific defect in grammar. There must be many such genes: after all, a 'language organ' could not be programmed by one gene. They may be hard to find, either because they are recessive (and so will not show up in parent and child), or because they have cognitive as well as linguistic effects. As an evolutionary biologist, I would be willing to place one bet. The language organ did not arise *de novo*. New organs usually arise as modifications of pre-existing organs with different functions: wings are modified legs, teeth are modified scales, jawbones are modified gill arches, and so on. It will be interesting to learn what the part of our brain now dedicated to language was doing in our simian ancestors.

The idea of a special language organ has led to the suggestion that there may be other modules specific to other cognitive abilities, for example for social interaction, the manufacture of tools, or the classification of living organisms (see Sperber 1994 for a general discussion). It has been obvious for some time that there are different modules for analysing different kinds of sensory input—for example, visual, auditory, tactile. It has sometimes been argued that the cognitive part of the brain, unlike the preceptual, cannot be modular because its whole purpose is to combine different kinds of information. But this does not really follow. There are separate perceptual modules for visual and auditory inputs, but their outputs can be combined in deciding that an animal is a dog because it looks like a dog and barks like a dog. In the same way, the existence of separate cognitive modules, for example for social and technological functions, would not prevent the conceptual output from such modules being combined.

Mithen proposed an ingenious application of the modular theory of the mind to explain the cultural revolution which marked the transition between the extreme technical and social conservatism of the lower palaeolithic and the inventiveness, cultural diversity and rapid change of the upper palaeolithic. His idea is that the earlier period saw the separate evolution of modules for social interaction and for the manufacture of tools, and that the cause of the revolution was the development of communication between these modules. The result was a symbolic revolution, in which the products of technology were used to carry social information.

Aiello discusses the cognitive and linguistic evolution of modern *Homo* from ape ancestors, using primarily anatomical evidence as a guide. As a non-anatomist, I was impressed by how much can be deduced about behaviour from fossils. How, one might ask, could one possibly deduce anything about the linguistic ability of an individual by examining his thoracic vertebrae? If you read Aiello's article, you will find out.

Mellars discusses the same revolution from a different standpoint, emphasizing the dramatic nature of the transition in Europe, and reviewing its social and demographic consequences. As an outsider, I am left with some sense of puzzlement. It seems natural to seek for a biological underpinning for the cultural revolution that took place some 50,000 years ago. In the absence of biological change, it is hard to explain why almost nothing happened for the preceding half a million years, and why so much has happened since. Most biologists (and, I suspect, all sociologists) would accept that the causes of the later neolithic and industrial revolutions were cultural rather than genetic, but the change that Mithen and Mellars discuss looks different, if only because it was preceded by such a protracted period of conservatism. But if the change was genetic, what was its nature? Which of the cognitive differences between apes and humans—in imitation, intentionality, social skills, linguistic competence, the integration of toolmaking and social modules, or in some combination of several of these— underlay the transition between early and late palaeolithic? Equally puzzling is the fact that the transition is not associated in an obvious way with a change in physical type. It is true that, in Europe, the change is roughly contemporary with the replacement of Neanderthals by modern *Homo sapiens*. But modern *H. sapiens* appear first in Africa some 100,000 years ago, and their appearance is not associated with any obvious cultural change. In Europe, there is some evidence associating upper palaeolithic technology with men of Neanderthal type. We badly need to know more about what was going on outside Europe.

REFERENCES

Brockmann, H.J. & Dawkins, R. 1979: Joint nesting in a digger wasp as an evolutionarily stable preadaptation to social life. *Behaviour* 71, 203–45.

Chomsky, N. 1957: Syntanctic Structures. Mouton, The Hague.

Dickemann, M. 1979: Female infanticide, reproductive strategies and social stratification: a preliminary model. In *Evolutionary Biology and Human Social Behaviour: An Anthropological Perspective* (ed. N. Chagnon & W. Irons), pp. 321–67. North Scituate, Mass. Duxbury.

Irons, W. 1979: Natural selection, adaptation, and human social behaviour. In N. Chagnon & W. Irons ed. op.cit.

Sperber, D. 1994: The modularity of thought and the epidemiology of representations. In *Mapping the Mind* (ed. L.A. Hirschfeld & S.A. Gelman), pp. 39–67. Cambridge University Press.